Man, Agriculture and the Tropical Forest
Change and Development in the Philippine Uplands

Editors:
Sam Fujisaka
Percy E. Sajise
Romulo A.del Castillo

Winrock International Institute for Agricultural Development
Bangkok, Thailand, 1986

WINROCK INTERNATIONAL 1986

Asia Regional Office:

Winrock International
Box 1172, Nana P.O.
Bangkok 10112, Thailand

Headquarters:

Winrock International
Route 3, Petit Jean Mountain
Morrilton, AR 72110-9537, U.S.A.

In the U.S.A. and Europe,
this book is available from:

Agribookstore
Winrock International
1611 North Kent St.
Arlington, VA 22209, U.S.A.

ISBN 0-933595-12-3

The publication of MAN, AGRICULTURE AND THE TROPICAL FOREST: CHANGE AND DEVELOPMENT IN THE PHILIPPINE UPLANDS has been made possible by a grant from the Ford Foundation's Southeast Asia Regional Office in Jakarta, Indonesia. The support of the Ford Foundation is gratefully acknowledged. Its role and support do not necessarily mean an endorsement of the materials chosen or the views presented.

CONTENTS

FOREWORD

For the past several years, the Agricultural Development Council (A/D/C) has been concerned with the degradation of forests and other natural resources in the Southeast Asia and the need to improve our ability to correctly manage these resources. "Man, Agriculture, and the Tropical Forest: Change and Development in the Philippine Uplands" grew out of this concern. In July, 1985, A/D/C merged with IADS (International Agricultural Development Services) and WILRTC (Winrock International Livestock Research and Training Center) to form Winrock International Institute for Agricultural Development. The new institute attached equal importance to this topic.

A focus on the Philippines is appropriate because of the experiments in natural resources management that have been conducted there over the past several years. These have mostly been in the hills and mountains where social or community forestry projects had taken place. These projects have been in response to the depletion of resources and sufferings of displaced people caused by logging, dam construction, and various forms of commercial exploitation. The proliferation of these projects was such that it became obvious that the time had come for documenting these field experiments. And the results have been mixed.

In 1982 Samuel Fujisaka, a Rockefeller Foundation Postdoctoral Fellow, joined the Program on Environmental Science and Management (PESAM) of the University of the Philippines in Los Banos which was headed by Percy Sajise. The association of a young anthropologist with an eminent and experienced Filipino biological scientist was productive. It led to more field research and a host of cooperative ventures, one of which was funded by A/D/C supported by a grant from IDRC. An important result was increased communication and collaboration among the many scientists in the Philippines who are interested in managing natural resources.

Samuel Fujisaka and Gerard Rixhon, Director of Regional Research and Training Program of A/D/C, thought that a book focusing on hill and mountain farmers as seen through a multidisiplinary group of scientists would be worthwhile. Tom Kessinger, Regional Representative (for Southeast Asia) of the Ford Foundation, agreed and the Foundation provided financial support to publish such a manuscript.

I believe this book is a welcome addition to the fast growing literature on natural resources and forest management. The authors and editors richly deserve our appreciation for a thorough description of what has been happening to the Philippines in its attempts to rehabilitate hill agriculture and make the upland areas more habitable for people. The authors' perspectives are based on much research and field observations as they look, at times a bit timidly, to the future. As the authors will themselves admit, theirs is not a final or even definitive statement as much room is left for further research. Theirs is an honest, empirical contribution to the understanding of a problem which goes beyond the Philippines itself, one which will welcome constructive criticisms, complementary analyses and further research and documentation.

David F. Nygaard, Director
Human Resource Development Division
Winrock International

December 1986

LIST OF MAPS

1
CHANGE AND DEVELOPMENT IN THE PHILIPPINE UPLANDS: AN INTRODUCTION

Sam Fujisaka

Upland and forest ecosystems of the Philippines and the resident users of these areas are currently in a state of change and transition. Changes have resulted from dynamic interactions between man and ecosystem, and considerable recent attention has been focused upon apparent problems stemming from such interactions. Deforestation, increasing populations, shifting cultivation, and watershed degradation with negative "downstream" effects have been viewed over the past several years as interrelated factors that together mean the destruction of Philippine upland forest ecosystems and further hardship for already poor upland populations.

Deforestation has resulted from the combined effects of logging and expansion of shifting cultivation often into areas made accessible by logging. Damage to watersheds has received more attention, as flooding and drought are thought to have become more common in lowland rice-producing areas and as reservoirs built for irrigation and hydroelectric power generation suffer from siltation. Problem-solving efforts — i.e., "upland development" — include attempts at controlling and guiding the directions of current and future change in the uplands in order to benefit both upland ecosystems and the uplanders as well as society in the larger sense.

This volume seeks to gather, analyze, discuss, and synthesize some of the current available information concerning the uplands and the uplanders, their dynamic interactions, the directions and magnitudes of systems and subsystems change, perceived problems, and attempts at problem-solving. Although existing knowledge gaps are admittedly considerable, relevant information is now available from the research of various academic disciplines — from agroecosystems research to law and land policy to anthropology — and the "hands on" experiences of development agencies and institutions. Strategies to improve conditions in the uplands are being designed, tested, and implemented. Ongoing work described in the chapters includes basic and applied research and the set of multifaceted tasks comprising upland development.

Chapters 2, 3, and 4 present, discuss, and synthesize basic research dealing with different interacting systems — the tropical forest (Sajise), upland cultures and societies (Russell), and population and migration (Cruz). These chapters consider both naturally-occuring change and change as a response to human disturbances. Chapters 5 (Samson) and 9 (Torres and Raintree) detail and illustrate the development of strategies and technologies for introduced change and development, while Chapter 6 (Segura-De Los Angeles) examines the economics of such technologies and strategies. Different types of case materials are presented in Chapters 7 (Aguilar, Jr.), 8 (Fujisaka and Capistrano), and 9 (Torres and Raintree). Laws and policies, especially regarding upland tenure, are considered in Chapter 10 (Lynch, Jr.). Change agents are discussed in Chapter 11 (Del Castillo and Castro). Social and community forestry in the regional context is considered in Chapter 12 (Kirchhofer and Mercer). A final chapter (Chapter 13 by Fujisaka and Sajise) synthesizes findings and implications for change and upland development.

THE ECOSYSTEM

The undisturbed tropical forest is a dynamic, but relatively productive and stable ecosystem. Few areas of the world's tropical forests, however, remain undisturbed. As a result, the functioning and dynamics of the ecosystem and its resilience in the face of human usage are currently being examined, although much more work remains to be done. Originally, most of the Philippine uplands were covered by tropical forest. Currently, of a total land area of 30 million ha, 16.7 million ha (56%) of the Philippines are classified as public forest, about five million of which are characterized by the national government as "open, denuded, and unproductive".

Observations show that degradation of tropical forest ecosystems is a dynamic process that often starts with timber cutting and continues as thin topsoils and nutrients are lost through erosion and leaching once the forest cover is removed. Losses can continue as lands are used for agriculture and become more serious as populations and competition for resources increase, productive lands become scarcer, and the fallow cycles of shifting agriculture are reduced. Lands may eventually become unproductive, lack potentials for rehabilitation, continue to erode, and be converted into fire-prone grasslands of questionable productivity, while, at the same time, the uplanders become poorer and have fewer alternatives for improving their circumstances.

MAP 1. THE PHILIPPINES

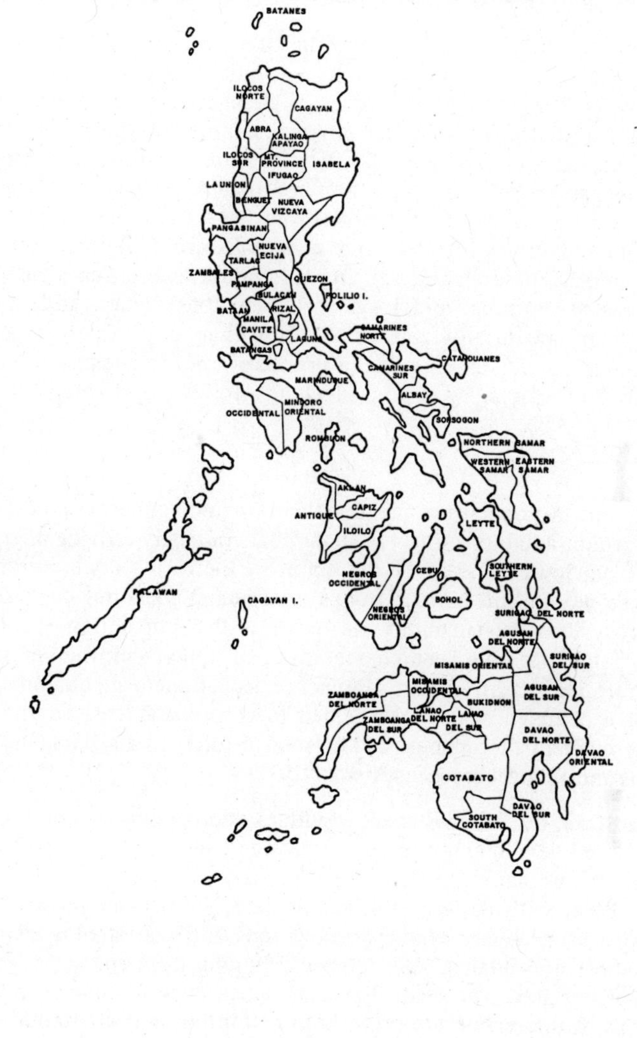

The upland tropical forest ecosystem is heterogeneous. Some forest areas or subsystems have been degraded, while other areas may be more resilient in the face of human utilization and perhaps more suited in terms of productivity and sustainability to the introduction of new forms of resources use. Such "new" systems of resources use involve both social system and ecosystem and can include technological, institutional, economic, cultural, political, and biophysical change.

Chapter 2, by Percy Sajise, explains the structure and function of the tropical forest ecosystem and current human resource use patterns. The chapter first considers estimates of global, regional, and Philippine tropical forest destruction. The physical factors of climate, location, water balance, and elevation classifications are described. Productive, protective, and regenerative functions are then examined in some detail, providing a clear view of the tropical forest as a dynamic, complex system. The sometimes intrusive and system-disturbing human activities are discussed. These include shifting cultivation, plantation agriculture and forestry, burning, and logging. Forest conservation and upland development are briefly considered. This chapter provides a starting point for the understanding of the productive but fragile physical environment and ecosystem within which the following chapters are set.

THE UPLANDERS

The human populations in the uplands include indigenous tribal and ethnic groups and increasingly large numbers of newly arriving migrants, many from lowland areas of the country. Increasing competition for resources within and among groups and greater pressures on resource and ecosystems sustainability accompany the population increases. While change has long been characteristic of upland cultures and societies, more recent shifts from subsistence to commercial economies are part of a complex set of interrelated local and non-local factors that have brought further change and often disruption to social and cultural systems and to local agroecosystems.

Chapter 3, by Susan Russell, examines ethnographic research in the uplands and details how upland cultures and societies are considerably diverse and are naturally changing, especially with population increases and contact with national Filipino society. Subsistence practices and strategies, social and economic organizations, and associated perceptions and beliefs are detailed and compared among different societies and types of cultures. The transition from subsistence to commercial economies and the related effects — both disruption and adaptation — are given particular attention. Changing patterns of access to resources — changing tenure patterns, systems of inheritance, resources sharing, credit, markers, information systems, labor organization, and agricultural practices — are detailed.

DEMOGRAPHIC CHANGE

Increasing population underlies all other types of change in the uplands. Current population increases and migration trends are associated with factors ranging from emerging forms of social and economic organization to decreased fallow periods and the adoption of new farming technologies. Demographic trends and the interactive effects of resource use patterns and technologies must be a starting point for considering and planning for upland development.

Chapter 4, by M. Concepcion Cruz, discusses population pressure and migration and emphasizes the need for more reliable data. Population pressure is first examined in terms of carrying capacity and changes in cropping patterns. Major migration flows and patterns are described, and some of the general characteristics of the migrants are discussed. The heart of the chapter, and the part that will probably generate the most discussion, is an estimate of upland population which, according to the author, ". . . is taken from a sample of 1179 barrios listed in the Bureau of Forest Development (BFD) social forestry program. The estimate reveals that about 11 million people currently reside in the uplands, representing 19.2% of the total Philippine population and over 50% of the total migrant population in the country". This chapter provides a needed update on population trends and an implicit warning as to the magnitude of the population "problem" in the uplands.

TECHNOLOGY GENERATION FOR UPLAND DEVELOPMENT

Existing local resource use patterns and technologies can adaptively change in response to population changes and resource competition. However, indigenous or in-place technologies and knowledge that are well adapted to local and current conditions can be inadequate in the face of new challenges and pressures. As one result — and as a part of what we are calling upland development — "improved" technologies are being developed, screened, tested, and evaluated at universities and research stations and on experimental plots and farmers' fields. Such technologies are ideally superior to in-place technologies in terms of profitability, sustainability, equitability, and ecosystems effects. Experience has shown that, at best, improved technology development for the uplands is a slow and painstaking process, requiring, among others, focused initial community appraisal, careful diagnosis of problems, rigorous screening and testing of technologies, and carefully planned and implemented work with potential adoptors. Improved technologies for upland development include those falling under the rubrics of "social forestry", "agroforestry", "farming systems", "polycultural systems", and others.

In Chapter 5, by Benjamin Samson, upland development technologies are described and evaluated. The chapter builds further upon Sajise's description of the tropical forest and tropical forest dynamics. Most upland development technologies are directed towards maintaining or improving the productivity or sustainability of the tropical forest agro-ecosystem. The chapter first examines a number of technologies that each address specific problems. Detailed are both ecosystemic problems and the dynamics of those problems — i.e., soil erosion, decreasing soil fertility, and deforestation — and technologies being developed to address those problems, such as bench terracing, vegetative soil erosion control methods, soil-conserving tillage methods, contour farming, green manuring, mulching, and the use of mycorrhiza. Samson then discusses what he calls "integrated technologies", or the multiple-faceted strategies now being tested and applied in upland development projects. Examined are different multiple cropping methods and the functioning of polycultural systems, different agroforestry systems, and different regeneration technologies for denuded and marginal uplands. Current directions and considerations in the development of technologies for use in the uplands and associated research gaps are discussed in a concluding section of Samson's chapter.

THE ECONOMICS OF TECHNOLOGIES AND PROJECTS

Samson describes the generation of technologies for the adoption and use by upland agricultural populations. Adoption of technologies depends upon their productivity, sustainability, and equitability. Unfortunately, technologies that are both individually profitable and socially beneficial have been difficult to find. Development projects in the uplands have faced a wide range of problems, perhaps the most basic of which has been a lack of concrete payoffs to participating uplanders.

Economic analysis of upland development technologies and projects are discussed in Chapter 6 by Marianne Segura-de los Angeles. She examines the economics of upland use by forest occupants, subsistence farming, economic impact analysis, and ex-ante and ex-post project evaluation. Research, actual project evaluations, impact analyses, and other available case materials completed by a range of investigators are described and evaluated, and various methodologies utilized by economists working in upland development are presented. Research gaps and policy-related issues stemming from the work presented are identified in a concluding section. This and the previous chapter provide needed assessments of the feasibility, including from the uplanders' point of view, of the various strategies being attempted in upland development.

SOME SPECIFIC CASE MATERIALS

The uplanders, whose numbers and resource use practices that contribute to observed problems, are also sources of adaptive change. Indigenous practices and technical knowledge, experimentation and innovation by pioneer migrant settlers in adapting to new agroecosystems, and evolving local social systems all reflect "natural" and often adaptive sociocultural responses to change. Ethnographic research on such phenomena is proving to be essential for the development of appropriate and improved technologies and strategies. Also from ethnographic studies, knowledge about many of the indigenous groups and their practices is fair. Knowledge about pioneer upland settlers is less complete.

Similar to the tropical forest, upland groups are diverse. The circumstances of different groups vary in terms of respective particular agroecosystems and related resource use practices, corresponding social and economic organization, types and degree of involvement in Philippine national culture, numbers, and past and present directions and degrees of change. Potentials for effective participation in "upland development" efforts thus vary among different groups.

The three case study chapters deal with different groups and with both naturally occuring change and introduced or guided for upland development. In Chapter 7, Filomeno Aguilar, Jr. presents findings from studies of eight "successful" ongoing social forestry projects. The author, interested mainly in levels and types of "popular participation", describes projects carried out by a variety of public and private agencies in diverse areas showing different ethnic and cultural mixtures.

Aguilar's studies describe the projects and local biophysical conditions, present data from samples of respondents on socioeconomic conditions and local participation in the respective projects, and offer lessons learned. Discussion reveals the difficulties involved in developing appropriate and effective strategies for working with upland communities and in guiding change in ways desirable to outside change agencies.

Two case study chapters describe necessary research carried out in local communities **prior to** the design and implementation of any introduced "development" efforts. Chapter 8, by Sam Fujisaka and Doris Capistrano, looks at evolving in-place practices and then discusses interactions among agroecosystem, resource use, social system, and national policy. Described are ecosystem and systems of forest resource use, the emergence and structuring of competing factions of settlers, and non-local factors that exacerbated local problems. Major constraints to local upland development included, first, ambiguities and confusion concerning national policies affecting the area and the people, government agency jurisdictions, and local and provincial land claims and, second, the community level effects of such confusion as expressed though emerging social and economic organization and resource use strategies. Current project work — the design, testing, and implementation of improved appropriate resource use strategies — may prove to be less problematical than the initial institutional problems. Other research data about the pioneer migrant community was used to screen possible improved local resource use patterns and to evaluate local suitability of government "integrated social forestry" policies.

Chapter 9, by Filemon Torres and John Raintree (with M. Dalmacio and T. Darnhofer), presents a case of more rapid and focused initial diagnosis and design work followed by further field experimentation and testing of "best-bet" technologies in an upland land reform community. The chapter, received as a working paper and somewhat revised for inclusion here, provides detailed appraisal information for screening and testing of potential technologies. The chapter describes upland farming and land reform in the Philippines, examines the study area in terms of physical variables, population, economy, agroecological zones, land tenure, farm size, and farming and production systems, and provides a diagnosis of major local land use problems and potentials. This latter section looks at the problems in household supply system and corresponding causal factors. The authors then explain the derivation of design specifications, discuss agroforestry and non-agroforestry alternative technologies, provide an ex-ante evaluation of the proposed system, and submit a detailed research proposal for field experiments to be carried out on farmers' farms, at a local field station, and at a research institution.

As in the previous chapter, the pre-project research and appraisal or diagnosis are interdisciplinary, with social, economic, agroecological, and institutional factors being of interlocking importance. In both cases, research is utilized for the design of the next stage of upland development activities.

NATIONAL POLICY AND UPLAND DEVELOPMENT

In spite of their locally adaptive characteristics mentioned above, different upland cultures and societies remain for the most part populations of poor farmers with little political power at the national level. Access to basic resources, especially land, is a major issue facing all uplanders — both tribal and migrant. National land tenure laws and policies have evolved over time, with changes reflecting the political and institutional climates of the different times. Laws and policies have variously granted forest lands to colonial powers, agencies of the national government, private interests, long-term indigenous users through the recognition of ancestral land rights, and even migrant settlers. In simplistic terms, however, "bad" policies have often led to short-sighted and destructive resource exploitation, while "good" policies have often lacked the necessary concomitant "political will" necessary for implementation.

Relatively recent national attempts to manage and maintain upland resources and environments included the control and regulation of logging and the banning of settlement and shifting agriculture in forest areas. Banning settlement by swidden farmers has been unsuccessful in both forests and in the more denuded areas made more accessible by commercial timber harvesting. Efforts also include reforestation and watershed or land rehabilitation projects.

The national government is more recently attempting "upland development" through programs that grant or recognize land rights of upland and forest occupants and that work with specific upland populations to improve local resource management practices while generally encouraging "popular participation" in farming systems improvement, cropping intensification, or adoption of more permanent crops and agroforestry technologies. The Integrated Social Forestry Program of the Bureau of Forest Development is granting limited land tenure rights in the belief that land tenure insecurity is a basic constraint to successful participation by upland farmers.

In spite of such policies, the government granting of commercial timber concessions and the use of destructive extractive technologies also continue, however, as do encroachment on tribal or ancestral lands, deforestation and resources depletion, and poverty among upland groups.

Owen Lynch, Jr., in Chapter 10, examines traditional practices and national laws and policies regarding land tenure in the Philippine uplands. The paper is concerned largely, on the one hand, with existing legislation that should, if implemented, protect the traditional land claims of tribal and indigenous groups, and, on the other, with the legal and policy processes that are now working to displace upland tribal or ethnic groups from traditionally held lands. The chapter discusses the colonial foundation of current laws, increasing jurisdiction among executive branch agencies over so-called public lands, the private property rights of many uplanders, and local or customary law regarding lands in traditional communities. The chapter concludes with recommendations on how to improve the situation faced by tribal and other uplanders.

THE AGENTS OF INTRODUCED CHANGE

The leading role in Philippine upland development is held by the Bureau of Forest Development (BFD), which has jurisdiction over the country's public forest lands. As mentioned, the integrated social forestry program of the BFD is now implementing projects that grant 25-year renewable individual or community land leases while encouraging adoption of non-shifting and agroforestry-based resource management systems. While the multi-agency and multi-institution Upland Working Group of the BFD and academically based institutions, such as the Program on Environmental Science and Management at the University of the Philippines at Los Baños are known throughout the country, the range of other public and private entities actively involved in upland research development is wide and varied.

Chapter 11, by Romulo Del Castillo and Charles Castro, shows that upland development is a broad field of research and applied field intervention, and that there are many participating agencies and institutions. The authors classify agencies and institutions according to mandate and development priorities. The five resulting categories are commodity-oriented institutions, service delivery institutions, community organizations, research and educational institutions, and funding institutions. For each category and for several sub-categories the authors describe organizations and their respective interest, goals, activities, and methods in Philippine upland development.

Particular attention is given to programs and institutions aided by the Ford Foundation, a private funding institution, and to the institutions represented in the BFD Upland Development Working Group. Concluding sections discuss working towards effective inter-agency collaboration and the activities of the working group. The chapter includes a useful directory of upland development agencies and institutions in the country.

THE REGIONAL CONTEXT

A penultimate chapter, Chapter 12 by James Kirchhofer and Evan Mercer, extends beyond the Philippines by considering social and community forestry in the Asia-Pacific region. The paper looks at activities having a "social" approach to forestry and describes the impetus and institutional responses to, and the appeal of, social forestry. Programs are classified as those requiring collective action and those requiring individual action. Agroforestry and social forestry are distinguished. Most of the chapter then deals with the subset of community or village based forestry programs requiring collective action. Potentials and limitations of community forestry, community forestry in rural development, and community forestry "in perspective" round out the piece. The chapter, in supplying information about different efforts in "social" forestry in the region, provides an implicit comparison between the problems and approaches of the Philippines and neighboring countries.

SYNTHESIS, IMPLICATIONS, AND DATA GAPS

A concluding chapter, Chapter 13 by Fujisaka and Sajise, integrates and builds upon lessons learned in the individual chapters, introduces additional materials for consideration, (re) evaluates problems facing the uplands and uplanders, examines unresolved issues, considers implication of the lessons learned and knowledge gaps for "upland development", social forestry policy, and change and adaptation, and presents a simple model of relevant human-ecosystem interactions discussed in this volume.

This volume discusses results from past and ongoing research. Some of the problems facing the uplands and the uplanders have been clearly identified; other problems and the magnitudes of their effects require much further research. Information is still needed for understanding the perceived problems and for addressing the problems with the development of productive, sustainable, and equitable technologies and related strategies of "development" project implementation. Estimates of the sizes of affected upland areas and rates of deforestation have often been informed guesses at best, as are many of the estimates of the sizes and rates of increase of upland populations. Forest and upland ecosystems are being converted and sometimes destroyed. The rates and seriousness of such destruction are not really known. Legal and illegal logging continues but numbers of trees and areas involved are not well known. Agricultural use of the uplands has been described as stable under some conditions and destructive under others. Knowledge about indigenous and tribal groups such as the Ifugao, Bontoc, and Mangyans is considerable. But, again, knowledge about the newer groups of migrant settlers — who are also very much "uplanders" — is lacking. Overall, the interplay between specific ecosystemic dynamics and particular human practices are still imperfectly understood, although much is being learned.

Both in spite of and because of such data gaps, it is the hope of the editors that the following chapters will present many of the relevant aspects of what is currently known about change and development in the Philippine uplands, and that such information will be readily applicable to the next "best bet" stages of research, data analysis, technology generation, project implementation, and policy-making.

2
THE CHANGING UPLAND LANDSCAPE

Percy Sajise

A tropical rainforest is a climax vegetation situated in the equatorial belt and bounded by the Tropic of Cancer (23° 27'N) and the Tropic of Capricorn (23° 27' S). The tropical zone accounts for about 40% of the earth's surface and fixes 25% of its total carbon dioxide. The tropical rainforest is one of the oldest ecosystems on earth (UNESCO/UNEP/ FAO, 1978). Fossil evidence suggests it has existed continuously since the Cretaceous period, more than 60 million years. Two hundred years ago, the tropical rainforest stretched almost unbroken over the lowland humid tropics of Central and South America, Africa, and Southeast Asia. Satellite photographs show that the tropical rainforest is no longer a continuous band, but a series of sporadic patches along this band (Richards, 1973). Extensive areas of rainforest remain in three regions: the Amazon River Basin in South America, the Congo River Basin in equatorial Africa, and the Malay Archipelago in Southeast Asia. Smaller forests are found along the gulf coast of Central America, in West Africa, and in the humid areas of India, Ceylon, and Australia (table 1). Much of these smaller tropical rainforests have been cleared for agriculture, animal husbandry, and habitation.

Since the late 1970s, tropical rainforests have generated increasing interest and concern, especially in the possible consequences of deforestation. Considerable differences of opinion have been expressed on this issue. According to Sedjo (1983), the deforestation controversy stems from an overestimate by Norman Meyers, published by the Carter administration's Global 2000 report (1979). Meyers assumed that each of the 20 million families currently living in tropical rainforests would deforest one hectare per year, resulting by 2000 in 40% deforestation worldwide.

A more recent and relatively optimistic estimate, based on an FAO/UNEP study, which used satellite photographs and extensive consultation with experts from 78 countries, is deforestation of seven million ha per year. The estimate gives little cause for alarm (Sedjo, 1983).

Table 1. Total area, forest area and human population in continents containing tropical forest (FAO/UNEP, 1981).

Region	Land Area $\times 10^3$ km^2	Total* $\times 10^3$ km^2	%	Forest Closed $\times 10^3$ km^2	%	Population $\times 10^6$	%
Tropical Africa	22150	7340	38	2100	29	267	25
South and Central America	14890	8210	43	5900	72	218	20
Southeast Asia and Australia**	9050	3600	19	3000	83	584	55

*Closed forest, open woodlands and shrublands.
**Without Oceania.

A report of the **Carbon Dioxide Review** (1982) implies that deforestation does not seriously affect global CO_2 exchange. Salati and Vose contend, however, that, "continued large scale deforestation in the Amazon could likely lead to increased erosion and water runoff with initial flooding in the lower Amazon, together with reduced evapotranspiration and ultimately reduced precipitation. At present, on the average, 50% of the precipitation is recycled. Reduced precipitation in Amazon could increase the tendency toward continentality and adversely affect climate and the present agriculture in southcentral Brazil" (1984: 137).

Although this review is generally optimistic, deforestation is a problem (Sedjo, 1983). Nearly all the shrinking of areas of close forest trees (tree crowns that cover 20% or more of the ground) occurs in the humid tropics. In the mid-seventies, moist tropical forests covered more than 935 million ha. already they have been reduced from their natural domain by more than 40% (Eckholm, 1979). Following are observations concerning the tropical rainforests in Southeast Asia.

Although Indonesia has the highest deforestation rate in absolute terms, 550,000 ha per year, it loses only 0.48% per year of its total forest area. In Thailand, only 333,000 ha are deforested per year, but this constitutes 3.6% of its total forest area (FAO/UNEP, 1981). Foresters estimate that in Thailand, between 18 and 30 trees are cut for every tree replanted; the land under forest has shrunk from 58 to 38% in just 17 years (Borsuk, 1977). Forests covered 74% of peninsular Malaysia in 1957, but only 55% in 1977; some 2,850 square kilometers were cleared for agriculture during each of the last five years (Myers, 1980). Equally rapid rates apply to Vietnam. Only Burma and Laos have not significantly converted forest areas to farming (FAO, 1980).

Another factor in rapid deforestation is population pressure. In 1980, in Southeast Asia alone there were some 350 million people. Given the rate of increase of 2 to 2½%, this number is expected to more than double by the turn of the century (Rambo, 1984).

About 70 years ago, the Philippine archipelago was almost totally covered with forest (Brown and Matthews, 1914). Dickerson described its biogeography, flora, and fauna:

> "In the open country and in the secondary forests of the Philippines, specific species endemism is less than 10%; while in the primary forests, it exceeds 80%. This condition merely emphasizes the fact that the primary forests represent the indigenous flora of the archipelago; that a high percentage of the species found in settled areas, open grasslands, and secondary forests represent species of recent or comparatively recent introduction. It may, with reasonable safety, be assumed that the original vegetation of the Philippine Islands before man reached the Archipelago was continuous forest of one type or another" (1926: 144)

Today, of the 30 million ha constituting the Philippines, only 16.7 million ha (or 56%) is classified as forest (BFD, 1981). This consists of about 5.6 million ha of unclassified land and 11.1 million ha of forest land. Forest lands can be further broken down into: 5.53 million ha of timberland, 3.25 million ha of forest reserves, 1.59 million ha of national parks game refuges, bird sanctuaries, and wilderness, 0.129 million ha of military reservations, and 0.312 million ha of civil reservations. Of the 5.53 million ha of timberland, only 2.7 million ha is classified as old growth, the rest as second growth forest (table 2). Bonita (1977) concluded from aerial photographs that, instead of 56%, the Philippines now has only 30% forest cover in areas concentrated in some few provinces. There are presently about 4.5 million ha of denuded grasslands, the result of logging, shifting cultivation, overgrazing, and burning. The annual rate of deforestation in 1970 alone was 200,000 ha/year, and between 1971 and 1980, 1.24 million ha of dipterocarp forests was converted to less productive second growth stands (Reyes, 1983).

Table 2. Present land use types and area coverage, Philippines.

Land Classification	Total Area — M ha. —	% of Total Land Area
1. Forest Land (BFD, 1981)	16.7	56
A. Unclassified	5.6	
B. Forests	11.1	
B.1. Timber	5.53	
B.2. Forest Reserve	3.25	
B.3. National Parks, Game Refuge, Wilderness, Bird Sanctuaries	1.59	
B.4. Military Reservation	0.13	
B.5. Civilian Reservation	0.31	
2. Alienable and Disposable (BAEcon, 1982)	13.3	44
A. Agricultural	9.11	
A.1. Irrigated	0.91	
A.2. Non-irrigated	1.65	
A.3. Sugar Cane	1.74	
A.4. Coconut	3.58	
A.5. Other seasonal crops	1.23	
B. Industrial, Resettlement, Infrastructure, etc.		

The bulk of the current forest lands is in Luzon (40%) and Mindanao (37%); the remainder is in Visayas (16%) and Palawan (7%). The country's commercial forests are located in Mindanao (57%), Luzon (26%), the Visayas (9%), and Palawan (8%). The causes of forest destruction are the same in the Philippines as in other tropical rainforest areas of the world: short-sighted timber harvesting, the spread of agriculture, and firewood collection. Behind these causes are the more basic issues such as agricultural stagnation, insecurity of land tenure and the "open access" nature of the forest resources, rising unemployment, rapid population growth, and incapacity to regulate private enterprise to protect the public interest (Eckholm, 1979, Sedjo, 1983).

PHYSICAL DESCRIPTION

Climate

Daily temperature changes in the tropics are greater than the very small monthly changes; available radiation is great; the long growth period spans the whole year. Day length varied little; temperatures in the lowlands are usually greater than 22°C, while in the high areas temperatures are about 18-22°C. Rainfall occurs throughout the year, there are not more than two-and-a-half dry months (where evaporation exceeds rainfall). There is, however, variability within the tropics, particularly among tropical islands: "Southeast Asia and the islands of the Western Pacific fall entirely within a broad environmental zone of the world referred to as the tropics. But islands maintain a range of ecological conditions of their own which depend on island size, topography, distance from a continental land mass, and so forth, that causes a continuous range of variability" (Hutterer, 1984: 81).

Location

Geography describes 3 major tropical rainforests: the American, African, and Indo-Malayan. The American extends east to the Guianas and west to the Andes. The African extends west into French equatorial Africa, Gabon, Cameroons, and French Guiana, east to the Great Lakes, and south toward Rhodesia. The Indo-Malayan tropical rainforest includes Malaysia, adjoining parts of Southeast Asia, and the Pacific Islands. The largest continuous rainforest of the Indo-Malayan group is found in the Malay Peninsula, Sumatra, Borneo, the Philippines, and New Guinea.

Water Balance

There are two systems of water balance: the per-humid and the seasonal. The per-humid type has no significant period of water stress, has an average annual rainfall of at least 100 cm, and has a predominantly evergreen forest. The area covered by per-humid forest mostly belongs to Western Malaysia, including the Malay Peninsula, the Philippines, New Guinea, and Indonesia. The seasonal tropical rainforest has annual distinct dry and wet seasons and is highly influenced by the tropical convergence zone.

Elevation Classifications

Tropical rainforest also varies with elevation. Generally, two types are distinguished — lowland and montane. Lowland tropical rainforest occurs up to 800 meters above sea level; montane above that.

TROPICAL RAINFOREST FUNCTIONS

Production Functions

The tropical rainforest ecosystem is a climax successional stage and has three distinctive but interdependent functions — productive, protective, and regenerative.

Although tropical rainforests make up only one-third of the world forests, they contain four-fifths of the earth's vegetation. A hectare of primary forest may support a plant biomass of 300 to 800 tons and have a net primary productivity of as much as 70 t/ha per year (Myers, 1984). The tropical forest is a very productive ecosystem, as high-yielding as such agricultural crops as sugarcane. Sugarcane, however, requires massive inputs; the tropical rainforest depends only on solar energy and highly efficient internal cycling of materials.

Estimates of the earth's primary production range between about 120 and 160×10^9 tons of dry organic matter per year, of which 45% is fixed by forests and woodlands, 37% by phytoplankton, 5% by agricultural crops, and 13% by other vegetation. Tropical rainforests, which cover 4% of the earth's surface, fix at least 25% of the earth's terrestrial carbon (table 3, Longman and Jenik, 1974).

Table 3. **Net primary production estimated for different types of vegetation (from Longman and Jenik, 1974).**

VEGETATION TYPE	NET PRIMARY PRODUCTION		
	Area (108 ha)	Per unit area (dry tons/ha/yr)	In world basis (109 dry tons/ha/yr)
Tropical Forest	20	20 (10 — 50)	40.0
Savanna	15	7 (2 — 20)	10.5
Temperate Forest	18	13 (6 — 30)	23.4
Boreal Coniferous Forest	12	8 (4 — 20)	9.6
Tundra and Alpine grasslands	8	14 (1.5 — 15)	1.1
Steppe and other temperate grasslands	9	5 (1.5 — 15)	4.5
Agricultural land	14	6.5 (1 — 40)	9.1

The age of the stand influences the total net production and relative amounts of stems, leaves, and roots. The Net Primary Productivity (NPP) in a mature forest should be less than in a young forest since most of the Gross Primary Productivity (GPP) would be spent in maintenance. Seventy percent of the GPP in mature forest and 4.0% in immature is used for respiration. Respiration at 25°C in Pasoh Forest was reported to be 54.5 tons/ha per year (Kira and Ogawa, 1969), indicating that much of GPP is spent in maintaining a large volume of biomass.

Many essential mineral nutrients are primarily bound in the tropical forest biomass, especially in the wood and leaves (table 4). The nutrient pool in oxisols and ultisols, the dominant soils in the tropics, is relatively small. Potassium, calcium, and magnesium are primarily bound in the biomass. On the other hand, more than 70% of the ecosystem's nitrogen and phosphorous reserves are in the top soil (Sanchez, 1979).

Table 4. Range, nutrient content in total biomass in mature forest in Zaire, Ghana, Panama, Puerto Rico (Sanchez, 1976)

Element	Range (kg/ha)	Element	One Observation (kg/ha)
Nitrogen	701 – 2044	Sulphur	196
Phosphorus	33 – 137	Iron	43
Potassium	600 – 1017	Zinc	13
Calcium	563 – 2760	Manganese	5
Magnesium	381 – 3890	Copper	3

In addition to high biomass, the tropical rainforest yields a wide array of chemical compounds and other products. Fruit trees such as *Durio, Garcinia, Eugenia,* and *Metroxylon* (sagu palms), provide food. Many tropical rainforest species are sources of valuable pharmaceuticals. Compounds extracted from other plants serve as chemical intermediates to produce useful compounds such as the steroids. Various species of *Discorea* . are steroid precursors.

The increased interest in natural rather than synthetic pest control agents in recent years has resulted in efforts to find sources of natural insecticides such as rotenone (from *Derris* sp.), repellents such as oil of citronella, and antifeedants such as those from the neem tree (*Azadirachta indica* A. Juss). Diosgenin, a sex hormone, is extracted from Derris sp. Illipe nuts; seeds of several species of *Shorea* are a source of oil and are an export of Borneo. At least 10 species of *Shorea* are used for oils in foods, soaps, and candle wax. A number of resins are gathered from forest trees. Gum damar, produced by various species of Dipterocarpaceae and Burseraceae, and copal from *Agathis alba* are among the more important. Rubber, gutta, and caoutchouc are produced by several plants, mostly in the families Sapotaceae, Apocynaceae, and Euphorbiaceae.

Several genera of rattan or climbing palms are used for cordage, weaving, and furniture. Bamboo is used for construction, and its young shoots are eaten. The bark of several species, especially the Urticaceae and Malvaceae, provides fibers. Many understory forest species and epiphytic plants are cultivated as ornamentals. Families with many ornamental species include Orchidaceae, Zingiberaceae, Palmae, and Gesneriaceae.

Protective Functions

The tropical rainforest effects soil protection by absorption and deflection of radiation, high rainfall, and strong wind, maintenance of high humidity and carbon dioxide, and maintenance of a diverse gene pool.

The tropical rainforest is characterized by both vertical and horizontal stratification. Three strata are generally recognized: "A" is characterized by most of the tree crowns being fairly well separated from each other and raised well above the very dense stratum; "B" is characterized by an almost continuous layer, with each crown usually in lateral contact with others; and "C" is characterized by a continuous stratum of young individuals of species which reach strata A and B when mature. Additional strata of understory vegetation include the shrub, herb, and root layers (Longman and Jenik, 1974).

The dome-shaped crowns of the emergents shade the main canopy and are themselves partically in sun and partially in shade at all times of the day, which gives them longer periods for photosynthesis. At midday when heat is highest, heat is dissipated, resulting in reduced transpiration.

Leaves of the emergents are generally scloerophyllous with thick cuticles, and stomata are mostly confined to the lower sides. These are adaptations that reduced transpiration rates. Scales or hairs also reduce water vapor diffusion rates. Most leaves have high reflectance (shiny, waxy surfaces) and are held at an angle that reduces the solar radiation per unit area. The canopy as a whole, then, has high reflectance, but much light is still transmitted, allowing growth beneath the canopy.

The canopy and the individual crowns permit moderate light penetration, moderate shade, and moderate sunfleck distribution on the forest floor during the hours around noon. Light intensity inside the forest is often greater than is generally supposed. It is reflected by the leaves and branches, transmitted through the leaves, or it passes between the leaves as sunflecks. Sunflecks may account for the higher light intensity during the 4 to 5 hours around mid-day, while reflected light accounts for the comparatively strong diffuse lighting.

Horizontal stratification is exhibited as a mosaic of light and dark patches due to variation in vegetation density. The dark patches correspond to areas with thick and tall vegetation; the light gaps or "chablis" are formed by trees falling from old age or other causes and include the clearings around the fallen trees.

It is the special structure of the tropical rainforest that is largely responsible for its protective characteristics. The structure influences solar radiation use, rainfall interception, nutrient cycling, and water and energy balance.

The tropical rainforest is known for floral and faunal richness. In Southeast Asia, from 100 to 600 species of higher plants per hectare are found (UNESCO, 1978). This richness may have been brought about by long evolution with minimal climatic change and stable growing conditions. The stable growing conditions may have led to the habitation of sites by species most adapted to them, thereby promoting optimum site utilization and dry matter production.

Species diversity means longer food chains, and more cases of symbiosis, mutualism, parasitism, and commensalism. Plants and animals in the tropical rainforest associate harmoniously by occupying specialized niches, thus creating environmental conditions favorable for one another (Odum, 1971).

Allard (1961), working with agricultural crops, showed that genetic diversity enhances constant productivity. He reported that mixtures of 30 F_4 families exceeded mean yield of the same families grown singly or in simple mixtures. This suggests that single or simple mixtures contain too few genotypes for efficient ecosystem productivity.

The tropical rainforest has properties that enhance ecosystem productivity. These properties (high water-holding capacity, high cation exchange capacity or CEC, stable soil structure, and porosity) are brought about by the high organic matter content. The total biomass of mature tropical rainforests ranges between 200 and 400 tons/ha dry matter. Approximately 75% of the biomass consists of branches and trunks; 15% to 20%, roots; 4% to 6%, leaves; and 1% to 2%, litter (Sanchez, 1976, see table 5.) Nutrients such as nitrogen (N), potassium (K), calcium (Ca), magnesium (Mg), sulfur (S) and phosphorus (P) accumulate in the biomass. Within an 8-year period, the biomass absorbs over 500 kg/ha of N, K, Ca and Mg as well as considerable quantities of S and P (Sanchez, 1976). Of the total biomass, about 40% to 50% is added to the soil as leaves, small branches, and roots. Fortunately, the nutrient accumulation for these is faster than for the remaining vegetation parts (Sanchez, 1976; see table 5). Thus, when these parts fall on the forest floor and decompose, substantial amounts of organic matter and nutrients are transferred to the soil.

Table 5.   Annual nutrient additions from the mature forest growing in an alfisol in Ghana (Sanchez, 1976).

Transfer Pathway	Dry Matter kg/ha	Nutrient (kg/ha)				
		N	P	K	Ca	Mg
Rainwash		12	3.7	220	29	18
Litterfall	10,528	199	7.3	68	206	45
Timber fall	11,200	36	2.9	6	82	8
Root decomposition	2,576	21	1.1	9	15	4
Total	24,304	268	15.0	303	332	75
Annual turnover %	7	13	11	33	12	19

The return of the organic matter from the vegetation to the soil occurs in litterfall, timber fall, and root decomposition. Annual rates of litterfall range from 5.5 tons/ha to 15.3 tons/ha in the tropics; in temperate areas, rates range from 1.0 tons/ha to 8.1 tons/ha (Ewell, cited by Sanchez, 1976). Studies at the Philippines National Botanic Garden (in Siniloan, Laguna) over a thirteen-year period showed that litterfall is continuous throughout the year with peak production in November, December, and June. Annual production rates ranged from 8.27 metric tons/ha to 18.63 metric tons/ha, with a mean annual value of 11.82 metric tons. Cuevas and Sajise (1978) observed annual litterfall is a function mainly of wind speed and is affected strongly by typhoons.

During decomposition, a large proportion of the soluble and dispensable decomposition products (humus) infiltrates the mineral soil. The rate of humus decomposition inside mineral soil is slow, however. The organic matter content of the soil then is high, resulting in large amounts of soil N, high CEC, high soil water holding capacity, and high stability of soil aggregates.

The tropical rainforest exists on a very small nutrient budget and survives only by maintaining an almost closed nutrient cycle. The nutrient cycle has two main storage compartments: the biomass and the top soil, which are connected by several nutrient pathways. The recognized mechanisms for nutrient transfer are litterfall, timber fall, root decomposition, and nutrient excretion from roots and rainwash. A direct nutrient transfer from decomposing litter to roots via fungal hyphae connection or mycorrhizal association has also been reported in the tropical rainforest (Richard, 1973). Nutrients accumulated in the biomass are transferred to the soil as roots and litter on the forest floor decompose. Litter decomposes at about 1.3% per day (Nye, 1961). The decomposition rate in the Philippines' Mt. Makiling forest is 48% per year (Cuevas and Sajise, 1978).

Nutrients and minerals are also released from decaying litter because of the interacting activities of soil biota. Enzymatic breakdown of organic matter by litter and soil organisms, mostly bacteria and fungi, can result in the release of nutrients, which can then be taken up by plants and recycled. Besides, through litter and root decomposition, nutrients are also transferred from vegetation to soil when precipitation passes through the forest canopy. This method is a direct transfer of elements to the nutrient pool: the nutrients are removed from the canopy and transferred to the forest floor by throughfall and stemflow. The transfer by throughfall and stemflow includes not only nutrients leached from vegetation, but also those nutrients washed from the vegetative surface and those contained in the incident precipitation.

The nutrient transfer from vegetation to soil is balanced by the nutrient uptake by the vegetation from the top soil. Thus, although the total stock of nutrients is not large, recycling is rapid and efficient. The great efficiency of the forest mineral cycle is indicated by the low concentration of mineral ions in rivers that drain forest areas (Richards, 1973). Without human intervention, tropical rainforests remain luxuriant because of high soil organic matter and the closed nutrient cycle.

Regenerative Functions

Tropical rainforests have adequate natural self-sustaining and regenerative capacities. Endemic species are capable of reproducing sufficient numbers of sustain biomass. This characteristic can be explained by discussing reproductive phases: phenology, pollination, fruit biology, fruit dispersal, seed germination, seedling establishment, and seedling physiology. (See Ashton, 1969).

Phenology. Mature phase species, including dipterocarps generally begin flowering only after reaching canopy height. Leaf changes and reproductive phases of the forest as a whole are often weakly seasonal, and most individuals of a population flower together. Canopy species do not lose and produce leaves continuously. Leaf abscission is usually associated with drought; in the seasonal tropics, so is bud break (Longman and Jenik, 1973). The young leaf flushes emerge almost simultaneously before leaf fall. Many species flush more than once a year; very few flush branch by branch (Koriba, 1958). Young leaves (of the canopy of the emergent and main canopy species) are pendant and bright anthocyanin-red. The function of the color is unknown. The young leaves' higher temperature may result from the color or from greater stomatal activity. Most canopy leaves live less than 18 months. Flowering is regular, but varies greatly in intensity from year to year. A few understory genera appear to flower continuously. In Dipterocarpaceae, slight rain following a drought has been found to induce flowering. Heavy rain promotes flowering of *Dendrobium crumenatum* (Coster, 1957) *Eusideroxylon zwageri* in Sumatra and Kalimantan was observed to flower irregularly from place to place, usually after dry periods.

No detailed comparative phenological observations exist for forests of different soils. Contrary to expectation, deciduous species increase significantly on well-watered fertile sites (Ashton, n.d.). The shortly deciduous *Firmiana, Pterocymbium, Alstonia scholaris, Octomeles,* spp. are confined to well watered sites in Malesian perhumid climates. Evidence suggests that leaf longevity is shorter on the mesic sites; no differences in flowering phenology are known.

Pollination. Among the emergents, flowers are copious and are usually small and inconspicuous, though scented and frequently having prominent nectaries. The flowers are usually hemaphrodite; less than 10% have morphological modifications favoring outcrossing. Pollinators are small insects (flies, thrips, beetles, polylectric bees). Self-pollination appears to be the rule among the emergents. The South American *Bertholletia excelsa* (Brazil Nut) is pollinated by large insects such as Bombus (a bumble bee), that have sufficient strength to raise the androphore covering the stigma of the hermaphrodite flower. The Brazil nut flowers only at certain times of the year. The pollinator (Polylectric bee) must depend on other species for food when the tree is not flowering.

In the understory, over 40% of the trees are either dioecious or otherwise morphologically adapted to out-cross. The main canopy species are intermediate in this respect; others are similar to the emergents. Floral morphology and size is most variable. Many families have large flowers, these are often few or they may smell like offal or rotting fruit. This suggests pollinators are more diverse and vectors more specialized (moths, beetles, flies, and other insects have also been identified) and that panmixis prevails.

The floral biology of woody climbers conforms to that of the emergent trees, but that of epiphytes and parasites is conspicuously different. The epiphyte's and parasite's bright and highly diversified large flowers, which are continuously produced, suggest a high vector specificity. The conspicuous and colorful labellum in many Amazon orchids is adapted for alighting insects. A very similar structure is exhibited in the terrestrial genus *Costus*.

Among canopy species, a few of the deciduous ones growing in mesic soils (soils with mean temperature between 8° and 15^OC) bear brilliant flowers during the deciduous phase and are often bird — (for *Firmiana*) and bee — (for *Wrightia*) pollinated. The largely wind-pollinated conifers and Casuarinaceae are gregarious and confined to skeletal and podsolized lowland soils.

Fruit biology. Emergent species may produce large quantities of fruit at frequent intervals, but the fruit is produced by a minute proportion of the original flowers. This has been variously attributed to failed pollination, rain during flowering, intrinsic physiological reasons, and predation. Understory species frequently produce a few, very large fruit (e.g., *Pangium edule*). A very few main canopy or emergent tree and climber genera produce light-winged fruits or seeds (e.g., *Kompassia, Engelhardtia* in Southeast Asia). Heavy fruiting occurs each year in the forest as a whole; occasionally it includes taxa that fruit at longer intervals. Related dipterocarps with staggered flowering times appear to have differential rates of fruit development leading to synchronous fruiting (Holmes, 1958; Ashton, 1964). Such periodicity must sustain a low carrying capacity of fruit predators, however generalized, and thus increase the proportion of seed that survives. At the genetic level, biochemical specialization may maintain high predator-host specifity and further lower predator carrying capacity (Janzen, 1970).

Dispersal. Many mature phase species posses no known means of fruit disperal. The fruit simply falls beneath the perimeter of the parent crown. For instance, the winged fruit of the Southeast Asian dipterocarps gyrates into the main canopy and then falls randomly but vertically below. The pattern can lead to aggregated clumping of seedlings and repetition of floristic composition. By favoring temporary inbreeding among a few individuals sharing parentage, such clumping may also be conducive to rapid ecotypic diversification (Ashton, 1964). In some species, however, differential predation has been found to favor survival of the seedlings farthest from the parent (Janzen, 1970).

Many propagules are dispersed by animals — particularly by monkeys, other arboreal mammals, and birds. Such fruits are usually conspicuously colored when ripe — most often red, but also black, white, or blue. They mostly have fleshy pericarps, mesocarps or endocarps, or have arils or arillodes. Those with arils or arillodes often have a dry protective pericarp which dehisces to reveal a colorful juicy interior (Comer, 1952).

The seeds are frequently toxic and are either rejected or pass unharmed through the vector. Such wider dispersal may be associated with lower rates of speciation (Ashton, 1969), and with changing local patterns of forest floral composition.

In general, the seeds of mature trees have no dormancy period. Dipterocarp seeds are viable for only four weeks when dessicated and cooled. The African *Terminalia ivorensis* and *Triplochiton scleroxylon* seeds lose viability if stored. The dormancy of 350 woody species from various habitats in the Ivory Coast ranged from less than two weeks to three years (Tang, 1971; Tang and Tamari, 1973).

Germination and establishment. Conditions for germination and establishment of mature phase tree species are nearly always highly specific. This prevents the return of such species after forest clearance (Gomez-Pompa et al., 1972). Inter-specific variation in moisture require-ments for successful germination is considerable. This may help explain both intrinsic spatial pattern and extrinsic spatial variation in forest flora. The West African *Aucoumea klaneana* seeds can be stored for several years if both temperature and humidity are lowered. But Bunck (in De La Mensbruge, 1966) found that seeds became inviable if relative humidity fluctuated more than 8%, although The Malayan ridge-top dominant *Shorea curtisii* required moister conditions to germinate than did *S. Ieprosula* and *S. parvifolia* growing on hillsides (Burgess, 1968). The *S. curtisii* was more tolerant of low light following establishment and was, thus, well-adapted to regenerate under the dense ground cover of the stemless palm, *Euqeissona tristis*, which abounds in the same habitat and which provides moisture and light conditions in which *S. curtisii* alone among its congeners can survive and become established.

For successful establishment, mature phase species on submesic sites require constantly moist conditions. Low light intensity is not limiting, and shade is needed to prevent the drying-out of the seedlings.

Predation is a major problem. Studies carried out by Synnot (1973) in the Budongo forests of Uganda examined the regeneration of *Entandrophragma utila* under a mature forest canopy of *Cynometra alexandri*, *Entandrophragma ssp.*, and *Khaya ssp.* Under simulated natural seed fall, about 40% of the seeds were eaten by rodents before and immediately after germination. And nearly 30% were killed within two years by antelope browsing. Other losses were from seed rot, insect and fungal attacks, and drought. Chan (unpub.) recorded 83% abortion in *Shorea ovalis* Korth., possibly because pollination failed. Over 90% of the sur-vivors were killed by a single insect predator before they fell, and 16% of these were killed by three others on the ground. Predation decreased with distance from the tree. On infertile and particularly xeric sites in Southwest Asia, especially ridge crests and podzols, mature species with dry, dehiscent, small-seeded, and wind-dispersed propagules (e.g., *Cratoxylon, Austrbuxus, Ctenolophon, Allantospermum, Axinandra, Agathis,* and *Casuarina*) have increased. This might be expected for environments that are frequently windy and adequately lighted on the forest floor — allowing for the rapid initiation of photosynthesis.

It is now fairly well established (Edmiston in Odum and Pigeon, 1970) that many rain forest families are mycorrhizal (e.g., Dipterocarpaceae, some Tiliaceae, Sterculiacease with ectotrophic, and Myrtaceae with vesicular-arbuscular mycorrhizae). Apparently, mycorrhizae become more wide-spread and physiologically important in infertile soils and determine species composition at the stage of seedling establishment.

Physiology of Regenerating Trees. Because of their generally plentiful seed resources and periodic heavy fruit production, perhumid rain forests are characterized by a dense, fluctuating, and often clumped distribution of seedlings and saplings on the ground. This silviculturally important fact, demonstrated in the West Malaysian dipterocarp forest (Wyatt-Smith, 1963) and in the mixed lowland forest in Irian (Loekito and Hardjono, 1965), has served as a basis for the practice of uniform cutting in logging.

The Malayan uniform system of regenerating lowland dipterocarp forest relies on such an assumption. Nevertheless, the vast majority of seedlings die in the first few years following fruit fall. Survivors (recruits) may be those that survive initial root competition. Increases in light intensity through partial canopy openings are known to lead to rapid height gains in seedlings, although many turn yellow and stagnate if complete canopy clearance is undertaken. The effects of light intensity on the growth of seedlings of some tropical forest tree species were investigated by Wadsworth and Lawton (1968). They concluded that *Khaya grandifoliola* was the only species with a significant growth rate at 1% daylight and that *Pinus caribaea* was the least adaptable to shade.

A suppression period has been recognized in the relatively young phase of Philippine species (Brown and Matthews, 1914). The crowns of understory trees and saplings are diffuse, taller than broad, and thus allow a high leaf area index. The leaves are mostly horizontal, thin, and have long narrow acumina and petioles. Leaf change is gradual and leaves apparently live longer in the crown than in the canopy. Continuous shoot extension is infrequent, but more common than in the canopy.

After the sapling stage, the growing trees gradually cease to respond to changes in the canopy above and around their crowns. This has been observed in the Solomon Islands (Whitmore, 1974) and explains why successful selective feeling systems and improvement thinnings have never been demonstrated.

It is clear that the tropical rainforest is endowed with basic structural and functional attributes reflected in its productive and regenerative properties. Management of the tropical rainforest and the determination of the impacts of human activities must necessarily take these attributes into consideration.

HUMAN ACTIVITIES IN THE TROPICAL RAINFOREST

When ecologists call the tropical rainforest stable, they do not mean static, but continually adjusting to maintain essentially the same overall structure, constitution, and function. The system's stability depends on two provisions: 1) that the forces acting on the system not exceed certain threshold values, and 2) that the environmental conditions be relatively constant. The system remains stable, then, if left essentially undisturbed. Such disturbances as those generated by man will result in succession that may resemble the original system. Deforestation and excessive removal of biomass have brought about ecological and socio-cultural disruptions.

Poorly planned forest clearance alters hydrological cycles, reduces absorptive capability of upland catchment areas, and increases the magnitude of downstream flooding following heavy storms. Water availability in dry periods may also be reduced, although this effect is less well-documented (Hamilton, 1981). Much upland agriculture seriously erodes the soil, which increases the rate of siltation of dams and lowland irrigation works (table 6). This destroys animal habitats; many little-known wild species are becoming extinct and their generic resources lost (Myers, 1979). The availability of protein for human consumption is also lowered thereby.

Large-scale settler movement into deforested upland areas traditionally inhabited by cultural minorities may cause conflicts. Examples include conflicts between ranchers and Mangyans (a Filipino cultural community group in Mindoro) and, over the past fifty years, those between ethnic Vietnamese moving into forested uplands and the Montagnard tribes who traditionally used them (Rambo, 1984). Even where expansion has been peaceful, as in Indonesian transmigration projects and the FELDA schemes in Peninsular Malaysia, social costs have often been high. These include heavy settlers dependency on the government (Rokiah, 1978), increased vulnerability to market fluctuations, and serious lowering of health and nutritional status (Meade, 1976).

Table 6.   Effects of forests covers on soil loss rates (Kellman, 1969; Veracion and Lopez, 1979).

Cover	Erosion Rate (T/ha/yr)
A. (Kellman 1969)	
Primary Forest	0.09
Softwod Fallow	0.13
Imperata	0.18
New Rice Kaingin	0.38
12-Year Old Kaingin	27.60
B. (Veracion and Lopez 1979)	
Pineapple	308.0[1]
Coffee	318.0
Tiger grass	396.00
Castor bean	360.0
Banana	414.0
Banana/Coffee/Pineapple Inter	421.0
Undisturbed Area	251

[1]Converted from cm/yr by assuming soil bulk density = 1.

By far the most serious disturbance of the tropical rainforest ecosystem has been by man. The degree of disturbance varies widely among the three great world tropical rainforests. The American forests are relatively undisturbed, while in Asia and Africa, man's influence is very conspicuous.

Shifting Cultivation

Man has effected the most destruction through shifting cultivation and shortened fallow periods (often due to population pressure). Shifting cultivation is practiced by nearly all native peoples of the tropics. Exceptions include a few regions in tropical India and in the lowlands of Java where intensive, sedentary cultivation systems are practiced. The world forest area destroyed annually by shifting cultivation is estimated at 30 to 50 million ha (UNESCO/UNEP/FAO, 1978). The 1978 Philippine forest destruction was estimated at 65,958 ha, with 54% attributed to shifting cultivation (BFD, 1978).

Shifting cultivation entails the cutting and burning of trees. Since humus is rapidly destroyed by exposure to the sun and since the ground generally receives little cultivation or manure, the soil becomes impoverished and infertile. The inherent fertility of the soil and the extent of leaching, erosion, and weed problems may dictate that the plot be abandoned. Then, if fire is prevented, second growth develops (with fire, grasslands persist as a fire disclimax).

Plantation Agriculture and Forestry

Included here are large artificial forest plantations and commercial vegetable or horticulture crops. When a plantation becomes unprofitable because of soil exhaustion, erosion, plant diseases, or economics, it is abandoned and secondary succession begins. Such long lived crops as coffee or rubber may temporarily survive in competition with the second growth vegetation, but succession will eventually proceed as in an abandoned shifting cultivation plot.

Burning

Shifting agriculture uses burning and, at times, so does grazing. Fire is a dominant factor in the development of the terrestrial ecosystem in the Philippines. Burning is a traditional pasture and grassland management practice and is used in preparing hunting grounds. Other reasons for fires are man's carelessness and lack of social responsibility, man's intentional burning, and nature. Burning changes the physical and chemical properties of the soil by releasing nitrogen into the atmosphere and into the soil water. Eventually, part of the latter will enter the drainage system and is lost by overland flow. Fires at the beginning of the dry season generally have a lower impact than those occurring toward the end (UHP, 1980).

Logging

The upper canopy is opened by logging, and this heavily influences the vegetation and soil. Primary rainforest may thereby become depleted of vegetation. Gaps left by the removal of timber are colonized by secondary forest species. The community thus comes to consist of irregular patches of primary and secondary forest. Such depleted forests can be large, such as the mahogany-producing areas of Nigeria (Richards, 1973). Sometimes depletion causes a vigorous outburst of growth in the undergrowth already present without much invasion of outside species. The treatment the area receives after logging and before secondary succession may vary greatly. It will depend on size of the cleared areas, distance from the uncleared areas, the local climate, topography, and soil, and the degree of humus destruction and soil erosion that have taken place.

Many forms of human activities and their disruptive effects on the tropical rainforest have clearly impaired its hydrology, nutrient cycling, biotic stability, and productivity. These biological and physical impacts, in turn, have been responsible for the poor socio-economic circumstances of the human communities in such upland areas. The development of these marginal areas is currently a major concern in many tropical countries.

THE UPLANDS

In the 71 years since Brown and Matthews (1914), all but 30% of the Philippine tropical rainforest has been destroyed. That 30% is concentrated in a few provinces (Bonita, 1977). The forest's transformation into degraded, unproductive, and unstable areas is a result of ecological disruptions. The upland tropical rainforests are important in three ways. First, hydrologically, especially in watersheds: the transformation of the tropical rainforest into grasslands and intensively cultivated areas has produced changes in surface runoff and ground water flow. These changes are manifested by more frequent floods and drought. Second, physically and technically: degraded areas lack vegetation cover and have highly eroded soils. Third, socio-economically: resource utilization often exceeds carrying capacity because of the marginal condition of the area, improper management, overpopulation, or a combination of causes.

The Pantabangan Watershed illustrates the seriousness of deforestation. It totals 82,900 ha and has a forest area of 36,915 ha covering 44% of the watershed. Grasslands dominated by *Imperata* and *Themeda* comprise about 43%, and crop land makes up about 12% of the area. Soil erosion in the watershed is a serious problem (NIA, 1978). According to studies conducted prior to the construction of the Pantabangan dam, sediment yield for the watershed was estimated at 2,020 tons/km^2/year.

After 20 years, the Agno River Irrigation System (ARIS) now operates at only 25% capacity because of the silting of the intake and canal system. ARIS can irrigate only 9,546 ha instead of the original 17,500 ha intended. In 1976, losses from desilting costs and crop failure amounted to $4.4M (UNEP, 1977; Coloma, 1984). Siltation is largely due to mine tailings discharged by nine mining companies in the watershed.

Of the Magat Watershed, with an area of 414,300 ha, 83% is affected by severe to excessive erosion. The sediment yield at the dam site was estimated at 2,050 tons/km^2/year during the planning of the Magat Dam (Coloma, 1984).

Upland areas are estimated to cover 31% of the Philippines' area. Rainfed areas range from 54% to 95% of different regions. The highest proportions occur in Western and Eastern Visayas, Northern and

Western Mindanao, Ilocos, Southern Tagalog and Bicol (Librero, 1978). These areas are planted to export crops such as coconut, abaca, fruit trees, vegetables, coffee, cacao, as well as to upland subsistence crops such as rice and corn. Approximately 5 million ha of these upland areas is covered by native grass species such as cogon *(Imperata cylindrica)* and bagokbok *(Themeda triandra)*. The target for upland development is comprised of 5M ha of denuded forest land, 1.65M ha of non-irrigated areas, and 3.0M ha of coconut lands and other lands that can be inter cropped. This may total 10 million ha, or 1/3 of the country's area.

The low productivity and instability of the upland resource base is reflected not only in the poor economic status of the approximately five million uplanders, but also in the lives of adjacent lowland communities affected by floods, drought, and siltation.

Gwyer (1977) reported that upland settlers or the "informal agricultural sector" of Philippine society have an average income of less than P3,000 per annum. He also said that in the next 21 years there will be an additional three million Filipino job-seekers. If even one third can be absorbed by the non-agricultural sector, two million remain jobless. Agricultural areas must be expanded since agricultural intensification can only absorb two percent of labor. Arable lands suitable for agriculture have been estimated at one million ha. However, since one million ha of the lowlands will be lost to urban and industrial settlement in the next 21 years (Gwyer, 1977), it is certain that attempts will be made to convert more uplands to the needed agricultural land.

The energy crisis will also exert pressures on upland resources. Firewood extraction will increase. Revilla (1984) estimates that our fuelwood requirements for household and industrial use will be 63 m^3 to 106 m^3 in the years 1990 and 2,000.

Recently, Noorgard (1984) proposed a conceptual plan for a development program that takes into account the relationship between the natural (biophysical) and the social (human) systems in identifying and designing development projects. Norgaard and Dixon (1984) later identified the following goals: 1) sustain system productivity and diversity, 2) start small and try many things, 3) learn from experience, 4) remain flexible, 5) reduce vulnerability, and 6) avoid big plans.

The human community determines upland technologies that are viable and culturally acceptable, the manner of introducing them, and the manner of acquiring, distributing, and managing material inputs for them. Experiences in Antique, Pantabangan, Bicol, and Makiling, Philippines have shown that while it is easy to conceptualize an ideal upland technology, the socio-cultural environment will determine its rate of adoption.

Use of the resource base is influenced by the socio-cultural environment. In Buhi, Camarines Sur, the Agta people plant sweet potato and upland crops for subsistence. The people of Puting Lupa, Mount Makiling, Laguna, on the other hand plant garlic and ginger for cash because they have access to a market. Land tenure also affects upland resource use.

A balance of subsistence and cash production from upland farms is needed to meet the needs of human communities. Nguu and Corpuz (1979) have shown that a diversity of upland crops (combining annuals and perennials) ensures efficient labor distribution and stable monthly cash income for upland farmers.

Ultimately, sustained growth of upland communities requires the proper channelling of materials and information from the different agencies of the government. This will have to be synchronized with the bio-physical setting. For example, the flow of planting materials from BFD and MAF to upland communities must take precipitation patterns into account. Animals from BAI must come when feeds are available.

Upland soils are generally acidic and low in nutrients, especially nitrogen and phosphorus. These are important constraints in vegetation establishment and crop production. Undesirable characteristics of upland soils also impair hydrological characteristics. Marginal upland areas have excess water during the wet season and little or no water during the dry season. Since water is needed for the household as well as for crop and animal production, its lack is a very important factor in upland development, as are water impoundment and conservation.

Biological constraints are imposed by the flora and fauna in the upland ecosystem. These are manifested in plant competition and the effects of vertebrate and invertebrate pests. While moisture is the limiting factor for crops in grass-covered hilly lands during the dry season, low light intensity almost limits growth during the wet season. A thick growth of grass two to three m tall can substantially shade shorter or newly established crops. Some important crops such as the legumes are not usually shade-tolerant.

The method used to relive light competition by brasses depends on cost and effectiveness. For example, if cogon is cut, it produces more tillers or young shoots (Sajise, 1972); cutting it is laborious and expensive. More effective would be lodging the grass using a log or similar heavy object for a roller.

FOREST CONSERVATION AND UPLAND DEVELOPMENT

Conservation

Three broad strategies can be identified: 1) conservation of natural forest ecosystems, 2) improved management of modified natural forest systems, and 3) proper management of plantation forestry.

Since there is still much to learn about the tropical rainforest ecosystem, some representative forest ecosystems must be preserved. These can be as national parks, nature reserves, wildlife refuges, and protected forests. Such areas would produce energy, biomass, and effect gas exchange. Some forests could be preserved untouched, and others could be managed to sustain yield.

As of 1981, there were 59 Philippine national parks. These cover 381,549 ha. Most of them are in a poor state owing to lack of maintenance, shifting cultivation, and other forms of deforestation.

Apart from the preserved areas, most of the natural forests will eventually be harvested as timber. With harvesting, natural forest ecosystems will change. The efficiency of the modified forests will depend on management practices adopted.

Harvesting timber removes much of the biomass. Nutrients tied up in the wood are removed, the remaining nutrients in the soil are leached if too much to the forest cover is removed. Growth rates may slow because of the depletion of nutrients in regenerating sites. Forest managers should not limit logging damage studies to seedlings and the standing regenerating crop, but should consider the environment as a whole.

Regeneration depends upon availability of seedlings and pole-sized trees and their growth after logging. For most dipterocarps, flowering and seed production are not a regular, yearly events. As discussed, dipterocarps tend toward gregarious flowering and mass seeding. If logging is carried out after a good seed year, there may be plenty of seedlings. If logging is carried out a few years after a good seed year, the seedling crop may not exist.

Enrichment planting, a current practice in areas poor in seedlings, involves the collection of seeds wildlings, establishment of nurseries, and subsequent planting. The recent trend toward utilization of less-known timber species often means greater damage and the removal of even more biomass. While plantation forestry is an option, it is not always efficient in the tropics.

Upland Development

A main objective of upland development is to improve the conditions of farmers and to enhance the protective character of the upland ecosystem. Needed are productive and ecologically and socially based farming system. The variability of upland biophysical and sociocultural situations throughout the country requires site-specific analysis. Aguilar (1982) described a situation in Buhi, Camarines Sur in which seedlings raised at a project nursery were not acceptable to farmers because the materials available did not meet their needs.

Farmers in Villarica, Pantabangan consider typhoons in choosing their crops. We mistakenly convinced a farmer to plant pole beans as source of cash instead of gabi (Colocasia). When he was about to harvest, a typhoon damaged his crop, leaving him worse off than his neighbors who had planted gabi.

The World Neighbors Project at Guba, Cebu, was successful because its strategies were based on a correct analysis of the needs and conditions of the farmers. Soil conservation work was organized around the family-kinship unit, the "alayon", and was hooked to vegetable and crop production, which increased farmer income in a relatively short time. Land tenure security is satisfactory with the cooperation of BFD. Needed inputs were available to the farmers.

The Antique Upland Development Program was successful in a cattle-fattening program for several reasons. Livestock constitute a major source of income to the farmers. Credit support was provided by the project. Extension support in the form of farmers' training was supplied. The project staff was credible to the community. Livestock could be raised on a cut-and-carry basis, thereby circumventing land tenure problems Household labor was available. There is a good market for livestock. Adapted forage species were available.

These experiences point out to the need for a quick but reliable upland community appraisal (Fujisaka and Capistrano, 1984; and this volume) to identify the following: key determinants of the present upland farming system practices, viable entry point or intervention strategies for an upland development project, and an appropriate administrative system for the project.

Alvarez (1983) pointed out other general lesssons relevant to the development of upland farming systems. Information provided by respondents is often questionable, and upland farmers can be conservative and suspicious of outsiders. The University of the Philippines at Los Baños Program on Environmental Science and Management (PESAM) is assessing a new step-wise community appraisal technique.

CONCLUSION

The uplands of the Philippines and many other parts of Southeast Asia are rapidly changing. The former stable, diverse, and productive tropical rainforest vegetation cover has been transformed into fragile, less productive, hydrologically unstable shifting cultivation areas, vast monocultures of plantation forest or agricultural crop, degraded pastures, and inadequately stocked secondary forests. These changes have been brought about by short-sighted timber harvesting, agricultural expansion, and firewood collection and have been exacerbated by unfavorable factors such as agricultural stagnation, land tenure insecurity, rising unemployment, rapid population growth, and the government's incapacity to regulate private enterprises to protect the public interest.

Regenerating the uplands must involve both the biophysical and the social components of the ecosystem and must recognize the high diversity and variability of conditions in the uplands. Specifically, upland development strategies should include: 1) conservation of some natural forest ecosystems, 2) proper management of secondary forests, 3) proper management of plantation forest and agriculture, 4) and upland farming systems development. We suggest adopting the watershed as a planning unit. An appropriate institutional administrative and support system should also be considered. Most important, a stable, productive and socially acceptable upland development program can only be attained if there is adequate community participation.

REFERENCES   CITED

Alard, R.W. 1961. Relationship between genetic diversity and consistency of performance in different environments. Crop Science 127 : 133.

Alvarez, J.B., Jr. 1983. Forest resources management and social forestry in the Philippine. Paper for Seminar, Management of Forest Resources: Issues of Forest Policy in the Developing Countries in Asia. U.P. Los Banos, Philippines, 11-15 July 1983.

Ashton, P. 1964. Ecological studies in mixed dipterocarp forests of Brunei State. Oxford Forestry Memoirs 25.

Ashton, P. 1969. Specification among tropical forest trees: some deductions in the light of recent evidence. Biol. J. Linn. Soc. 1 :155-196.

Bonita, M.L. 1977. Philippine forest resources: Their present critical state and their conservation through rational utilization. Guillermo Ponce Professorial Chair lecture. UPLBCF, College, Laguna.

Borsuk, R. 1977. Illegal loggers wreck havoc on Thai forest. Kuala Lumpur, Malaysia: Malaysian Business Times, 22 June: 4.

Brown, W.H. and D.M. Matthews. 1914. Philippine dispterocarp forest. Phil. J. Sci. Sec. A: 413-61.

Bureau of Forest Development. 1978. Philippine Forest Statistics. 1981. Philippine Forestry Statistics.

Burgess, P.F. 1968. An ecological study on the hill forests of the Malay Peninsula with special reference to the regeneration of tree species of economic importance: an assessment after nine months study of the project. Malayan Forester 31 (4): 314-325.

Coloma, A.G. 1984. Management of sedimentation problems in irrigation systems in the Philippines. East-West Environment and Policy Institute Paper, July, 1984 (Mimeographed).

Corner, E.H.J. 1952. Wayside Trees of Malaya. Second Edition. Singapore

Coster, C. 1957. Wortelstiidian in the Tropen. Tectona 25:828-872.

Cuevas, V.C. and P.E. Sajise. 1978. Litterfall and leaf litter decomposition in the Philippine secondary forest. Kalikasan, Philipp. J. Biol 7(2): 99-109.

De La Mensbruge, C. 1966. La germination et les plantules des essences arbores de la foret dense humide de la cote d/Ivoire. CTFT. 389 p.

Dickerson, R.E. 1926. Distribution of the Earth: Land Reform and Sustainable Development. Worldwatch Institute, Washington, D.C.

Eckholm. E. 1979. The dispossessed of the Earth: Land Reform and Sustainable Development. Worldwatch Institute, Washington, D.C.

Edmiston, J. 1970. Survey of mycorrhiza and nodules in the El Verde forest. In: Odum, H.T., Pigeon R.F. (eds.) A Tropical Rainforest. A Study of Irradiation and Ecology at El Verde, Puerto Rico. Division of Technical Information, U.S. Atomic Energy Commission.

Food and Agriculture Organization (FAO). 1980. FAO Production yearbook, Vol. 33. Rome: Food and Agriculture Organization of the United Nations.

FAO/UNEP. 1981. Tropical Forest Resources Assessment Project: Forest Resources of Tropical Asia. Rome.

Gomez-Pompa, A., C. Vasquez-Yanes, and S. Guevarra, 1972. The tropical rainforest: a non-renewable resource. Science 177: 762-756.

Gwyer, G. 1977. Agricultural employment and farm incomes in relation to land classes: A regional analysis. Proceedings of the International Workshop on Hilly Land Development, Albay, Philippines.

Hamilton, L. 1981. Upper Watershed Land Use: A research prospectus. Honolulu. East-West Environment and Policy Institute (Typescript).

Holmes, C.H. 1958. The broad pattern of climate and vegetational distribution in Ceylon. In: Proc. Kandy Symposium, Paris, UNESCO, 226 p.

Hutterer, K. L. 1984. Ecology and Evolution of Agriculture in Southeast Asia. In: An Introduction to Human Ecology Research on Agricultural Systems in Southeast Asia, A. Terry Rambo and P.E. Sajise (Eds.) (In Press).

Janzen, D. 1970. Herbivores and the number of tree species in tropical forests. American Naturalist 104 (940): 501-528.

Kellman, M.C. 1969. Some environmental components of shifting cultivation in upland Mindanao. J. Trop. Geog. 28.

Kria T. and H. Ogawa. 1969. Assessment of Primary Production in tropical and equatorial forests. In: Productivity of Forest Ecosystems (Proc. Brussels Symposium): 309-321. Paris: UNESCO.

Koriba, K. 1958. On the periodicity of tree growth in the tropics with reference to the mode of branching, the leaf fall and the formation of the resting bud. Gardens Bull. Sing. 17(1): 11-81.

Lewin R. 1984. Parks: How big is big enough? Science 225 (4662): 611-612.

Librero, A. 1978. Socio-economic constraints in technology transfer in rainfed crop production. Proceedings National Workshop on Development and Management of Rainfed Crop Production (Mimeographed).

Leokito and Hadjono. 1965. Survey regenerasi hutan bekas tebangan di Orasbari Utama. Rimba Indonesia. 10: 265-274.

Longman, K.A. and J. Jenik, 1974. Tropical Forest and its Environment. London: Longman Grp. Ltd.

Mende, M. 1976. Land development and human health in Malaysia. Annals Association American Geographers 66(3): 428-439.

Myers, N. 1979. The Sinking Ark: A New Look at the Problem of Disappearing Species. New York: Pergamon.

1984. The Tronsy Source: Tropical Forests and Our Future. W.W. Norton and Co. N.Y. & London.

1980. Conversion of Tropical Moist Forest. Washington, D.C.: National Academy of Sciences.

Nguu, N.V. and E.B. Corpuz. 1979. Resources, production activities and financial status of kaingin farm. Proceedings of the Vth International Symposium of Tropical Ecology, April 16-21, 1979, Kuala Lumpur, Malaysia.

Norgaard, R.B. 1984. Coevolutionary development potential. Land Economics 60(2): 160-173.

Norgaard, R.B. and J.A. Dixon. 1984. Project design and evaluation using economic and coevolutionary criteria. East-West Center Environment and Policy Institute (Mimeographed).

Nye, P. H. 1961. Organic matter and nutrient cycles under a moist tropical forest. Plant and Soil 13: 333-346.

Odum, E.P. 1971. Fundamentals of Ecology. Philadelphia: W.B. Sanders and Co. 574 p.

Rambo, T.A. 1978, No Free Lunch: Fire as an energy input in swidden agriculture (Mimeographed).

1984. Human ecology research on tropical agroecosystems in Southeast Asia. In: An Introduction to Human Ecology Research in Southeast Asia, Rambo, A.T. and P.E. Sajise (eds.). (In Press).

Reyes, M.R. 1983. Conserving the Philippine Dipterocarp forest. Canopy 9(2): 14-15.

Richards, P.W. 1973. The tropical rainforest. Scientific American 229 (6): 59-67.

Rokiah Binte Talib. 1978. Two models of land development: Malaysian experience: Paper presented at the Xth International Congress of Anthropological and Ethnological Science, Post Plenary Session, Hyderabad, India, 19-21 December. (Mimeographed).

Sajise, P.E. 1972. Evaluation of cogon (*Imperata cylindrica* (L. Beauv.) as a seral stage in Philippine vegetational succession. II. Auterological studies on cogon. Ph.D. Thesis, Cornell University, Ithaca, N.Y.

Salati, E. and P.B. Vose. 1984. Amazon Basin: A system in equilibrium. Science 225 (4658): 129-137.

Sanchez, P.A. 1976. Properties and Management of Soils in the Tropics. New York: John Wiley and Sons.

_______ 1979. Soil fertility and conservation considerations for agroforestry systems in the humid tropics of Latin America, pages 79-124 in H.Q. Monzi and P.A. Huzley, eds. Soil Research in Agroforestry. International Council for Research in Agroforestry, ICRAF: Nairobi, Kenya.

Sedjo, R. 1983. The situation of the worlds tropical forest. Paper presented in a seminar on Management of Forest Resources: Issues of Forest Policy in Asia sponsored by the Agricultural Development Council and the Japan Center for International Exchange, Los Baños, Philippines.

Segura-De los Angeles, M. 1980. Agroforestry development project: Economic and social impact analysis. ESIA-WID, Publication, Philippine Center for Economic Development Diliman, Quezon City.

Synnott, T.J. 1973. Seed problems. In: IUFRO International Symposium on Seed Processing (Bergen).

Tang, H.T. 1971. Preliminary tests on the storage and collection of some *Shorea spp. eeds*. Malayan Forester 34: 84-92.

Tang, H.T. and C. Tamari. 1973. Seed description and storage tests of some dipterocarps. Malayan Forester 36: 38-53.

United Nations Environment Program. Report to the Government of the Philippines on Mine Tailings Disposal in the Baguio Area-Agno and Bued River Systems. October 26, 1977 (Mimeographed).

UNESCO. 1978. Tropical rainforest ecosystems. A state of knowledge report prepared by UNESCO/UNEP/FAO. Natural Resources Research XIV. Paris 683 p.

Upland Hydroecology Program Annual Report. 1980. U.P. at Los Baños, College, Laguna, Philippines.

Veneracion, V.R., A.B. Lopez. 1979. Comparative effects of different vegetative covers on soil erosion rates in roads banks in Benguet. Annual Report, FORI, College, Laguna.

Wadsworth, R.M. and J.R.S. Lawton. 1968. The effects of light intensity on the growth of seedlings of some tropical tree species. J. West Africa Sci. Ass. 13 (2): 132-145.

Whitmore, T.C. 1974. Change with time and the role of cyclones in tropical rainforest on Kolombangara, Solomon Islands. Oxford Commonwealth Forestry Institute, Paper No. 46. 92 p.

Witkamp, W. and A. Crossley. 1966. Decomposition of litter in relation to environmental, microflora and microbial respiration. Ecology 47: 194-201.

Wyatt Smith, J. 1963. Manual of Malayan silviculture for inland forests. Kuala Lumpur, Malayan Forestry Res. 23. 400 p.

3
MOUNTAIN PEOPLE IN THE PHILIPPINES: ETHNOGRAPHIC CONTRIBUTIONS TO UPLAND DEVELOPMENT

Susan D. Russell

A crisis of deforestation and consequent soil erosion looms in many Southeast Asian upland ecosystems. Blame for this problem has been attributed to both slash and burn agriculturalists or timber extraction industries. To the extent that the actions of uplanders rather than legal and illegal logging are seen as being responsible for diminishing forest reserves, a growing demand exists among policy-makers and foresters for a more holistic, interdisciplinary approach to appropriate intervention strategies. Anthropologists are now frequently incorporated into technical programs and are given a growing recognition that economic change cannot be understood adequately in isolation from the social institutional framework. Recent efforts in social forestry, for example, suggest that anthropological perspectives have direct practical applicability to development programs.

An unfortunate truism in the region is that public interest in the uplands has arisen less from a concern to improve the livelihood of uplanders, than from a realization that lowland environmental and agricultural problems are directly related to upland deforestation and ecosystem degradation. Regional governments and development agencies are interested in the practical potential of involving upland communities in the development process by granting them a stake in resource management and preservation programs. Current program incentives for uplanders include material inputs for agroforestry and conservation projects, rather than the primarily punitive measures of unsuccessful past approaches. Other aspects of this new social forestry strategy in principle involve an institutional framework for increasing uplanders' access to improved technology, an emphasis on community participation in project management, and the provision of more secure claims to land (Aguilar, 1982; Bennagen, 1983).

As this paper illustrates, **anthropological perspectives** provide an understanding of the socioeconomic milieu of Philippine upland populations. Analysis is comparative, and generalizations are made concerning the social organization and forms of economic activity of upland Filipinos. Forming generalizations beyond typologies is hampered, however, by the great diversity among upland indigenous ethnolinguistic groups. Besides variations in subsistence strategies, there are major differences in cultural heritage, history, and the degree to which any single group has been incorporated into the regional or national society.

A second difficulty in generalizing is presented by the ethnographic literature. The times of field work range throughout this century. "Ethnography' refers to the analytic description of the ideas or cultural knowledge that produce group-specific behavioral patterns and social institutions. Early anthropologists working in the region were primarily concerned with recording ritual, kinship, economic, and political beliefs and practices that distinguished upland groups from each other. As a result, information often minimized variations within specific populations and gave an impression of unchanging, traditional societies. While upland Filipinos have experienced dynamic changes for centuries, increasingly disruptive impacts on livelihood strategies have been intensifying in the last few decades. While recent literature addresses factors currently transforming upland societies, many groups are represented solely by descriptive ethnographic accounts from the early half of this century. In numerous cases, the actual fieldwork done was limited, or was devoted primarily to examining specific limited topics. For these reasons, more attention is given in this paper to recent works that have built on earlier ethnographic accounts.

Due to the incomplete record of change for most upland groups, the examples selected were those that best illustrate certain themes in a dynamic content, or that represent different kinds of social change or adaptation. Specific cases should not be construed as reflecting the exact responses of upland groups in all places at all times. People are not passive subjects reacting in the same way to outside forces; instead, individuals and groups act and decide according to culturally meaningful systems of belief and culturally distinct standards of evaluation. A major contribution of ethnography to development is in the illumination of how belief systems generate and influence behavioral patterns in interaction with specific physical and social environments.

This paper first provides a brief sketch of the distribution and diversity among indigenous ethno-linguistic groups in the Philippine uplands. A second section outlines upland indigenous subsistence strategies. A third section discusses forms of social organization characteristic of upland populations, and a fourth part reviews on-going transitions to the market economy. The final section summarizes and assesses the implications of the ethnographic literature for the future of upland development.

ETHNO-LINGUISTIC DIVERSITY

Over 80 different languages are indigenous to the Philippines, and more than 70 of these are considered to be minor, owing to their comparatively smaller number of native speakers (Reid, 1971). Most of these minor languages originated in the uplands, and the complex diversity of subcultures that such variation implies makes comprehensive taxonomy of upland groups somewhat arbitrary (map 2). Islamic influences in Mindanao and parts of the Visayas varied but seem to have been confined largely to trade and tribute payments exacted by lowland sultanates. Upland populations in the southern Philippines are generally similar to each other in terms of social organization and economy, although the elaboration of warrior status among certain upland peoples in eastern Mindanao formerly distinguished them from groups in the western part of the island (Yengoyan, 1975: 50). In contrast, populations of the mountain chains of northern Luzon exhibit strikingly different socioeconomic and political adaptations to their more rugged physical environment.

Unlike the lowland experience from the mid-1500's to 1898, most upland areas successfully resisted Spanish subjugation (Scott, 1982). Many uplanders recognize this common historical theme, which to some extent bolsters a growing awareness of shared problems today. One of these shared problems is greater underdevelopment in comparison to much of the lowlands. Past isolation and low population densities mean that the uplands now generally have more limited access to health and educational services, fewer infrastructural advantages, and lower land productivity and annual income (Integrated Research Center, 1983). The inability of industry to absorb population increases and the shortage of unoccupied, level, and arable land elsewhere in the lowlands has led to increased migration to the uplands, a factor which exacerbates the pressure on upland resources.

Mountain agricultural systems are characterized by the multiple exploitation of diverse ecological zones. Upland environments vary in terms of slope, elevation, moisture regime, soil characteristics, and exposure to wind and sunlight. In a monsoon island country like the Philippines, latitude affects seasonal climatic differences. Mountainous environments in northern Luzon, for example, experience extreme

MAP 2. PHILIPPINE UPLAND (TRIBAL) COMMUNITIES

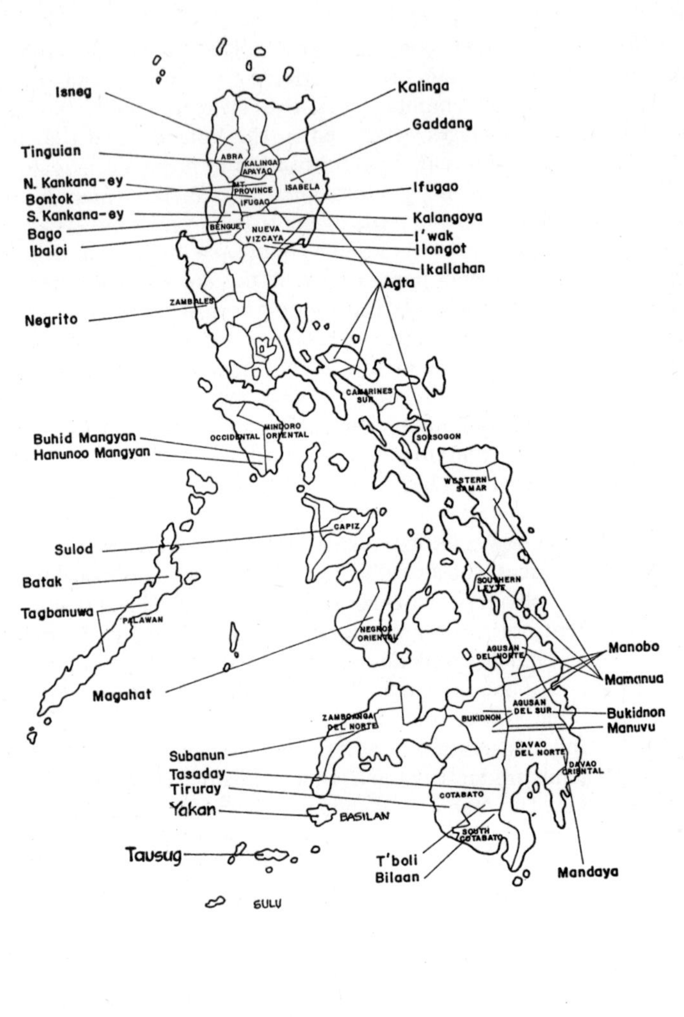

rainy and dry seasons, whereas rainfall in Mindanao is more evenly distributed throughout the year (Wernstedt and Spencer, 1967).

Despite environmental diversity, land use patterns and indigenous farming systems developed a number of similar characteristics. Recent studies of mountain environments usually stress that the most appropriate units of comparison are based on vertical land use patterns, rather than on elevation (Soffer, 1982; Guillet, 1983). The elaboration of strategies for exercising control over different vertical ecological zones in the Philippines has been addressed in the ethnographic literature in terms of the periodic or seasonal movement of groups between forest and coastal areas (Peterson, 1978; Pennoyer, 1981; Connelly, 1982). For ethnographic comparison, however, it is feasible to group upland Filipinos according to respective primary subsistence emphasis, that is, swidden horticulture, terraced wet-rice agriculture, and foraging.

Under ideal conditions, swidden horticulture can be highly productive compared to permanent field agriculture. Given low population density, swidden horticulture represents a rational adaptation to the tropical ecosystem. In the tropical forest ecosystem, nutrients are easily leached by heavy rainfall and hot climate (Spencer, 1966). Nutrients cannot be maintained for long in the soil itself, and fertility is derived from the vegetative cover. Although fertility decreases rapidly after the first year or two of cultivation, ashes produced by burning the site before planting provide essential nutrients, although not all upland swiddening groups burn their fields. In the eastern Mindanao uplands, for example, absence of a dry season precludes burning. In order to sustain acceptable yields, therefore, Mandaya continually make swiddens in primary forest with its naturally higher levels of soil fertility and farm steeper slopes so that gravity can aid in clearing (Yengoyan, 1971).

Whereas swidden horticulturists generally rely on burning to provide fertility, wet-rice agriculture depends on the creation of paddy soils, blue-green algae for nitrogen fixation, and irrigation to bring nutrients to the pond-fields. Greater labor coordination is required for wet-rice production than for swiddens. Hydraulic control, weeding, transplanting, and terrace and irrigation canal maintenance are complex and more time-consuming in the uplands due to the rough terrain. Production in wet-rice agriculture responds favorably, however, to increased labor intensity (Geertz, 1963).

Subsistence hunting and gathering has a long history in the Philippines but is now relatively rare as a primary subsistence emphasis. Hunter-gatherers may also engage in swidden horticulture, although this is a relatively recent transition for some groups (Estioko-Griffin and Griffin, 1981; Warren, 1984).

There are several caveats to a subsistence-based typology. First, most upland groups exploit multiple resource zones with emphasis depending on time of year and schedule of activities for the most important crops. Many groups engage in more than one of the major subsistence emphases. Also, in some areas upland agriculture consists of permanent hillside farming (Barnett, 1967; Olofson, 1980). Secondly, many, if not most, upland groups are involved at least peripherally in the market economy, either as purchasers of manufactured or processed goods, or as petty commodity producers and wage laborers. In Benguet, many Ibaloi and Southern Kankana-ey are involved in capital-intensive vegetable farming on permanent fields for Manila markets (Davis, 1973). While such heavy market dependency may be unusual for the uplands, partial cash crop production of fruits, coffee, and vegetables typifies many upland agricultural strategies. Even in remote areas, forest products or animals may be traded for consumer goods or grains even by hunting and gathering peoples (Peterson, 1978; Griffin, 1981). Perhaps even more typical is the permanent or seasonal entry of uplanders into the wage labor market, especially in construction, mining, and commercial agricultural labor. Finally, it should be emphasized that categorizing upland populations by predominant subsistence strategies is conventional but deceptive. Variation between swiddening groups often may exceed that obtaining between any single swiddening population and a wet-rice terracing group.

Migration between highlands and lowlands is common. Peace and order problems and declining forest cover have caused groups such as Gaddang (Wallace, 1970), Mandaya (Yengoyan, 1964), and Tiruray (Schlegel, 1979) to settle permanently in the lowlands. Migration to the uplands has been extensive in Palawan, Mindoro, and Mindanao, where much of the ancestral or communal lands claimed by indigenous groups has been usurped by outsiders holding legal titles (McDonagh, 1983). As a result, upland populations are mixed in certain regions. This population heterogeneity is not recent. Populations originally from the lowlands have resided in the uplands in many areas for generations (Vilar-Basco, 1956; Allison, 1963; Vander Meer, 1963b; Olofson, 1980; Connelly, 1982). The uplands often are an ethnic mosaic with blurred social, linguistic, and economic boundaries.

Some upland groups have preserved their ethnic and economic autonomy more than others, a factor helping to insulate them from external change (Aguirre, Beltran, 1967). Keesing (1962) argued that lowland Filipinos frequently escaped Spanish repression by moving to the mountains. Internal colonization from both national and international sources continue to push small groups into marginal areas (Atal and Bennagen, 1983). The political economy of Filipino nation-state expansion and the extensive highland-lowland interactions historically largely account for the current ethnic diversity in the uplands, despite the widespread misconception that mountain Filipinos are isolated and conservative cultural minorities.

INDIGENOUS SUBSISTENCE EMPHASES

Olofson (1981) reviewed recent research among Philippine swidden horticulturists. In this, he attempted to dispel some of the misconceptions that persist among government policymakers. One prevalent myth is that all indigenous groups had ecologically-sound systems of shifting cultivation, while all lowland Filipino horticultural practices in the uplands are ecologically destructive. Although this assessment is true in some cases, it is ethnographically untenable. As Ellen observed:

> As a description of empirical phenomena, the concept of ecological balance is at best highly relative and at worst misleading. The notion of a perfectly balanced ecosystem is incompatible with the empirical evidence for species extinction, population depletion, and undernutrition. While approximate balance may be the result of short-term adaptive processes, no biological mechanism is perfect. The isolated populations traditionally studied by anthropologists might seem to fit such a model rather well, but even in relatively closed small-scale populations, there is evidence that this balance is frequently upset - through innovations, migration, conflict and resource depletion. The more complex ecosystems become the less equilibrium assumptions hold true (Ellen, 1982; 187).

While many indigenous upland groups practice swidden horticulture within distinct territories, Mandaya traditionally expanded continually into virgin territory (Yengoyan, 1971). Evidence indicates that not all uplanders with lowland origins maintain ecologically disruptive cultivation techniques (Allison, 1963; Vander Meer, 1963a). Resource-conserving agricultural practices are not a monopoly of indigenous cultural minorities; they simply tend to characterize groups with low population densities who have been resident and dependent for many generations on the upland forest environment.

Unfortunately, long-term residence does not preclude less ecologically-stable cultivation practices when people experience severe subsistence pressure and demands for cash income. The current upland pattern of shortening fallows despite lower yields is testimony to increasing land-use pressure and restricted economic options. Ifugao, for example, are aware of the importance of maintaining the forest watershed for better hydraulic control in their rice terraces (Conklin, 1980:30). Yet swidden horticulture in combination with tree cutting for fuel and the woodcarving handicraft industry increased the dry season water shortage in Banaue, Ifugao, and are partially responsible for the abandonment of some terraces (Eder, 1982b:107).

A variety of methods are employed by swiddeners to control burning: clearing firebreaks, removing cut underbrush to protect domesticated trees, and sheathing vines by means of protective coverings (Conklin, 1957: 52, 65-66). While the ethnographic literature demonstrates the careful coordination of burning among many long-term residents, not every group is as conservation-oriented as were the traditional Hanunoo. Relatively few precautions are taken by some Gaddang to confine fire to single plots (Wallace, 1970: 61). Accidental burnings of nearby forest occasionally occurred, with the frequency perhaps depending on the length of the dry season.

Production Regimes

Philippine swiddens can either be devoted to a single staple such as upland rice, corn, or sweet potato, or may range from simple to complex multiple and intercropping (Scott, 1958; Rice, 1981). Hanunoo recognized 92 different varieties of rice and regularly intercropped several dozen of 280 recorded food crops in rice swiddens (Conklin, 1957: 30). The large number of crop varieties typically cultivated may reflect the continual experimentation necessary to coordinate plant varieties with the many different upland microenvironments (Warner, 1981). Perhaps the most common reasons for extensive intercropping are consumption variety and a more even spread of labor and harvests throughout the year. I'wak stagger planting to allow daily harvesting once the crops begin to mature (Peralta, 1982: 59). Ilongot plant both quick-growing and slow-growing rice varieties to reduce the risk of a typhoon or drought ruining all of their crops. This pattern also makes it possible for a single person to harvest (Rosaldo, 1981: 34).

Reliance on a variety of crops also diminishes the risk of pests and disease and is one of a number of farming practices farmers engage in to help spread risks. If pigs or deer are a threat to crops, for example, Ifugao construct trenches (Conkin, 1980); whereas I'wak (Peralta, 1982) and Kalinga (Lawless, 1977) may erect fences. Also, single plots are often quite small and geographically dispersed, reducing the likelihood of damage to all of a household's crops in the event of calamity. Field dispersal also increases the speed of natural reforestation during the fallow period. The upland custom (Frake, 1955; Fox, 1982; Prill-Brett, 1983; Lopez-Gonzaga, 1983) of considering certain forest reserves as sacred and off-limits to farming also promotes reforestation.

Uplanders frequently cultivate trees for food, animal feed, fuel, medicines, building materials, and cash. Trees are often planted in swiddens, thereby protecting crops from direct exposure to heavy rainfall and excessive sunlight. Partial replication of the layered structure of the natural tropical forest vegetation decreases soil erosion. Many uplanders maintain woodlots near their villages. Cultivated woodlots are often mentioned in the ethnography of wet-rice terracing groups in northern Luzon (Drucker, 1974; Lawless, 1977; Conklin, 1980), and may be individually owned. Ifugao woodlots are former swiddens managed both as tree farms and gardens and are usually less than a hectare. With up to more than 200 varieties of plants, carefully attended woodlots can reflect greater diversity than is present in the rest of the upland forest (Conklin, 1980: 31). Indigenous agroforestry extends beyond casual planting of trees and arboraceous plants to encompass thinning, weeding, pruning, and protection (ibid: 9).

Ifugao swiddens are an important and expandable subsistence option when wet-rice production is limited or threatened, and root crops rather than rice provide the bulk of daily diet in some areas (Barton, 1922; Conklin, 1980). Among groups such as Bontok (Drucker, 1974) and Kalinga (Takaki, 1977), rice is the main staple even if many families do not produce enough to last an entire year. For most wet-rice farmers with sufficient terraced land, however, primary attention is devoted to paddy-field agriculture rather than to swiddens. Among Kalinga (Takaki, 1977; Lawless, 1977) and Bontok (Drucker, 1977), swidden production has declined in favor of intensification of rice production. The labor required in wet-rice agriculture generally reduces time available for swiddening, and swiddens tend to be located near the village. The relative contribution of swidden crops to the total diet varies among and within upland populations in Luzon, depending on the amount of water available for wet-rice.

Rice is the preferred crop in most areas, whether cultivated in swiddens or in paddy. In highland Luzon, the agricultural calendar revolved around different phases of the rice cycle, which in turn depends on climate and environment. Most groups now plant two rice crops annually; in higher elevations such as in Sagada, only one crop is grown due to the longer growing period (Scott, 1969). Rice-growing has strong ritual significance in many areas, and special varieties are used for rice wine. Freshwater fish and other aquatic animals provide additional food sources from paddy agriculture. Finally, since rice is a storable staple, terraces are valuable assets distinguishing wet-rice farmers from swiddeners. Differential possession of such assets are status and prestige distinctions between and within groups in northern Luzon.

Cordillera inhabitants developed wet-rice terraces gradually several centuries ago (Scott, 1958; Breeman et al, 1970; Conklin, 1980).

Terraces exist in elevations as high as 500 feet above sea level and frequently are transformed out of 35° to 45° slopes. Sagada terraces range from 2.5 feet to 12 feet high, with angles of 45° to 80° (Scott, 1969: 222). Ifugao terrace embankments are mostly earthen with 30% to 35% stone walling (Conklin, 1980: 25). Bontok and Northern Kankana-ey terraces are constructed with even more stone. Wall height ranges from 1 m to 20 m in Ifugao, and annual maintenance and repair of terraces requires 200 to 1000 man-days of labor per ha (ibid: 37). Much labor goes into maintaining waterworks, although there are rainfed terraces (Lawless, 1977).

Hydraulic technology can be extremely complicated. The longest irrigation canal in Mountain Province is 25 km (Bacdayan, 1974). Some canals in Ifugao are 15 km, although larger waterworks are usually for rainy season drainage rather than irrigation (ibid:8). Terraces must be built close to rivers in villages without springs. Residents of Bontoc yearly rebuild their dam and waterworks system which is washed away during typhoon season by the Chico River (Jenks, 1905: 91).

Differences in population density between swiddening and wet-rice terracing groups are large. Although swiddening may support an upper limit of 50 people per square kilometer (Pelzer, 1945 : 24; cited by Frake, 1955: 264), density in Southeast Asia rarely exceeds 10 persons per square kilometer. Ifugao population density is 126 persons per square kilometer of total area and 230 persons per square kilometer of cultivable land. Population sizes in nucleated villages also vary. Tagbanuwa swiddeners live in large permanent villages of up to 200 individuals (Fox, 1982), whereas Bontok villages range from 800 to 3000 persons (Prill-Brett, 1983).

A major factor allowing such dense populations was the higher productivity of wet-rice agriculture. Yields in the uplands are considerably lower than in the lowlands due to soil and water differences, and because high-yielding rice varieties and seed-fertilizer technologies have not been adapted effectively to most upland conditions. It is also notoriously difficult to obtain reliable information on upland yields, since actual sizes of paddies and fields are quite variable. Because productivity is greatest in the paddy centers, multiplying yield of small terraces underestimates yield per ha for much larger terraces (Scott, 1969: 222).

Ifugao average per hectare yield over an eight year period was 2500 kg of unhusked rice (Conklin, 1980: 35). Shorter-term data from Kalinga, where sweet potatoes form a more minor contribution to the diet than in Ifugao, indicate higher productivity with more than 4500 kg per ha of unhusked rice (Takaki, 1977: 805-807).

Complementary Economic Activities

For swiddeners and mixed swidden and wet-rice farmers, domesticated animals, fishing, hunting, orchard cultivation, and the gathering of forest products can be significant subsistence supplements. Pigs are an essential sacrificial animal in Cordillera rituals, and a great deal of swidden labor is expended on raising food for domestic pigs. Carabao are important draft animals in gently sloping terrains, and their sacrifice at rituals contributes additional prestige to a feast-giver. Cattle ranching used to be a major part of the economy among the Ibaloi elite, where the wide pasture lands in southern Benguet favored large herds. Many of these animals were cared for by non-owner farmers in return for a share of the offspring (Pungayan, 1978), a practice which still exists elsewhere in the uplands (Eder, 1982a). There are fewer cattle herds in Benguet today due to population increases, the spread of commercial vegetable gardening, and the decline in power and wealth of the traditional elite (Tapang, 1982). In many upland villages, chickens, pigs, and dogs continue to be significant in the domestic household economy.

Another characteristic of the upland Luzon economy is the long tradition of independent lode and placer gold mining (Scott, 1974). Both Ibaloi and Southern Kankana-ey are involved in this occupation. Downturns in the economy coinciding with high world market prices of gold frequently herald an increase in the number of Benguet farmers who become temporary miners.

Foraging

Swiddening and wet-rice terrace cultivation are not the only forms of indigenous upland subsistence strategies in the Philippines. Hunting and gathering, an older and once more widespread adaptation in Southeast Asia, is still practiced by Negrito and Negrito-like groups (Garvan, 1964; Bennagen, 1976). Griffin (1981) emphasizes that "hunting and gathering" is a misnomer for the Agta of northern Luzon and suggests that "foragers" is more accurate. Whereas the Tasaday in Mindanao (Yen and Nance, 1976; Fernandez and Lynch, 1979) are a foraging group without sophisticated hunting technology, the Agta have a well-developed bow and arrow technology that, in rich environments, led to their provisioning of nearby agriculturalists with protein in exchange for grain and other consumer needs (Peterson, 1978). Northeastern Luzon Agta employed a diversified, opportunistic subsistence strategy based on hunting, fishing, and gathering of forest and riverine cultigens. This strategy was intimately bound to seasonal cycles of rainfall and the associated movement of game and the maturation of fruit and root crops. Rainfall patterns also influence expoitation of rivers, streams, and coastal waters for other protein sources (Griffin, 1981: 28-30).

As lumber companies and lowland migration increased strains on hunting territories several decades ago, some upland foraging groups began practicing swidden horticulture, especially in areas where they had been employed as part-time farm laborers. In the past 30 years, swidden horticulture has become the dominant subsistence activity of Palawan Batak. Despite difficulties, they learned to farm more or less efficiently (Cadelina, 1982; Warren, 1984). Unfortunately, debt bondage and stress have been among the costs of socio-economic transition for many Agta (Eder, 1977; Peterson, 1981).

Some Agta of northeastern Luzon practice swidden horticulture near lowland groups. They have experienced lackluster success so far, partly because swidden labor is still of secondary importance to foraging (Estioko-Griffin and Griffin, 1981: 66). Other factors explaining Agta difficulties in adopting settled agriculture, despite familiarity with gardening (Peterson, 1981: 47), include resource competition with other groups. The more successful agricultural Agta have fields remote from expanding lowland migration; the less successful have experienced land take-overs by non-Agta (Griffin, 1981: 39; Peterson, 1981: 51-54). Social discrimination experienced by upland foragers strongly affects their ability and willingness to defend lands from intruders. Vague conceptions of land ownership do little to prevent their relinquishing of cleared lands to outsiders, and many Negrito groups end up as laborers or tenants of lowland farmers (Maceda, 1974: 7).

The Agta lack social institutions that might bolster the transition to permanent agriculture. These include stable settlement patterns and strong leadership roles (Estioko-Griffin and Griffin, 1981: 71). Physical mobility is an important cultural value prominent in conflict resolution within and between groups (Peterson, 1981: 50). Difficulties experienced by foraging groups in radically altering their productive activities are partly organizational, rather than simply the result of resistance to change. The relative ease with which such groups enter into swidden horticulture, part-time wage labor, and trading reflects the innate flexibility and opportunism of their traditional economic strategies, even though certain newer economic lifestyles are not as feasible or preferred as others.

Implications

The diversity and multiple nature of upland food procurement activities is well-substantiated. Such diversity enables upland groups to compensate for short-run climatic fluctuations, thereby lessening the risk of natural disasters. On the whole, however, socio economic adaptations are not always capable of dealing with recent changes brought about by increasing population, expansion of commercial interests in the uplands, and consequent long-term changes in resource availability. Increased dependence on the commercial economy is a strategy open to upland farmer, but it is not a solution for people far from markets or roads. Desire for cash for modern goods or to send children to school is strong in the uplands, and short-run decisions to forego subsistence farming in favor of a cash income may have unanticipated consequences for community welfare and environmental preservation.

Social institutions of groups with different subsistence strategies influence how well and how quickly they adjust to new economic opportunities or pressures. Groups with private property or individual land ownership have an advantage over other groups without such traditions in confrontations with land-encroaching outsiders. Groups with cooperative labor arrangements have greater organizational capabilities for programs requiring similar coordination.

Problems facing uplanders are not due to cultural inabilities to adapt to change. Particular organizational forms encourage selective responses to changing opportunities and constraints. Problems of economic transition often reflect a lack of viable choices for people relatively neglected in the national system of resource allocation. To better understand ongoing socio-economic transformations, examination of indigenous forms of social organization complements the previous discussion of subsistence strategies.

FORMS OF UPLAND SOCIAL ORGANIZATION

Many aspects of social organization are influenced by the interaction between technology and environment. Labor requirements of different technologies can produce social institutional variability in similar environments, and every society establishes sanctions governing resource use according to culturally-specific notions of appropriate behavior. Noneconomic environmental factors influence social organization and generate unique patterns of behavior. Owing to the overlapping nature of social relations in small-scale societies, it is often impossible to separate strictly economic behavior from other ritual or kinship-oriented motivations. Cultural rules regarding land use, inheritance, and status mobility determine equitable or inequitable access to resources. These institutional arrangements, in turn, support individual or collective resource management. Each of these overlapping aspects of socio-economic organization are addressed in the following sections.

Property Control

While the ways by which groups distinguish types of property vary, indigenous classifications tend to distinguish between personal and public property, and between divisible and nondivisible property. Two essential and appropriate questions are: who has the right to use particular resources and who has the right to dispose of such resources? Frequently, the right to primary resources are invested in small kin groups rather than individuals, although a wider range of individuals may acquire temporary-use rights.

Until recently, many upland swiddening groups viewed land as a common resource not subject to permanent private ownership. People perceived themselves as having use rights to the land and ownership only of the products (Frake, 1955; Conklin, 1957; Yengoyan, 1971; Schelegel, 1981). Once a swidden reverted to fallow, use rights were abandoned by the cultivator, although rights to still productive trees or root crops on a fallowed swidden were often retained. Mandaya claimed future cultivation rights to forest lands in the linear trajectory of the field clearings. Other cultivators avoided crossing another's path in the opening of new swiddens (Yengoyan, 1971: 367). Kin groups also control specific territories within which any member may clear a swidden (Garvan, 1931; Manuel, 1973; Lopez-Gonzaga, 1982). Still, a kin group's relationship to ancestral property was perhaps more akin to stewardship than onwnership per se, especially if supernatural beings were considered to be the true owners.

Garden plots of foraging groups were traditionally not privately claimed after cultivation ceased (Peterson, 1978). Fishing and hunting territories, however, could be communally controlled with recognized borders (ibid: 25; Maceda, 1975: 30).

In terraced wet-rice areas, jural relations between individual house-holds, corporate kin groups, and village institutions are more complex. These communities are permanent rather than mobile, and their pro-ductive property reflects intensive labor investment. Rules developed in part to keep property intact across generations, thereby avoiding the division of land resources into smaller and smaller plots.

While rice terraces are private property in the Cordillera (Lawless, 1977; Drucker, 1977; Conklin, 1980), there are restrictions on rights to dispose of the land. Strong feelings exist that rice fields are held in trust for the next generation (Jenks, 1905: 164-165, Moss, 1920: 251; Borton, 1949: 95). Taboos against the sale of inherited property persist, except when sale is necessary to obtain items needed for rituals or to purchase better land elsewhere. In all land sales, close relatives must first be offered the purchase right. Only if they decline that option can other villagers buy the property. There is a general reluctance to sell land to outsiders, especially to those not married locally. Such social and physical boundary maintenance enabled many Cordillera groups to insulate themselves against possible conflicts arising with greater cultural heterogeneity. Indigenous religious beliefs further instill the desire to protect land from outsiders, since ancestors are believed to influence the fertility of land, animals, and kin groups. Selling inherited property to an outsider while not making provisions for one's offspring will surely incur the displeasure of one's ancestors and consequent misfortune.

The manner in which claims are established to uncultivated common lands illustrates the difficulty involved in trying to ascertain "ownership" in Western terms. Individual Kalinga (Takaki, 1977: 207-208) may privately claim unused land, but other village members retain rights to graze their water buffalo on private pastures or can request temporary use of claimed but uncultivated land. There are restrictions on owner-ship, therefore, in that co-villagers share access to such property. Access is allowed as long as it does not adversely impinge on the owner's personal use (ibid: 219-220; Barton, 1949: 91-102). Kalinga land claims are considered valid only if people actually use the property at the time the claim is made (Takaki, 1977). This perception that imme-diate use is necessary to fix claims may well underly uplanders' incom-prehension of government rights to exclude access to unused forest areas.

Indigenous rules preventing property use by outsiders are not well-developed outside the wet-rice terrace regions. In Mindanao, Mindoro, and Palawan, loss of indigenous territories is a dreary and growing trend. Land has been lost through debts to outsiders or through failure to register for titles. The latter problem is especially serious where people are less familiar with private property concepts (Maceda, 1974; Mc-Donagh, 1983). Government and private corporations bear heavy res-ponsibility for group territory losses in the uplands, as occurred in dam construction and granting of timber concessions (Krinks, 1974; Cordil-lera Consultative Committee, 1984).

Resource ownership is a very important political and economic issue in that institutional arrangements facilitate and restrict individual access in various ways. Property control vested in individuals, kin groups, or other social units offer significant incentives for increasing productivity compared to systems of relatively free access. The more productive wet-rice systems, therefore, have the most developed indi-vidual and group ownership rules. Swidden groups in turn have more definite conceptions of land use rights than do foraging groups. Since people in small-scale societies tend to be members of a resident kin group, however, no one is deprived of access to at least some form of productive resources.

Inheritance

Kinship structures many socio-economic processes, including the transmission of property across generations. Like land ownership, more complex kinship patterns characterize wet-rice terracing groups. These groups, for example, distinguish between private and conjugal property. Inherited properties of husband and wife are typically individually owned and jurally distinct from joint properties acquired after marriage. Conjugal property is almost exclusively reserved for offspring, although some land may be sold and animals sacrificed in rituals following the death of the parent.

Seniority by order of birth has been integrated into the inheritance patterns of several upland groups. Primogeniture, or being the first born, is significant among many wet-rice terracing groups, with the eldest children receiving the lion's share of the property regardless of sex (Eggan, 1960; Conklin, 1980). Bontok and Kalinga inheritance rules frequently allot the best rice fields to the eldest same-sex offspring by each parent (Jenks, 1905; Keesing, 1949; Lawless, 1977; Drucker, 1977). Less valued property tends to belong either to the corporate kin group or is considered communal village property. Bontok and Northern Kankana-ey villages, however, are divided into wards comprised of adult males. One function of the ward is to hold property corporately for its members (Drucker, 1977; Prill-Brett, 1984). While younger offspring in wet-rice areas may inherit few rice terraces, they, therefore, do have alternative avenues of gaining access to productive property.

In some groups, inheritance is more or less equal for all children (Moss, 1920; Bello, 1972), even though the eldest sometimes inherit more because of their earlier start in cultivation. Land inheritance occurs at the time of marriage in the Central Cordillera groups. Lands are not transferred formally until after the parents' death in Benguet, although property use-rights are granted at marriage. In commercial cropping areas, however, rising land values have led to property transmission before parental death to eliminate inheritance disputes and to provide offspring with collateral for production loans (Russell, 1983).

Among swiddening groups, property is generally inherited by offspring of both sexes. Some groups, however, emphasize primogeniture in dividing property (Jocano, 1968; Warren, 1975; Manuel, 1975). Foragers in the past had fairly incomplete notices of land inheritance (Peterson, 1978) but, in principle, appear to subscribe to equal inheritance (Warren, 1975).

Kinship Organization

Kinship is reckoned bilaterally through both parents in the Philippines. This pattern allows for maximum flexibility in establishing social relations instrumental for access to land, labor, and mutual aid. The most important social unit is the nuclear family, usually comprising an independent household, although elderly parents often reside with a married child. Another important social unit is the bilateral kindred. The significance of the kindred in upland social structures varies, partly depending on the mobility and size of local settlements. Settlement size and permanence are influenced by land use and resources dispersion (Yengoyan, 1973). Among Mandaya, the effective kindred consists of persons from whom an individual may request assistance in certain tasks. Group composition fluctuates according to the work concerned, but the number of individuals will be less than the total kinsmen to whom one is related, including people related through marriage (ibid: 168-169). Similar principles of kinship network formation characterize many upland swiddening and foraging groups.

In the wet-rice region, permanent villages, terrace agriculture, and high population densities have stimulated more formal rules for transmitting property and forming coordinated labor groups. Since a kindred potentially embraces all of an individual's bilateral relatives, any individual is a member of overlapping kindreds. This overlapping inhibits the kindred from functioning effectively as a corporate group with respect to land control, since conflicting loyalties may make alliances for joint cooperation unstable (Eggan, 1960: 46-47). The ward system of the northern Kankana-ey and Bontok offers more stability than a bilateral kindred and represents another social unit with functions similar to those of a kin group (Keesing, 1949; Eggan, 1960). As social units, wards are less susceptible to the demographic vagaries of bilateral kindreds.

Ecological factors influence but do not determine the pattern and intensity of social relationships in upland societies. Resource distribution influences the nature of subsistence strategies, which in turn affect settlement patterns. Yet no strong association exists between the type of settlement pattern and the ecosystem. Even warfare and the consequent advantage of nucleated villages for defense does not explain variations in Southeast Asian swidden settlement patterns (Frake, 1955: 272). Peaceful groups such as the Sindangan, Tabanuwa, and Hanunoo each have a different residential pattern; the peaceful Subanun have the same settlement pattern as warlike groups in Mindanao (ibid: 273). Ifugao, living in small dispersed hamlets, were similar in terms of warfare to Bontok and Northern Kankana-ey, both of whom live in nucleated villages.

Frake (1962) illustrated how cultural perceptions conditioned decisions affecting swidden location. Rather than living in nucleated villages, Subanun prefer homes close to their swiddens and far from each other so that family disagreements will not be overheard by neighboring households. They believe that a new swidden should be opened with a minimum of wild vegetation boundaries, a rule producing shared borders between new and previous fields. As Frake explains:

> The ecological rules. . . which determine Subanun swidden and household arrangements are explicitly geared to protection of swiddens from animal pests with a minimum expenditure of time and energy in such tasks as fence building, field-house construction, and travel to fields for daily watching. Yet the practice of clearing large areas adjacent to previously cleared areas increases, under certain conditions, the probabilities of succession to grassland instead of forest, thus removing the land from future swidden cycles. The Subanun emphasize the immediate returns of increased swidden protection and accessibility at the cost of some loss of control over the fallowing stages of the swidden cycle. Other swidden farmers of the same part of the world weigh the advantages and disadvantages of alternative techniques for controlling faunal and flora enemies differently with different consequences for swidden arrangements and settlement patterns. . . (Frake, 1962: 57-58).

Political Organization and Stratification

Societies develop institutions to help resolve difficulties arising from external threats, internal disputes, and the need to collectively organize. Political institutions vary depending on cultural history, interaction with other groups, population density, economy, and technology. The denser the population and the more complex the economic organization, the more complicated are the problems to be solved.

Political processes in many upland societies today reflect an ongoing struggle between indigenous, introduced, and external forms of authority. Leadership may not reside in elected officials where there continue to be other bases of respected power. Inequalities in the distribution of power are inherent in many indigenous systems, but recent rapid economic change has also expanded the sources of power available to individuals with experience and contacts in the larger society.

Swidden cultivators traditionally had decentralized political organizations, no formal leadership roles, and weakly-defined territorial political linkages beyond the local settlement. Dispersed hamlets often supported ad-hoc leadership based on oratorical ability or religious specialization. Egalitarian consensus predominated decision-making, although skills and bravery in battle formerly served as another basis for leadership among Kalinga, Isneg, Mandaya, and Manobo (Scott, 1979).

Hereditary status differentiation occurred among Tagbanuwa and other groups (Fox, 1982). Such status distinctions resulted from attempts by Muslims to instill more manageable political organization on swidden groups in order to exact tribute and to trade more effectively. Similarly, groups that recognized specific **datu** often allowed them jurisdiction rights over disputes and varying degrees of control over their territories (Cole, 1913, 1956).

Political variation among permanent wet-rice cultivators ranges from a single head spokesman for a large region to decentralized ward organizations. Kalinga developed elaborate peace-pact systems between geographical regions that centered on one individual administrator and enforcer, the **pangat** (Barton, 1949: 146-148). A pangat's reputation as a warrior was significant in determining community support. Now, wisdom, wealth, family connections, oratorical ability, and education comprise desired qualities (DeRaedt, 1969: 763-764).

Even among neighboring Cordillera groups there are great differences in socio-political organization. Ifugao have no formal indigenous political structure, and kinship serves as the primary organizing mechanism. Bontok village organization is decentralized and dispersed among wards within each village (Eggan, 1941: 13); although there were supra-ward organizations for matters affecting the entire village (Prill-Brett, 1977: 22). Each ward regulated the socio-economic and religious affairs of its members and established peace-pacts with other villages. Although warfare is no longer practiced, peace-pacts are still significant in settling disputes today.

Casual observers of upland society often assume that class differentiation is absent since people often have similar attire and lifestyles. Uniform appearances, however, often reflect attitudes against conspicuous consumption rather than for equality or homogeneity. Where lands are adequate and swidden horticulturalists still maintain communal-based rights to land (rather than private ownership) there may be little class differentiation. There is little stimulus for accumulation when possibilities for greater productivity or income through agricultural intensification are limited. The predominant political organization of such groups is more egalitarian than hierarchical.

In Luzon, inequality was based on differential possession of rice terraces. Ifugao status designations reflected the amount of rice (as opposed to sweet potato) that individuals consumed throughout the year (Barton, 1922: 416-418). The wealthy elite tended to arrange marriages among their children to keep their wealth intact (Barton, 1969; Moss, 1929). Class differentiation appears to have been greatest among traditional Ibaloi. Aristocratic kin groups dominated through control of land, gold mines, and cattle herds. Along with the elite in nearby Ifugao and Abra, wealthy Ibaloi frequently maintained usurious debt relations for rice or animals with large numbers of community members (Barrows, 1902; Cole, 1922; Barton, 1922; Keesing and Keesing, 1934). Power monopolies were prominent in Benguet, partly due to closer colonial contacts with Baguio City. Juridical arbitration by the elite was, however, somewhat tempered by consensus decisions of a council of community elders.

Upland class differentiation often allowed for influence to be exerted by community members through open debates. Elites had to solicit public support, and their ability to coalesce followers (especially resident kin members) was usually a necessary if not a sufficient condition for leadership. Permanent central authority or political offices such as tribal chiefs were largely absent in the uplands, except perhaps in Benguet and where the datu leadership complex was introduced by Muslims (Scott, 1979: 158-159). Difficulties that government officials frequently experience in eliciting cooperation among some upland communities attest practically to the diffused and limited forms of external authority recognized by uplanders.

A major motivation for accumulating rice lands and livestock in the wet-rice region traditionally was status enhancement. Status was achieved or maintained by sponsoring large, prestige-building rituals (DeRaedt, 1964). Redistribution of wealth to the community through rituals served marginally to offset some of the exploitative relations that benefited the rich in the more stratified areas. Aristocratic status in most upland wet-rice regions tended to circulate among a relatively small number of families from generation to generation. Some social mobility was possible, however, since higher status could be achieved in some groups through ritual sponsorship.

Status distinctions, inequality, and a well-developed political organization are absent among foraging groups. These peoples tend to group together during the rainy season, splitting into smaller groups of several families during the dry season to more efficiently exploit forest and riverine resources. No specific leaders exist among such groups. Resources, especially game, are shared.

Agricultural Labor

Stratification based on the sexual division of labor is weak in upland societies. Women are accorded similar economic rights as men, as generally is true elsewhere in the Philippines. The importance of women in agriculture has been noted by practically all anthropologists who have worked in the uplands (Olofson, 1981; Cherneff, 1981). While men generally contribute more of the heavy labor required in forest clearing, women do more of the weeding. Women not only dominate swidden root crop production in wet-rice areas, but also work in all phases of the rice cultivation cycle (Bacdayan, 1977). Among foraging groups such as Agta, women also have been reported to hunt independently (Estioko-Griffin, 1981). Women, therefore, make extremely important labor contribution to the household. Understanding institutional forms for mobilizing labor beyond the household is crucial for designing development strategies that increase popular participation in decision-making and encourage effective indigenous resource management. New crops or technologies often require the creation of new organizational forms. Capabilities for collective action vary. Upland agriculture ranges from systems in which most field labor is done by individual household members to efforts carried out by large-scale cooperative labor groups encompassing most of the adult village population. Typical labor arrangements beyond the household level include exchange labor in which households take turns working on each other's fields, feast labor wherein animals are consumed by the workers in return for a specified amount of labor, and hired labor in which workers are paid in either cash or in a certain share of the harvest (Frake, 1955:105-107; Cocklin, 1957:53-55; Lopez-Gonzaga, 1982:24).

Institutional coordination of agricultural tasks is generally weaker among swiddeners than among permanent farmers. Forms of mutual assistance exist among neighboring field owners or kin, especially if the activity is arduous or must be performed quickly to be effective. Reciprocal labor for clearing swiddens among Gaddang, for example, is important only when the trees are very large (Wallace, 1970: 60). Mandaya also rely on reciprocal labor for felling trees since frequent rain prevents them from burning their swiddens (Yengoyan, 1971:364).

The timing and nature of wet-rice agriculture result in a greater variety of extra-household labor practices, and cooperative labor groups are generally larger. Formation of labor groups among Bontok and Northern Kankana-ey depend on the task and may be organized by ward members or by individual field owners (Scott, 1969; Prill-Brett, 1983; Voss, 1983). Same-sex work teams are quite common (Reid, 1972), although there is a high interchangeability of tasks among men and women (Bacdayan, 1977). Among both of these Cordillera groups, reciprocal labor is obligatory for some tasks; and non-compliance necessitates payment of fines. Wealthy farmers may mobilize labor through payment in rice shares or, more recently, in cash. This arrangement often includes a free noon meal for the workers. In contrast, most wet-rice terrace labor in Ifugao is done by small groups or individuals, depending on household preference (Conklin, 1980). Emphasis on individual rather than cooperative labor arrangements may partially reflect the greater contribution of swidden crops to Ifugao subsistence.

Water Management

The most impressive upland cooperative labor organizations are those concerned with water management. While terrace building is usually a household activity, construction and maintenance of irrigation canals are usually the responsibility of associations of farmers with fields watered by a common source. Equitable water allocation, which can be a major factor in reducing yield and income disparaties among rural producers (Coward, 1980; Siy, 1982), requires complex decision-making, cooperation, and enforcement institutions.

Maintenance and repair of primary irrigation canals are organized by individual wards among Northern Kankan-ey in Sagada. Wards assess each household a number of participants to work according to land sizes: that is, the largest field owners provide three workers; medium-sized land owners send two; and smallest rice field owners send one individual (Voss, 1983; 158-159). Obligations are thus doled out according to benefits. Repairs and cleaning of canal divisions are handled by smaller groups of owners. Organization and leadership within these associations are usually informal, and the obligation of each owner is often determined through discussion. Members failing to show up for work are fined a certain amount of rice or the missed amount of labor (ibid: 155-157).

Ward institutions can organize large-scale tasks, such as the construction in 1954-55 of a 25 kilometer irrigation canal (Bacdayan, 1974). "Free riders" are a problem in distributing water to individual terraces, however, with some farmers trying to get more water than allotted. Individual distributors are paid to enforce equitable water allocation in some regions (ibid: 176). In much of Mountain Province, the threat and effect of water theft is so strong that nightly vigils are maintained at fields throughout the dry season. Arguments are common, even between relatives, and can lead to violence (Scott, 1969: 227). Unresolved difficulties are taken to the wards for adjudication (Voss, 1983) and may require elders from all the village wards if the dispute is between several groups (Prill-Brett, 1983).

Water distribution rules in the uplands vary, but fields are often ideally watered according to the time of terrace construction, that is, older fields receive water first (Dozier, 1966: 151-152; Voss, 1983: 162-163). In practice, fields closest to the water source frequently have priority. The dispersal of fields often means that many people have terraces both close to, and distant from, the water sources. This fragmentation of land helps ensure equitable water distribution, as has been observed in the Ilocos (Siy, 1982) and Bontok regions (Prill-Brett, 1983). The water distributor role is rotated among Bontok association members, a practice which minimized incidents of favoritism. Erring distributors cannot hope for future fair treatment once they are replaced by someone else. Water theft is not subject to fines, and withholding water as punishment is not done since every family depends on water access (Prill-Brett, 1983).

Ritual Regulation

While swiddeners reduce risk through straggered planting and dispersed fields, synchrony in the cropping cycle is ritually-enforced among some wet-rice groups (Prill-Brett, 1983). Agricultural activities in each Ifugao district, for example, are controlled by the owner of a ritual rice field. He or she is responsible for initiating planting, harvesting, and other phases of the rice cycle. The timing is coordinated both with the unique ecological features of the district and with the performance of rituals designed to ensure a prosperous harvest (Conklin, 1980).

The interweaving of ritual and agriculture strengthens commitment among irrigation association or ward members to fulfill their labor obligations through cultural sanctions. Ritual feasts, taboos on working certain days, competitive inter-ward relationships, and the festive nature of the harvest period all act as cultural incentives for compliance, thus lowering the cost of enforcement.

Implications

This review of upland social organization indicates the fallacy of postulating economic or ecological determinants of behavior. Upland populations responded to similar problems in ways cushioned by previous patterns and unique experiences influencing their perception of viable behavioral options. Social factors are relevant to almost all resource use decisions. Ritual ideology may affect the timing of the agricultural calendar or restrict the availability of land for religious purposes. The presence of communal village, corporate kin group, or ward property may greatly augment a household's access to land and forest resources, possibly excluding rights to outsiders. Leadership patterns can also serve to mobilize labor for large-scale projects or simply serve to settle inter-household disputes. Variations in the organizational capabilities of upland agricultural groups underscore the need for appropriate, socially-grounded development programs that consider such heterogeneity.

In summary swidden groups generally have a weaker institutional coordination of agricultural tasks than do permanent farmers. Cooperative groups tend to be ad hoc, small, and based on casual contracts that expire with the completion of the task. Work groups usually lack community sanctions, being based on voluntary commitments between neighboring field owners or co-residents. Leadership of labor groups is generally informal and seldom vested in a single individual. For upland wet-rice farmers, equitable sharing of water resources requires considerable individual household overseeing, regardless of whether water distributors are present. Even with competition between households, complex group coordination developed in some areas to prevent resource monopolization. Equitability increased by assigning individual labor obligations through group decision-making and public discussion.

Where there are indigenous labor organizations, externally-induced programs to improve resource management can be more successful and less costly if the community can organize such projects within their own framework of labor and resource allocation. Integrated pest management programs for new rice varieties, for example, depend on synchronous planting (Goodell, 1984), an organizational requirement already familiar to some upland groups. On the other hand, institution building may be necessary among groups that lack strong cooperative organization.

Social relations in small-scale communities vary in response to external and internal pressures. Little demonstrated correlation exists, for example, between collectivist traditions and success of modern communal agriculture programs (Wong, 1979). Culturally-determined perceptions of appropriate resource use must be understood to assess local impacts and responses to new programs. The heterogeneity of perceptions distinguish not only upland populations, but also different categories of individuals within populations, a theme explored further in the next section.

TRANSITION TO THE MARKET ECONOMY

Upland socioeconomic organization currently represents both short and long-term adjustments to different environments and economic pressures. Upland groups are *not* necessarily conservative, risk averse, or resistant to change. Many indigenous agricultural systems have fully incorporated new crops and technologies, with some new crops grown for the market and others supplementing subsistence. The current upland pattern is mixed cash and subsistence cropping, with some areas experiencing seasonal, temporary, and permanent out-migration for wage labor. Just as multiple exploitation of ecosystems optimizes subsistence sources, thereby reducing risks and spreading household labor more effectively throughout the year, diverse combinations of income-earning and subsistence-oriented strategies can enhance family survival. Population pressure on resources and the presence of market incentives for cash cropping or wage labor may either or both be sufficient to encourage transition to a commercial economy.

There are drawbacks to becoming totally dependent on the market in mountain environments (Kunstadter, 1978; Guillet, 1983: Brush, 1983). Given the often cooler climate, limited cultivable areas, and expensive transportation over poor roads, risks associated with commercial agriculture can be high. Exacerbating these problems are a lack of low-interest credit sources and opportunistic middlemen. Research designed to increase the productivity of small upland farms is needed. Ethnographic case studies of communities undergoing transformation into market economies illuminate the interaction between community members and the nature of changing social relationships affecting access to local resources and new opportunities. The beneficial or non- beneficial impacts of economic change are major concerns.

Returns to Labor and Land

There has been insufficient analysis of variations of upland labor patterns by sex, age, and background. An exception is Eder's study of work patterns among individuals of different status in Palawan. He concluded that ". . . lower-group farmers not only do less remunerative work than do other farmers, they do less work altogether; less work even where they possess the resources that would make work nominally possible" (Eder, 1982a: 105). Although arboriculture was more rewarding, poor farmers were reluctant to invest labor and capital in an enterprise having deferred returns, especially since tree planting conflicted with labor demands of subsistence rice swiddening. A similar preference among poorer farmers for activities that generate subsistence goods or more immediate returns, such as daily wage labor, has been noted elsewhere (Connelly, 1982: 58-60).

Tree crops provide favorable returns to land and labor if commercial demand is steady (Eder, 1981; Connelly, 1982). Tree crops can increase in importance as land becomes more intensively utilized (Lopez-Gonzaga, 1983). Coincidental with the transition from subsistence rice production to the cash cropping of maize, for example, the Buhid expanded production of crops such as cacao, coffee, coconut, and fruit.

In San Jose, Palawan, Eder (1981: 99-102) estimated returns per man-day for rice swiddens at 3.18 pesos, compared to returns of 15 pesos per man-day for tree crops (table 1). Tree cropping in Napsaan, Palawan yielded returns of 32 pesos per man-day, compared to returns of 6 pesos to 10 pesos per man-day for rice swiddens. Although comparing these two studies is somewhat arbitrary given different time

periods and data collection methods, evidence suggest that tree cropping yields higher returns to labor than other options. Yet both authors
stress problems preventing small farmers from pursuing this strategy
to the exclusion of other income or subsistence-oriented activities.

Table 1. Net returns to labor in Palawan, 1971 and 1981.

San Jose, Palawan, 1971*		Napsaan, Palawan, 1981-82**	
Activity	Returns Per Man-Day	Activity	Return Per Man-Day
Rice Swiddens	3.93 pesos	Rice Swiddens	6-10 pesos
Arboriculture		Arboriculture	
(bananas)	15.00 pesos	(cashews)	32 pesos
Day Labor	3.50 pesos	Day Labor	12 pesos
Vegetable Gardens	7.08 pesos		
Livestock Raising	15.00 pesos		
		Forest Gathering	
		(rattan)	16 pesos
		(almasiga)	20 pesos

*Data from Eder 1982a: 104.
**Data from Connelly 1982:56-59.

Permanent agriculture does not necessarily provide higher returns to
land than upland swidden horticulture. Schlegel's (1979) study in
Cotabato, Mindanao revealed a striking difference in the productivity of
traditional Tiruray rice and corn swiddens, as opposed to the plowed
fields of Tiruray peasant tenants (table 2). Although sample sizes were
small, data suggests that the swiddens were more productive. Tiruray
peasants derived a somewhat superior diet, but at the cost of almost
twice the annual labor expended by swidden farmers (Schlegel and
Guthrie, 1973). Under these conditions, Tiruray rationally oppose
commercialization which involves harder work for a fewer returns and
greater vulnerability to market forces (Schlegel, 1979:166).

Several points are suggested. It is not clear whether Tirurays are
"resisting" market agriculture or simply avoiding burdens of tenancy
farming and unfavorable profit-sharing arrangements. Voss (1983),
Russell (1983), Lopez-Gonzaga (1983) and Eder (1982a) suggest that
favorable returns to cash cropping and the people's perception of benefits frequently stimulates commercialization. The Tiruray experience

Table 2. Tiruray yields for rice and corn, Cotabato*

	Cavans Per Hectare		Seed/Yield Ratios	
	Rice	Corn (1st Crop)	Rice	Corn (1st Crop)
Figel Swiddens (1966)	51.9	19.6	1:40	1:124
Kaba Kaba Plowed Fields (1967)	38.2	18.7	1:28	1:195

*Data from Schlegel, 1979: 165.

underscores the importance of monitoring nutritional impacts when new crops or technology are introduced. Diet may or may not be enhanced by changing productive activities. A constant theme in these three studies is the danger of specializing too narrowly in one economic strategy, especially if subsistence activities are abandoned in favor of one or two market activites. Indebtedness for consumption needs among poorer farmers is sometimes the result of engaging in cash-oriented activities as opposed to subsistence production (Connelly, 1982; Russell, 1983). Diversified strategies may also be more economically stable, since various fall-back options are then available if problems arise in pursuing a primary activity (Yengoyan, 1974). Environmental heterogeneity, staggered labor demands for each activity throughout the year, and skewed land distribution in a community can be constraints, of course.

New crops, including aboriculture, must be considered relative to competing demands for labor, land, and capital. It is essential to understand, for example, who does or does not plant trees, and why (Brokensha and Castro, 1984: 11) Upper income groups and larger farmers often have the land and cash resources needed for tree farming, livestock raising, or cash cropping (Eder, 1982a). Such groups are also able to invest in activities with deferred returns, since they are most insulated from short-run subsistence demands.

Cash cropping usually decreases labor and land available for subsistence production. But market specialization is not necessarily harmful since productivity is often increased. Upland populations increasingly need cash to offset land-use pressure or enhance family well-being. Seasonally impassable roads, inadequate market intelligence, and an underdeveloped rural financial market often underly the risks of market specialization in remote mountain areas.

Marketing and Credit Relations

Local or outside traders are frequently the major credit source in the uplands, a situation that can adversely affect small-scale commercial farmers. The abaca market structure in eastern Mindanao reflects a typical pattern of indebtedness to lowland middlemen or local large producers. In return for no-interest loans of cash, livestock, or stripping machines, the buyer receives a farmer's harvest at whatever price he can set. Where a farmer lacks processing equipment, the arrangement can be like sharecropping in which net profits are divided equally (Yengoyan, 1966: 693).

There are similar marketing and credit arrangements in Benguet (Russell, 1980a and 1983; Voss, 1983), where credit is used to purchase fertilizer, seeds, and other inputs, Most middlemen in this vegetable growing region were formerly Chinese-Filipinos and lowland Filipinos (Davis, 1973). Southern Kankana-ey now act as middlemen, although many larger farmers are still financed by Manila Chinese-Filipino traders (Russell, 1980b). Agricultural loans from middlemen in Madaymen, Benguet are often accompanied by rice supplies, the cost of which is deducted from the harvest receipts, as is the "hidden interest" for the production loan. For example, if prices are low, the trader's profit may comprise 5% to 10% of the net harvest receipts. When prices are high, 20% to 35% is more typical. Traders consolidate ties with producers not only by extending credit, but by building social indebtedness through no-interest emergency loans, agricultural advice, free rides into town, and holiday gifts. This pattern is geared for long-term, rather than seasonal or even yearly gains. Farming vegetables in Benguet is a hit or miss activity marked by seasons of overall loss or extremely minor gains, but punctuated by occasional seasons of high profits. Traders manipulate the fluctuations in net gains and losses to ensure long term returns without allowing farmers to financially capsize during poor seasons. Extreme indebtedness for both production and consumption loans, however, are common among small producers in this region (Russell, 1983, and 1985).

Access to accurate market price information may be limited in the uplands if prices are volatile throughout the day, if farmers are far from the market, or if farmers cannot seek alternative outlets due to indebtedness to a single trader. Mandaya cultivators corroborate price information through local residents making frequent trips to town. Since creditors are aware of this check, they rarely engage in predatory pricing for fear of losing a producer's loyalty (Yengoyan, 1966: 698). Where traffic between the regional marketplace and the settlement is infrequent, however, the source of market reports are often limited to middlemen. Lower prices may be offered in this case, especially if the middleman has a monopoly. Even in areas of intensive production a producer may be unable to take advantage of higher prices offered by free agents owing to indebtedness to a particular middleman. When accounts are eventually settled, farmers can seldom challenge the prices recorded since the trader may claim an inability to obtain the free agent price when the crop was sold. Benguet farmers' inability to check selling prices is exacerbated by some middlemen's unwillingness to settle accounts more than once every year or two (Russell, 1985).

Market participation can be confounded not only by a lack of reliable price information, but also by the typical peasant cycle of having to sell harvests cheaply and buy consumer goods at higher prices due to remote location (Yengoyan, 1966: 700-710; Schlegel, 1979: 163). Early stages of commercial transition may be especially difficult if people lack accounting skills or are unfamiliar with money as a medium of exchange (Schlegel, 1979: 100-111; Lopez-Gonzaga, 1983: 166-167). Capital-intensive market cropping increases vulnerability to price fluctuations for both production and consumption. Cultivators may either become more indebted or have to invest more time in less remunerative activities such as wage labor (Wallace, 1970; Connelly, 1982). Class differentiation between wealthy producers or traders and the bulk of small producers often increases (Yengoyan, 1966).

Land Ownership and Inequality

Private property often increases at the expense of communal property in transitional commercial areas. Tiruray traditionally distinguished between swidden lands that reverted to public forest after cultivation and door-yard gardens subject to long term, continuous use and private ownership (Schlegel, 1981). Tiruray shifting to permanent plow agriculture expanded permanent possession to include plowed fields since these could no longer revert to public forest. Partial adoption of permanent property concepts has also occurred among Mandaya (Yengoyan, 1971: 370). Small abaca producers continue to see abaca swiddens as primarily subject to use rights, but they now construct fences and continually cultivate the fields in order to retain such use rights. Rising land values and production costs act as further incentives to titled land, but the costs of legal titling favor large producers with better non-local ties (Yengoyan, 1971; Russell, 1983).

Lowland migrants and decreasing forest lands were incentives to register land as private property among the Buhid Mangyan (Lopez-Gonzaga, 1983: 126, 185). Among the Manobo in Bukidnon, however, rather than individual land titling, in-migration of lowlanders resulted in the increased clearance and sale of forest swiddens to settlers (Allison, 1963: 168-170). Clearly, forms of externally-induced social change among upland groups are diverse. While swiddening groups such as the T'boli adopted private property concepts, such adaptation was sometimes too late to prevent land loss to more prosperous farmers or outsiders (McDonagh, 1983).

Wealth and land inequality in mixed subsistence and commercial agricultural areas can disrupt traditional leadership patterns. Individuals with outside contacts and knowledge can often challenge elders in decisions concerning resource use or ownership. Voss (1983: 159-160) described how a wealthy, educated Sagada farmer constructed terraces close to an irrigation canal, thus defying customary laws. The farmer relied on the Philippine court system rather than the ward council to successfully circumvent local opposition, although restrictions were enacted on his water allocation. In another case, traditional deference to order, wealthier farmers allowed some individuals to misuse funds and equipment of a Benguet marketing cooperative. Disenchantment over the abuses then led to a reluctance of other members to continue their support (Russell, 1980a and 1980b).

With wage labor opportunities or where the agricultural cycle has fewer peak labor demands, reciprocal or feast-labor arrangements may decline. Benguet vegetable farmers using skilled laborers stopped hiring poorer relatives for havesting (Davis, 1978: 66-67). The hiring of laborers has also replaced exchange labor, except among smaller farmers (Voss, 1980; Russell, 1980). The replacement of indigenous cooperative labor in mixed subsistence-cash cropping economies, however, is neither absolute nor inevitable. Exchange labor can adapt to new conditions, especially if participation is necessary to maintain community membership rights (Voss, 1980; Voss, 1983: 198-200).

New cash crops and technologies, and wage labor opportunities produce shifts in socio-economic relationships and land distribution. While results may include greater community income, benefits are seldom equally distributed. Land mortgaging to obtain production capital, for example, is increasing in Benguet and may lead to land consolidation (Russell, 1983: 215-218). Increasing land tenancy has also been observed in the cash cropping areas of this province (ibid: 214; Voss, 1983: 226). When new techniques or crops require larger amounts of land, capital, or information, such changing land tenure patterns indicate that better-off farmers are benefiting more than poorer members (Yengoyan, 1966; Eder, 1982a; Lopez-Gonzaga, 1983).

Implications

Land access inequality has a long history in the uplands, the basis of which may shift with rising land values, however, since the size of holdings in different ecological zones may not correspond to values. More important determinants of value, and therefore, inequality, may be nearness to urban centers, water, roads, possession of animals or vehicles, and differential access to credit and marketing channels (Brokensha and Castro, 1984: 35).

Commercial agriculture often stimulates the desire for greater land holdings and capital goods. Resulting changes in stratification do not necessarily lead to a replication of lowland Philippine class structure, however, since local histories, systems of resource allocation, and indigenous social institutions in the uplands are so diverse (Voss, 1983; Lopez-Gonzaga, 1983; Russell, 1983). Patron-client relationships between farmers, inter-household cooperative alliances, or indigenous traditions of wealth redistribution may offset market forces that otherwise increase stratification. The levelling influence of the ward institution, peripheral market position, and cultural attitudes toward capital accumulation, for example, impeded class differentiation in Sagada (Voss, 1983: 211). In any case, a landless class in the uplands has not yet emerged.

Replacement of a diversified subsistence cropping regime with specialized cash cropping can increase farmer indebtedness and vulnerability to price fluctuations, especially if the market crops are not consumable staples. While upland Filipinos willingly adopt commercial agriculture when they perceive that benefits outweight risks, the information and foresight with which such decisions are made are far from perfect or equally advantageous for all community members.

CONCLUSIONS

The cultural context of economic change can baffle non-anthropologists. Development planners often assume that cultural beliefs are either obstacles to be overcome or remnants of the past that will disappear as market forces transform material incentives into rationalized economic action. Implicit in such erroneous interpretations is that anthropological insights are overly cumbersome since they require too much information about a culture. This mistaken conception of the holistic ethnographic approach often serves as a rationale for retreating into number-crunching or model-mongering based on faulty assumptions that market forces ultimately influence everyone equally and at the same rate. Such aggregated data reduces upland social structural variations into homogeneous, presumably predictable trends that find little empirical support at the micro-level of any one population (Bennett and Kanel, 1983).

There is considerable socioeconomic heterogeneity within and between upland populations in lifestyles and responses to environmental constraints and opportunities. This diversity should not obscure that integration into national society has eroded self-sufficiency and isolation in these communities. Lowland Filipino and Western ideas and values have infiltrated many indigenous belief systems, albeit in varying ways. For development to occur, it is important to understand former and current forms of resource management and the factors influencing behavior. Technical solutions alone are unlikely to be widely adopted if local socioeconomic organization is incompatible or if only a few individuals are able to benefit. Projects or techniques found to be workable in one area may not be effectively transplanted to another seemingly similar area.

Upland populations include a diversity of ethno-linguistic groups, not all of whom are cultural minorities. Resource competition between resident groups, government agencies, and private corporations now form the context in which land use changes occur. Land pressure in the uplands has resulted in innovations and alterations in resource management and productive activities. Assumptions of socioeconomic equilibrium in most regions are now misplaced given increasing population pressure and economic demands. Mixed subsistence and commercial agricultural are common and should remain so given the advantages in such diversification.

Indigenous irrigation control, terrace construction, and agroforestry provide important information on effective upland resource management. Current productive strategies may or may not have desirable environmental impacts, and introducing inputs requires an understanding of peoples' perceptions.

Cultural perceptions of appropriate economic strategies and the social institutions governing access to resources vary greatly among upland populations. Redirecting change is difficult since pressures for change impinging on any single group and their respective histories are diverse. Needed is the monitoring of changes in land distribution and in rules controlling access to property and in social relations. Such information could help prevent the designing of programs in which certain groups, such as smaller producers or women, are neglected or adversely affected. Clear understanding of contemporary socioeconomic activities, their requisite labor demands, and contribution to subsistence will help ensure that benefits are not monopolized.

Rural social organizational capabilities of different groups are not identical. Participation in project design and implementation can strengthen existing institutions or can provide support when such institutional mechanism are lacking. Cultural norms that restrict the self-serving actions of a few may be weak when new sources of power or wealth are presented. Ensuing conflicts between different groups have often led to widespread disillusionment and reluctance to contribute support. Motivations underlying land use or project participation may be primarily either long-term subsistence or short-term profit; the resultant impact on ecological stability can vary accordingly.

Increasing local self-sufficiency is an important goal in the uplands. Owing to different abilities within the population to absorb capital costs, reliance on external inputs can create sharper inequalities within a community than would be use of locally produced inputs alone. Dependency on commercial inputs can increase vulnerability in fragile ecosystems to price fluctuations, leaving farmers indebted to middleman or wealthier community members. This condition may then force poor farmers to engage in ecologically damaging practices to survive. One way to prevent this outcome is to tap farmers' experience and cultural knowledge of resource management by involving them in experimentation and evaluation of local crop combinations or appropriate technologies. This interactive approach can also stimulate greater economic independence in the long run, thereby decreasing the need for further intervention in the future.

Raising living standards in ways that are ecologically and economically self-sustaining will go a long way in correcting the social injustices and disrupted local economies that competition between different economic interests have produced in the uplands. The rights of indigenous groups to maintain their cultural integrity and autonomy demand a morally responsible stance, especially if socioeconomic integration into the national society threatens local control over traditional resources and ways of life. Although it is still contentious as to whose interests upland development program are intended to serve, lessons learned from previous failures alone argue for the necessity of decentralized, flexible, and innovative efforts if success is to be achieved. If environmental preservation and the well-being of upland peoples are not both truly desired, then the alternative outcome is likely to be even more costly in social, economic, and political terms.

REFERENCES CITED

Aguilar, Filomeno V., Jr., 1982. Social Forestry for Upland Development: Lessons from Four Case Studies. Quezon City . Institute of Philippine Culture, Ateneo de Manila University.

Aguirre Beltran, G., 1967. Regiones de Refugio. Mexico . Instituto Indigenista Interamericano, Ediciones Especiales: 46.

Allison, William W., 1963. A Compound System of Swidden ("Kaingin") Agriculture. Philippine Geographical Journal 7 (3): 159-172.

Atal, Yogesh and P.L. Bennagen, 1983. Introduction. In: Swidden Cultivation in Asia, v. 1. Bangkok .UNESCO.

Bacdayan, Albert S., 1974. Securing water for drying rice terraces; Irrigation, community organization, and expanding social relationships in a Western Bontoc group, Philippines. Ethnology 13: 247-260.

1977. Mechanistic cooperation and sexual equality among the Western Bontoc. In: Alice Schlegel (ed.), Sexual Stratification: A Cross-Cultural View, New York: Columbia University Press.

Bernett, Milton L., 1967. Subsistence and transition of agricultural development among the Ibaloi. In: Mario Zamora, (ed.), Studies in Philippine Anthropology, Quezon City: Alemar-Phoenix.

Barrows, David P., 1902. Field notes . Benguet and other parts of the Luzon Cordillera. (Unpublished diary). Quezon City. Igorot Study Center, St. Andrew's Seminary.

Barton, Roy F., 1922. Ifugao Economics. University of California Publications in American Archeology and Ethnology 15 (5):385-446.

1949. The Kalingas . Their Institutions and Custom Law. Chicago . University of Chicago Press.

1969. Ifugao Law. Berkeley:University of California Press.

Bello, Moises C., 1972. Kankanay Social Organization and Cultural Change. Quezon City: Community Development Research Council.

Bennagen, Pons L., 1976. Kultura at kapaligiran : Pangkulturang pagbabago at kapanatagan ng mga Agta sa Palanan, Isabela. M.A. Thesis. University of the Philippines, Quezon City.

1983. Philippines. In: Swidden Cultivation in Asia, v. 2. Bangkok. UNESCO.

Bennett, John W. and Don Kanel, 1983. Agricultural Economics and Economic Anthropology: Confrontation and Accommodation. Topics and Theories, ed. Sutti Ortiz. Lanham, Maryland: University Press of America.

Breeman, Nico Van, L.R. Oldeman, W.J. Plantinga, and W.G. Wielmaker, 1970. The Ifugao rice terraces. In: N. Van Breeman, A.J.F. Heydendael, D. Hille Ris Lambers, H.C. Molsters, L.R. Oldeman, W.J. Platinga, and W.G. Wielmaker, (eds.), Aspects of Rice Growing in Asia and the Americas, Miscellaneous Papers 7. Wageningen, Netherlands: Landbouwhogeschool.

Brokensha, David and Alfonso P. Castro, 1984. Fuelwood, Agro-Forestry, and Natural Resource Management : The Development Significance of Land Tenure and Other Resource Management/Utilization Systems. Binghamton, New York. Institute for Development Anthropology.

Brush, Stephen B., 1983. Traditional agricultural strategies in the hill lands of tropical America. Culture and Agriculture 18 (Winter). 9-16.

Cadelina, Rowe V., 1982. Batak interhousehold food sharing: A systematic analysis of food management of marginal agriculturalists in the Philippines. Ph.D. dissertation, University of Hawaii.

Cole, Fray-Cooper, 1913. The wild tribes of Davao district, Mindanao. Field Museum of Natural History, Anthropological Series 12:49-203.

1922. The Tinguian: Social, religious ,and economic of a Philippine tribe. Field Museum of Natural History, Anthropological Series 14:231-493.

1956. The Bukidnon of Mindanao. Chicago Natural History Museum, Fieldiana: Anthropology 46.

Conklin, Harold C., 1957. Hanunoo agriculture . A Report on an Integral System of Shifting Cultivation in the Philippines. Rome: Food and Agriculture Organization.

1980. Ethnographic Atlas of Ifugao. New Haven, Connecticut . Yale University Press.

Connelly, W. Thomas, 1982. Economic adaptation in an upland environment in Palawan . A preliminary summary of field research. Philippine Sociological Review 30: 51-62.

Cordillera consultative committee, 1984. Dakami Ya Nan Dagami: Papers and Proceedings of the 1st Cordillera Multi-Sectoral Land Congress, 11 to 14 March 1983, Baguio City.

Coward, E. Walter, Jr., ed., 1980. Irrigation and Agricultural Development in Asia. Ithaca, New York: Cornell University Press.

Davis, William G., 1973. Social Relations in a Philippine Market. Berkeley: University of California Press.

_______ 1978. Anthropology and theories of modernization: Some perspectives from Benguet. Papers in Anthropology 19(2).59-72.

DeRaedt, Jules, 1964. Religious representations in Nothern Luzon. Saint Louis Quarterly 2(3):245-348.

_______ 1969. Myth and Ritual: A relational study of buwaya mythology, ritual and cosmology. Ph.D. dissertation, University of Chicago.

Dozier, Edward P., 1966. Mountain Arbiters . The Changing Life of a Philippine Hill People. Tucson, Arizona: University of Arizona Press.

Drucker, Charles B., 1974. Economics and social organization in the Philippine highlands. Ph.D. dissertation, Stanford University.

_______ 1977. To inherit the land: Descent and decision in Northern Luzon. Ethnology 16 (1).1-20.

Eder, James F., 1977. Portrait of a dying society: Contemporary demographic conditions among the Batak of Palawan. Philippine Quarterly of Culture and Society 5:12-2.

_______ 1981. From grain crops to tree crops in a Palawan swidden system. In: Harold Olofson, (ed.), Adaptive Strategies and Change in Philippine Swidden-based Societies, Los Banos: Forest Research Institute.

_______ 1982a. Who Shall Succeed? Agricultural Development and Social Inequality on a Philippine Frontier. New York: Cambridge University Press.

_______ 1982b. No water in the terraces . Agricultural stagnation and social change at Banaue, Ifugao. Philippine Quarterly of Culture and Society 10:101-116.

Eggan, Fred, 1941. Some aspects of culture change in the northern Philippines. American Antropologist 43:11-18.

1954. Some social institutions in the mountain provinces and their significance for historical and comparative studies. Journal of East Asiatic Studies 3: 329-335

1960. The Sagada Igorots of northern Luzon. In: George P. Murdock, (ed.), Social Structure in Southeast Asia, Viking Fund Publications in Anthropology, No. 29. New York. Wenner Gren Foundation.

Ellen, Roy F., 1982. Environment, Subsistence and System: The Ecology of Small-Scale Social Formations. London: Cambridge University Press.

Estioko-Griffin, Agnes, 1981. Woman the hunter. In: Frances Dahlberg, (ed.), Women the Fatherer, New Haven, Connecticut: Yale University Press.

Estioko-Griffin, Agnes and P. Bion Griffin, 1981. The beginnings of cultivation among Agta hunter-gatherers in northeast Luzon". In: Harold Olofson, (ed.), Adaptive Strategies and Change in Philippine Swidden-based Societies, Los Banos: Forest Research Institute.

Fernandez, Carlos A and Frank Lynch, 1979. "The Tasaday, a food-gathering forest band". In: Mary Racelis Hollnsteiner, (ed.), Society, Culture and the Filipino, Quezon City: Institute of Philippine Culture, Ateneo de Manila University.

Fox, Robert B., 1982. Religion and society among the Tagbanuwa of Palawan island, Philippines. Monograph No. 9. Manila: National Museum.

Frake, Charles O., 1955. Social organization and shifting cultivation among the Sindangan Subanan. Ph.D. dissertation. New Haven, Connecticut: Yale University.

1962. Cultural Ecology and Ethnography. American Anthropologists 64:53-59.

Garvan, John M., 1931. The Manobos of Mindanao. Washington, D.C.: U.S Government Printing Office.

1964. The Negritos of the Philippines. Herman Hochegger, ed., Weiner Beitrage Zur Kulturgeschichte und Linguistik 14.

Geertz, Clifford, 1963. Agricultural Involution. Berkeley: University of California Press.

Goodell, Grace E., 1984. Bugs, bunds, banks, and bottlenecks: organizational contradictions in the new rice technology. Economic Development and Cultural Change 33(1):23-41.

Griffin, P. Bion, 1981. Northern Luzon Agta subsistence and settlement. Pilipinas: A Journal of Philippine Studies 2-26-42.

Guillet, David, 1983. Towards a cultural ecology of mountains: The central Andes and the Himalayas compared. Current Anthropology 24(5):562-567.

Integrated Research Center, 1983. Dimensions of Upland Poverty: A Macro-View and Micro-View. Manila: De La Salle University.

Jenks, Albert E, 1905. The Bontoc Igorot. Manila: Ethnological Survey Publications, v.1.

Jocano, F. Landa, 1968. Sulod Society. Quezon City: University of the Philippines Press.

Keesing, Felix M., 1949. Some notes on Bontok social organization, northern Philippines. American Anthropologist 51:578-601.

1962. The Ethnohistory of Northern Luzon. Stanford, California. Stanford University Press.

Kundstadter, Peter, 1978. Alternatives for the development of upland areas. In: Peter Kunstadter, E.C. Chapman, and Sangha Sabhastri, (ed.), Farmers in the Forest: Economic Development and Marginal Agriculture in Northern Thailand, Honolulu: East-West Center.

Lawless, Robert, 1977. Society ecology in northern Luzon: Kalinga agriculture, organization, population and change. Papers in Anthropology 18(1):1-36.

Lopez-Gonzaga, Violeta, 1982. Bolo agriculture and subsistence production in upland Mindoro: The case of the Buhid Mangyan. Contributions to Contemporary Southeast Asian Ethnography 1 (September):19-34.

1983. Peasants in the Hills: A Study of the Dynamics of Social Change Among the Buhid Swidden Cultivators in the Philippines. Quezon City: University of the Philippines Press.

Maceda, Marcelino N., 1974. A survey of landed property concepts and practices among the marginal agriculturalists in the Philippines. Philippine Quarterly of Culture and Society 2(1-2):5-20.

1975. Mamanua. In: Frank M. LeBar, (ed.), Ethnic Groups of Insular Southeast Asia, v. 2, New Haven, Connecticut: Human Relations Area Files.

Manuel, E. Arsenio, 1973. Manuvu's Social Organization. Quezon City: Community Development Research Council.

1975. Upland Bagobo (Manuvu'). In: Fronk M. LeBar, (ed.), Ethnic Groups of Insular Southeast Asia, v.2, New Haven, Connecticut: Human Relations Area Files.

McDonagh, Sean, 1983. Modernization, multinationals and the tribal Filipino, Solidarity 4:73-82.

Moss, Claude R., 1920. Nabaloi law and ritual. University of California Publications in American Archeology and Ethnology 15(3):207-342.

Olofson, Harold, 1980. Swidden and kaingin among the southern Tagalog: A problem in Philippine upland ethno-agriculture. Philippine Quarterly of Culture and Society 8:168-180.

Ed., 1981. Adaptive Strategies and Change in Philippine Swidden-based Society. Los Baños: Forest Research Institute, U.P.L.B.

Pelzer, Karl J., 1945. Pioneer Settlement in the Asiatic Tropics: Studies in Land Utilization and Agricultural Colonization in Southeast Asia. American Geographical Society Special Contribution No. 29.

Pennoyer, F. Douglas, 1981. Shifting cultivation and shifting subsistence patterns among the Taubuid of Mindoro. In: Harold Olofson, (ed.), Adaptive Strategies and Change in Philippine Swidden-based Society, **op cit.**

Peralta, Jesus T., 1982, I'wak Alternative Strategies for Subsistence. Anthropological Papers No. 11. Manila National Museum.

Peterson, Jean T., 1978. The Ecology of Social Boundaries: Agta Foragers in the Philippines. Urbana: University of Illinois Press.

Peterson, Warren, 1981. Recent adaptive shifts among Palanan hunters of the Philippines. Man (N.S.) 16:43-61.

Prill-Brett, June, 1977. The Bontok Peace Pact Institution. Quezon City: Social Sciences and Humanities Research Committee, University of the Philippines.

1983. The social dynamics of irrigation in Tukuran society. Cordillera Studies Center Monograph Series 2, Baguio City: University of the Philippines.

1984. Bontok land tenure. Paper presented at the Conference on Ancestral Land Law, University of the Philippines College of Law, Baguio City.

Pungayan, Eufronio L., 1978. A sociolinguistic analysis of the kinship structures among Benguet Ibalois in the Philippines. M.A. thesis, Saint Louis University.

Reid, Lawrence A., 1971. Philippine Minor Languages: Words Listed and Phonologies. Honolulu: University of Hawaii Press.

1971. Ward and working groups in Guinaang, Bontoc, Luzon. Anthropos 67:530-551.

Rice, Delbert, 1981. Upland agricultural development in the Philippines: An analysis and a report on the Ikalahan program. In: Harold Olofson, (ed.), Adaptive Strategies and Change in Philippine Swidden-based Societies, **op cit.**

Russell, Susan D., 1980a. A Socio-Economic Analysis of Cooperatives and Marketing in Buguias, Benguet Province. PHI/76/001, Manila: Food and Agriculture Organization.

1980b. Motivating development: Alternative strategies and long-range planning in Benguet. Aghamtao 3:88-98.

1983. Entrepreneurs, ethnic rhetorics, and economic integration in Benguet province, highland Luzon, Philippines. Ph.D. dissertation, University of Illinois.

1985. Final Report: Urban Food Markets Project. (Unpublished ms.) New York: Rockefeller Foundation.

Schlegel, Staurt A., 1979. Tiruray Subsistence: From Shifting Cultivations to Plow Agriculture. Quezon City: Ateneo de Manila Press.

1981. Tiruray gardens: From use-right to private ownership. Philippine Quarterly of Culture and Society 9:5-8.

Voss, Joachim H., 1980. Cooperation and market penetration: transformation in indigenous forms of cooperation on the Cordillera central of the Philippines. Aghamtao 3:42-50.

1983. Capitalist penetration and local resistance: continuity and transformations in the social relations of production of the Sagada Igorots of Northern Luzon. Ph.D. dissertation, University of Toronto.

Wallace, Ben J., 1970. Hill and Valley Farmers: Socio-Economic Change Among a Philippine People. Cambridge, Massachusetts: Schenkman Publishing Company.

Warner, Katherine, 1981. Swidden strategies for stability in a fluctuating environment: The Tagbanwa of Palawan. In: Harold Olofson, (ed.), Adaptive Strategies and Change in Philippine Swidden-based Societies, **op cit.**

Warren, Charles P., 1975. Palawan. In: Frank LeBar, (ed.), Ethnic Groups of Insular Southeast Asia, v.2, New Haven, Connecticut: Human Relations Area Files.

1984. Agricultural development of the Batak of Palawan, Philippiees: A case study. Crossroads: An Interdisciplinary Journal of Southeast Asian Studies 2(1):1-11.

Wernstedt, Frederick L. and J.E. Spencer, 1967. The Philippine Island World. Berkeley: University of California Press.

Wong, John, ed., 1979. Group Farming in Asia. Singapore: Singapore University Press for A/D/C.

Yen, Douglas E. and John Nance, 1976. Further studies on the Tasaday. Panamin Foundation Research Series No. 2. Manila: Panamin Foundation.

Yengoyan, Aram A., 1964. Environment, shifting cultivation, and social organization among the Mandaya of eastern Mindanao, Philippines. Ph.D. dissertation, University of Chicago.

1966. Marketing networks and economic processes among the abaca cultivating Mandaya of eastern Mindanao, Philippines. In: Raymond E. Borton, (ed.), Selected Readings to Accompany Getting Agriculture Moving, v.2, New York: Agricultural Development Council.

1971. The effects of cash cropping on Mandaya land tenure. In: Ron Crocombe, (ed.), Land Tenure in the Pacific, Melbourne: Oxford University Press.

84 MAN, AGRICULTURE AND THE TROPICAL FOREST

Schlegel, Stuart A. and H. A. Guthrie, 1973. Diet and the Tiruray shift from swidden to plow farming. Ecology of Food and Nutrition 2:181-191.

Scott, William H., 1958 A preliminary report on upland rice in northern Luzon. Southwestern Journal of Anthropology 14(1):87-105.

______ 1969. Growing rice in Sagada. In: On the Cordillera: A Look at the Peoples and Cultures of the Mountain Province. Manila: MCS Enterprises.

______ 1974. Discovery of the Igorots. Quezon City: New Day Publishers.

______ 1979. Class structure in the unhispanized Philippines. Philippine Studies 27:137-159.

______ 1982. The Spanish occupation of the Cordillera in the 19th century. In: Alfred M. McCoy and Ed. C. de Jesus, (ed.), Philippine Social History: Global Trade and Local Transformations, Honolulu: University of Hawaii Press.

Siy, Robert, Jr., 1982. Community Resource Management: Lessons from the Zanjera. Quezon City: University of the Philippines Press.

Soffer, Arnon, 1982. Mountain geography: A new approach. Mountain Research and Development 2:391-398.

Spencer, J.E., 1966. Shifting Cultivation in Southeastern Asia. Berkeley: University of California Press.

Takaki, Michiko, 1977. Aspects of exchange in a Kalinga society, northern Luzon, Ph.D. dissertation, Yale University.

Tapang, Ben J., 1982. Innovation and economic change: A case history of the Ibaloy cattle enterprise in Benguet. M.A. thesis, Center for Research and Communication.

Vander Meer, Canute, 1963a. Agricultural rituals for corn crops on Cebu island, Philippines. Philippine Journal of Science 96(3):305-318.

______ 1963b. Corn cultivation in Cebu: An example of an advanced stage of migratory farming. Journal of Tropical Geography 17:172-177.

Vilar-Basco, Carmen, 1956. Two bago villages. Journal of East Asiatic Studies 5:125-212.

1973. Kindreds and task groups in Mandaya social organization. Ethology 12(2):163-177.

1974. Demographic and economic aspects of poverty in the rural Philippines. Comparative Studies in Society and History 16:58-72.

1975. Davao gulf. In: Frank M. LeBar, (ed.), Ethnic Groups of Insular Southeast Asia, v.2, New Haven, Connecticut: Human Relations Area Files.

4

POPULATION PRESSURE AND MIGRATION IN PHILIPPINE UPLAND COMMUNITIES

Ma. Concepcion J. Cruz

A major requirement for the appraisal of upland development is reliable information on the rate of growth of the upland population. A profile of population numbers and distributions is important in pinpointing priority areas for development. As population increases, the relationship between the population growth rate and the long-term stability of the upland environment is threatened as population increases and as man-to-land ratios decline. Over-exploitation of forest resources arises as migration into upland areas exceeds available resources. Migrants may have farming experiences different from those required in the uplands, leading to the use of destructive cropping systems.

The need for a reliable population estimate can also be seen in the conflicting figures currently available, most of which appear to underestimate the problems of population pressure. Rough estimates include 379,372 (Llapitan, 1977), 400,000 (Spencer, 1966), and 600,000 (Pollisco, 1978). Recently, the Bureau of Forest Development (BFD) Upland Development Working Group estimated that 15% of the national population or 7.5 million people live in the uplands (Lynch, 1984b). The BFD reported that only 800,000 persons were living in the uplands as of 1980. This means that less than 1.7 percent of the national population occupies about 55% of the nation's land (Lynch, 1984a). This seems unrealistic considering that in 1960 alone, Wernstedt and Simkins (1965) reported that 716,183 persons from Luzon and the Visayas moved to the frontier areas of Bukidnon, Surigao, Agusan, and Cotabato provinces in Mindanao.

This chapter examines population pressure in Philippine upland communities with respect to conditions affecting migration. Specifically, the chapter analyzes (1) the relationship between population growth, migration, and population density, (2) the effects of migration on the long-term stability of the upland environment, (3) the historical conditions and government policies which encouraged population movement into the uplands, and (4) the direction and composition (characteristics) of the major upland migration streams. The chapter will provide policymakers and researchers with basic information on population movements in the Philippine uplands. However, the evaluation of reasons for moving and selection of settlement sites is tentative and suggestive of the need for more comprehensive data gathering and research.

The chapter is divided into four parts. The first provides a brief overview of theoretical models dealing with the relationship of population pressure on the use of forest resources. Then Philippine microlevel studies of migration to the uplands are reviewed in terms of the general characteristics of migrants and the motives, circumstances, and effects of such movements. The chapter further examines upland migration using National Census and Statistics Office (NCSO) data for 1970 for a sample of 486 municipalities and 1,179 barrios covered by the Integrated Social Forestry Program of the BFD. A brief description of the implications on the development of policies for social forestry is presented in the last part.

POPULATION PRESSURE IN THE UPLANDS

Upland communities respond to population pressure in terms of changes in the basic production system and in patterns of cooperation in the use of forest resources. A major concern of researchers has been the conditions under which population pressure appears and is sustained, and the corresponding shifts to more intensive production systems leading to shortened fallow periods and resource degradation (Flieger, 1977; Gourou, 1966).

Boserup (1965) argued that, when population density is low, only a small portion of the total area available is cultivated, and land reverts to long periods of natural fallow. As population density rises, the cropping period is increased and fallow years are decreased. As a final stage, continuous cropping takes place; the transformation from shifting cultivation to sedentary agriculture is completed. In this sequence, the shift to more intensive cropping occurs as population growth demands greater total output. The transition from shifting cultivation to bush fallow and, finally, to multiple cropping is prompted by greater demand for food availability and falling crop yields.

An assumption of the Boserup model is that population pressure demands a technical response, such as more intensive cropping or the adoption of new cultivation techniques. Grigg (1979) has shown that intensification is not the only possible response to demographic stress.

Communities faced with diminishing returns to labor may migrate to uncolonized areas. When the alternative of moving is impossible, the population may also turn to cottage industries (handicrafts and trades) or engage in off-farm work (Levi, 1976).

Population Pressure and Carrying Capacity

Allan defines the carrying capacity of a resource as "the maximum population density the system is capable of supporting permanently in the environment without damage to the land " (1965). The concept of carrying capacity implies a fixed limit to the productive activity that can be supported by a given environment. Production below this limit can continue indefinitely, but attempts to exploit the environment beyond this limit lead to a breakdown of the system.

Carrying capacities have been measured for shifting cultivation systems (Conklin, 1961; Street, 1969; Brookfield, 1963). Using case studies of tropical upland communities, Gourou (1966) estimated that a decrease in the fallow period from 24 to 9 years increased the potential population that could be supported from 12 to 30 persons per square kilometer.

Estimates of population density of indigenous shifting cultivation communities in the Philippines show that even in the remote forested areas, the man-land ratio has increased, making current subsistence patterns of production inadequate. Eder (1982) reported that the pioneer Cuyunon swidden farmers in Palawan had up to 24 ha of land for cultivation in the early periods of settlement. These farmers now have only an average of 3.3 ha. Excluding land owned by outsiders (which is about half of the land area of Palawan), population density is now close to 200 persons per square kilometer. Intensification of agricultural land use has taken place so that income from subsistence production accounts for only 29% of total income, while income from sale of agroforestry products represents a larger share (71%) of household income (Eder, 1981: 93-94).

Changes in Cropping Patterns Due to Population Growth

When population growth depresses average output sufficiently to lower current living standards, a community switches to a new technique that improves average output despite a reduction in the fallow period. In most cases, this would mean introducing higher-yielding varieties and fertilizers or using new and shorter crop rotation schemes. Based on research on the Mindoro Hanunoo farming practices, Conklin (1961) estimated that about 8 to 15 years of natural regeneration are needed to sustain tropical forest productivity. This type of long-fallow cultivation cycle has changed dramatically even among traditional shifting cultivation communities. Among the Ikalahans in the mountains of Nueva Vizcaya, the mean fallow period is only 5.8 years in a swidden (or *uma*) cycle of 10 years (Aguilar, 1982). In the Lake Buhi watershed in Camarines Sur, which is less remote and more accessible to a lowland market, that fallow period is 2.2 years in a cultivation cycle of 4.4 years (Aguilar, 1982). The trend is generally towards a cultivation cycle based on the annual cropping calendar of the dominant subsistence crop. In the upland rice-growing areas of Hamtic, Antique, land is allowed to rest only during May and April (Tapawan, 1981).

Intensification also can be accomplished by changes in farm implements induced by the shift to commercial farming. The influx of Visayan migrants from nearby Panay, Romblon, and Masbate provinces into the Batangan Buhid territory in Mindoro Oriental reduced ricecrop production in favor of maize. Using the surplus income from the sale of maize, a significant number of the Buhid were observed to have acquired Visayan tools, such as the plow and carabao (Lopez-Gonzaga, 1983).

Furthermore, intensification often followed diversification, or the broadening of alternatives for income generation. In the Mountain Province, such responses include seasonal migration of Bontok males to nearby mines (Prill-Brett, 1982), the adoption of the *loktanon tangdanan* (or lowland wage labor system) by the traditional Batangan Buhid in Mindoro Oriental (Lopez-Gonzaga, 1983), or the planting of vegetable cash crops in certain "upland zones" of the Lake Balinsasayao swidden farmers in Negros Oriental (Cadelina, 1983). As environmental and economic problems increased with population stress on resources, communities have expanded livelihood alternatives. If the alternatives are "adaptive", then the costs for restoring flexibility in the system are reduced (Slobodkin and Raport, 1974; McCay, 1978).

MIGRATION IN PHILIPPINE UPLANDS

Studies of internal migration in the Philippines have concentrated on national or macro-level analyses of the actual number of migrants before 1960 (Nava, 1959; Kim, 1972) and during the period 1960 to 1970 (Smith, 1976; Perez, 1978). Other migration studies attempt to determine the major correlates of migration movement on the inter-provincial level (Zosa, 1973; Smith, 1974) and on specific flows, such as the rural-to-urban one (Pernia, 1975). The characteristics of all types of migrants have been examined by Smith (1975) and Zachariah and Pernia (1975). Lopez and Hollnsteiner (1974) discussed the adjustment process which migrants undergo at their destinations. Studies on migration policy can be found in Laquian (1970) and Carino (1974). Good summaries and reviews of these studies are found in Flieger (1977) and Herrin (1982).

Specific cases of rural-to-rural migration have been documented for a few migrant settlement areas, especially in Mindanao. The works of Carabaran et al. (1975) in Misamis Oriental, Ulack (1972) in Iligan City, and Wernstedt and Simkins (1971) in the Digos-Padada valley, Davao Province, provide observations on characteristics of migrants moving into frontier settlement areas. The studies indicate that a similar pattern exists between rural-to-urban flows and rural-to-rural upland migration streams. Movement, once established, tends to be self-reinforcing as migrants send back information and assistance for the transfer of relatives, friends, and other migrants.

Migration occurs before half of the presumed population limits are reached, and factors other then economic hardship enter directly into the decision process. Two stages of population movements in the Philippines are consistent in completed studies. First, a birth-to-1960 phase was identified solely for rural-to-rural migration. This phase was characterized by the dominance of frontier destinations. The UNFPA-NCSO (1976) study estimated that about 33% of rural-to-rural migration flow in the period 1960-1970 ended up in frontier upland areas.

The second stage (1960-1973) was characterized by the sustained heavy influx of rural migrants to urban centers (mainly Manila) and a significant reshuffling of population among the Southern Tagalog provinces. In-migration to upland areas continued during this period but a lower proportion to the total migrant population than in the pre-1960 period.

As in 1970, there were 132 identified migration streams in the country. Only 12 are significant (involving 50,000 or more migrants). The largest rural-to-rural upland migration steams originate from the Central and Eastern Visayas area and move towards Mindanao. Western Visayan migrants have a preference for Southern Mindanao. The major receiving upland areas in Luzon are in Palawan, Mindoro, and Benguet (Perez, 1978).

Counter-flows — return migration movements from the place of destination to the place of origin — have been insignificant in the upland migration streams. The Central Visayas-Southern Mindanao migration stream, for example, has a net transfer turnover ratio of 68 to 1. Perez (1978) concluded that for every ten migrants moving to the uplands, less than three engage in a counter flow. While the focus of measurement is on lowland-to-upland movement, migration from upland to upland areas also occurs.

General Characteristics of Upland Migrants

Studies of upland migration indicate that for all destination areas, migrants tend to be of two groups. The first group consists of young males between ages 21 and 40 years, travelling as singles or as sets of brothers. The second consists of older males, 50 years and above, travelling with their families. Females who migrate are between 11 and 25 years and single. Unfortunately, historical accounts are limited to the Luzon central plain.

There are no distinct age-sex differentials among migrants who engage in long-distance movements. In Mindanao, there was a marked increase in the male population of frontier provinces within the 20-35 year range (Wernstedt and Simkins, 1965). Female migrants travel short distances and are younger, between 11 and 25. The 1982 study by Madigan et al. of migration in northern Mindanao showed a similar trend where migrants were mostly middle-aged males. In Diadi, Nueva Vizcaya, 60% of the migrants were male, 40 years and above (Duldulao et al., 1981).

McLennan (1973) and Anderson (1974) note that the majority of upland migrants were lowland rice farmers and tenants before moving. Other migrant occupations include logging operators, lumberjacks, prospectors, travelling merchants, landless workers, adventurers, missionaries, and town officials, such as those found among the *loktanon* (lowland) migrants into the Buhid frontier territory in Mindoro Oriental (Lopez-Gonzaga, 1983).

Expected income at destination is higher for migrants with previous access to agricultural lands, although expected occupation mobility at destination is significant for migrants without land (Pernia, 1977). Land ownership and access to resources are more important for long-distance, interregional migration than for interprovincial movements (Zachariah and Pernia, 1975).

Frontier migrant households in Pangasinan that do not have access to land and that fail to get regular employment migrate periodically and "float around the hillsides doing occasional jobs, planting and harvesting, and drawing on the support of their kinsmen and neighbors" (Anderson, 1974; see also Mozo, 1763 as cited in Doeppers, 1968). These migrants are, however, depicted as quick to respond to changes. Although the previous upland crop of native tobacco was less laborious, the migrants readily adopted the planting of Virginia tobacco. Within two years, virtually all migrants in the uplands were selling high-grade tobacco to a government subsidized market (de Salazar, 1742, as cited in Blair and Robertson, eds., 1903-1909).

Female Visayan migrants in Mindoro Oriental spend a greater amount of their time in income-generating activities such as trading. Several of these Visayan merchants (or *biyaheras*) created a market for agricultural produce by establishing steady trade partnerships with local Buhid cultivators. The Visayan males, on the other hand, cultivate profitable upland crops (such as maize) and engage some Buhid as paid laborers, or *tangdanan* (Lopez-Gonzaga, 1983).

Patterns of Upland Population Movements

There is no Philippine counterpart of the Buginese migration organizer, or *pengurun perantun,* in Indonesia who arranges travel and accommodation of migrants to Jakarta (Vayda, 1980). Most Filipino migrant families arrange for their own transfer and go through a two-stage process where a family member (usually a male member or pair of brothers) goes to the destination areas to explore possibilities for movement. Then the whole family makes the move, and, by word of mouth, neighbors and kinsmen are told of opportunities at the new site. For family migrants in Sisya, Pangasinan, an initial investment in land and sufficient savings to support a family for the first few years are required for moving (Anderson, 1974). In this two-stage process only three families with adequate provisions have returned to Pangasinan (Wernstedt and Simkins, 1965).

Some movements occur after a reallocation of resources. Real location can occur after a change in tenure status (Pernia, 1977). In Bohol, changes in income following a bad crop or disease also initiated out-migration (Zosa, 1973). Reallocation of resources in Pangasinan took the form of a death in the kin group, leading to land fragmentation from inheritance. Emigration occurred after the lands were subdivided (Anderson, 1974).

A description of the migrants' mode of travel is found in McLennan (1973). Ilocanos travel along waterways in wagon caravans (bullcarts), indicating the transportation available at the time of movement. Long-distance migrants travel by sea, disembarking at ports where bullcarts are purchased for overland travel. The migrants bring provisions for food and shelter; and for some, even "whole houses which are dismantled" are loaded into the cart (BFD, 1980). Dried fish and the fish-shrimp condiment, *bagoong,* are usual food supplies. Household size restricts the distance of movement (see also Pernia, 1977).

Changes Introduced by Migrants

Areas occupied by migrants are close to national roads or at the fringes of interior settlements. As more migrants begin to establish residence, a slow expansion process occurs. Such expansion has had three major impacts on the traditional Buhid society in Mindoro (Lopez-Gonzaga, 1983). First, indigenous Buhid settlers congregated in small settlements — rather than in their usual arrangement of scattered houses — in order to protect their lands more effectively from migrant landowners. Secondly, cash crop production was stimulated as *loktanon* (lowland) trading increased, and a new class of Buhid traders *(ahentes)* emerged. These *ahentes* specialized in dealing with Visayan *biyaheros* (or merchants) and acted as middlemen for the Buhid. Some *sari-sari* (sundry) stores selling kerosene, soap, and salt were later set up in interior Buhid settlements. Third, Buhid lands were slowly sold to lowland merchants. Lands were no longer communal but were treated as a commodity to be bought and sold. Some Visayan merchants acquired lands through token cash payments for debts incurred by the Buhid in the purchase of exhorbitantly-priced goods. As Lopez-Gonzaga (1983: 131) explains:

> . . . The land-oriented migrants supported themselves by cultivating short-term crops, such as maize and sweet potatoes, on the land borrowed from the Buhid cultivator, but they eventually achieved direct control over this land by filing a land declaration to the local land bureaus, a procedure whose legal implications often remain unknown to the original Buhid holders. . . The Bisayan migrant eventually secures the borrowed land with a token cash payment.

In Mindanao, the influx of Visayan and Ilocano migrants reduced the proportion of the local Muslim population from 31% in 1948 to less than 20% in 1960. The overall effect of migration in Mindanao, however, has been the rapid increase in total area cultivated from 2.4 million ha in 1948, to more than 5.04 million ha in 1960 (Wernstedt and Simkins, 1965).

Changes in cropping patterns also developed after a period of sustained in-migration. Among the Buhid in Mindoro Oriental, a shift to more profitable maize occurred. Some upland migrants to Pangasinan were found cultivating high-grade tobacco and cutting and selling first-class acacia wood to Benguet wood carvers (Anderson, 1974).

Settlers were drawn into heavier work schedules in their fields as migrants became strong competitors for local resources. The average labor input of one Buhid was 2,700 man-hours per crop season, 35% to 40% of which (or 700 to 900 hours) were devoted to the cultivation of maize (Lopez-Gonzaga, 1983).

Places of Origin and Destination

The traditional out-migration groups have been the: 1) Visayans, mostly from the eastern, central, and western provinces of the Visayan region, 2) Ilocanos, from the nothern Ilocos coast of Luzon, and 3) Tagalogs, mostly from the coastal towns of Southern Luzon. The Visayan frontier migrants seem to prefer Mindanao as a destination. Pascual's

1966 analysis of rural-to-rural migration in 1960 shows that 60 percent of Visayan-born migrants end up in Mindanao, the largest group coming from Cebu. The Visayans also form distinct ethnolinguistic sub-regions at their places of destination. The Central Visayan (Cebu, Bohol, and Negros Oriental provinces) migrants settled in Davao and Zamboanga del Sur; the West Visayan (Negros Occidental, Aklan, Antique, Capiz, and Iloilo) migrants went to Cotabato. Leyte province migrants showed a preference for Surigao (see map 3; Perez, 1978).

Ilocanos were not as selective as Visayans concerning destination. The first wave of Ilocano migrants concentrated in the Central Plains and surrounding Caraballo, Zambal, and Sierra Madre mountains. Using civil and ecclesiastical enumerations for the period 1887 to 1905, McLennan (1973) estimated Ilocano emigration at 1,000 to 2,000 per year. After the American homestead program, the number swelled to 10,000 in a year. By 1920, according to McLennan's (1973) overview of Ilocano migration based on Zuniga's census of the population in Ilocos, a total of 45,000 migrants had moved out of the area.

Keesing (1962) argues that the Ilocano reputation for wet-rice cultivation may have attracted the Spanish missionaries. The missionaries opened up trails to the uplands to teach pagan swiddeners (the Tinguians, Negritos, and Igorots) fundamentals of the cross and the plow. Regardless of motivation, the Ilocano movement was massive, especially in the Central Plains and highlands. In Diadi, Nueva Vizcaya, Ilocanos comprised 39% of all migrants. The pioneer Ilocano migrants were mostly experienced wet-rice farmers (Duldulao et al., 1981). They also settled in the uplands of Mindoro Oriental and Palawan. It should also be noted here that the Ilocanos preferred Cotabato province, Mindanao, as a destination, accounting for 90% of Luzon-born migrants and 11% of total migrants to Cotabato in 1960 (Wernstedt and Simkins, 1965).

The Tagalogs have not been as mobile as the Visayans and Ilocanos. Most Tagalog migrants come from the coastal towns of Luzon and move short distances, such as from Batangas province to Mindoro Oriental and Palawan. The Tagalogs who moved to Mindanao comprised only 7% of total migrants in Mindanao in 1960 (Wernstedt and Simkins, 1965). Within Luzon, the usual destinations of Tagalogs were the frontier areas of Western Pangasinan (Caraballo mountains), Zambales, Nueva Ecija, Isabela, Cagayan, Quezon (in particular, Baler), Mindoro Oriental, and Benguet provinces (Anderson, 1974).

Distance was not the crucial factor in the Pangasinan residents' decision to migrate (Anderson, 1974). Increasing land pressures and social tension in the Central Plain, non-availability of lands for cultiva-

MAP 3.: Location of Integrated Social Forestry Project Communities Included in the Demographic Study (1975).

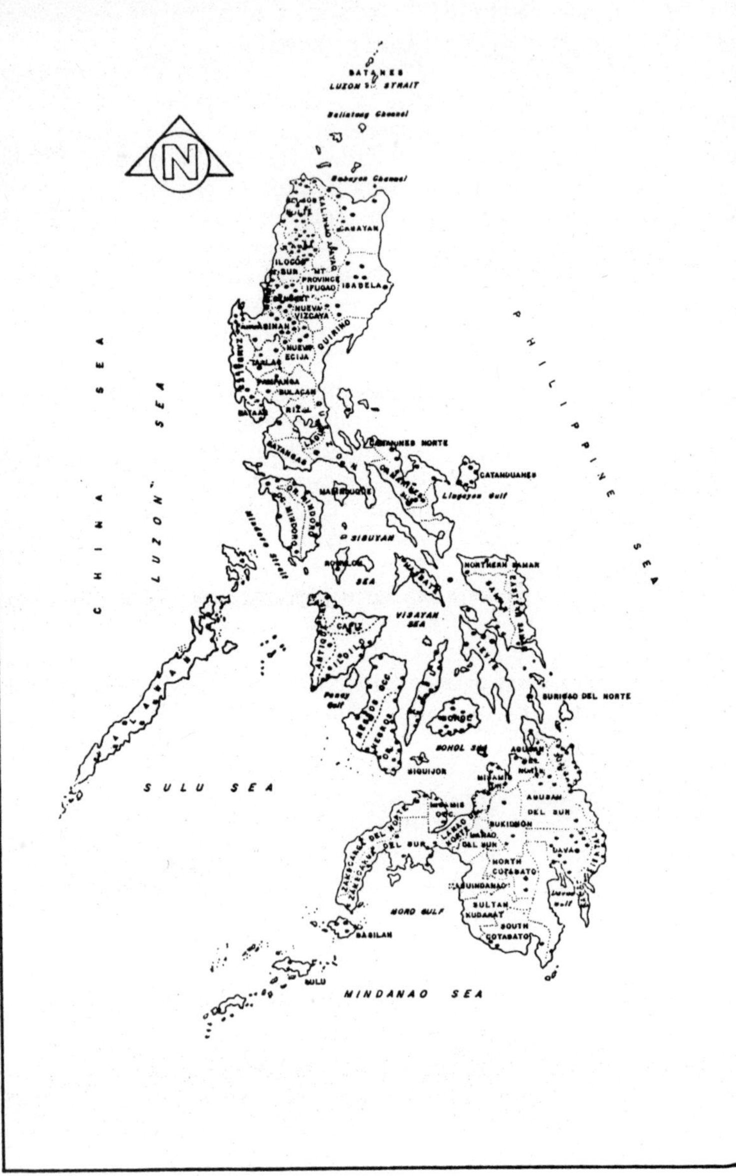

tion, and lack of employment opportunities were the more important factors. The income prospect (or "better living conditions") appeared to be the primary motivating force among migrants in Diadi, Nueva Vizcaya (Duldulao et al., 1981). Of the migrants surveyed, 71% replied that the prospects for owning land and getting higher incomes were their reasons for moving. Around 70% of the migrants were lowlanders without access to land, and 30% were former kaingineros. About 13% of the kainginero migrants said that the reason for moving was lack of kaingin (land).

Historical Upland Movements

Three events encouraged massive migratory movements to the uplands: the opening of trails by Spanish missionaries, changes in peace-and order conditions, and the government's planned resettlement schemes.

Prior to the 19th century, upland migration was essentially an expansion of the proselytizing work of missionaries. Doeppers (1968) noted that the new missions after 1700 were located along the western and eastern margins of the Central Plain and were oriented to upland groups. Mission trails were established which later led to new settlements. The earliest trail, the Balete Pass from Pangasinan to Cagayan, was established in 1685 by the Dominicans. The trail paved the way for the growth of the upland town of San Luis Beltran (present-day Asingan). Keesing (1962) discusses development of a similar trail in 1739 connecting the Igorot villages to Asingan via Cagayan and San Jose. Fray del Rio describes how Igorot Christians later travelled this route to carry vegetables and upland rice (Keesing, 1962).

Spanish missionaries forcefully transferred Christian farmers in order that they could instruct swidden cultivators in the use of wet-rice farming. These actions accelerated the opening up of the uplands to lowland merchants. Salazar noted that in 1680 ". . . salaried Indians (Filipinos) were taken from other provinces so that they might cultivate the land, and so that the Zambals might learn of them" (1742: 93-94). Integration relative to ethnic affiliation and language took effect as the small heterogeneous groups of converts (the indigenous upland settlers) eventually resettled and intermarried with lowland migrants. The lowlanders who migrated gradually occupied the interior upland areas (Doeppers, 1968).

Government Resettlement Programs

The government's resettlement schemes started as early as the Tobacco Monopoly in 1781. McLennan (1973) notes that because of Spanish control over the tobacco plantations, mass emigration and settlement in upland areas "to escape the hardship of the monopoly" occurred. In 1930, several groups of migrants were attracted to U.S.-sponsored sugar plantations. The rates of migration "fluctuated with the futures of the sugar industry" (McLennan, 1973). During this period, around 25,000 Ilocanos emigrated yearly to Mindoro's sugar plantations. The rinderpest epidemic in the 1880s deprived farmers of work animals and later led to massive emigration to Mindoro.

With the American-sponsored homestead program, emigration shifted from the Central Plains to Mindanao. Public alienable and disposable lands were sold or leased at the rate of as much as 1,000 hectares per individual in1903. Agricultural colonies were established in 1913 through a subsidized migration plan in seven areas in Mindanao, six in Cotabato, and one in Lanao. In 1937, the National Land Settlement Administration (NLSA) was established. By the end of the year, 30,000 to 35,000 families had resettled in Mindanao. Other areas, such as the Koronadal and Allah valleys were opened up (Wernstedt and Simkins, 1965). After the Second World War, the NLSA was replaced by the National Resettlement and Rehabilitation Administration (NARRA), which extended direct financial support to migrants through procurement of seedlings and credit for equipment. By 1948, total government involvement in migration in Bukidnon represented over one-fifth of total migration to the province.

Government settlement programs continued after 1965, primarily to alleviate congestion in Metro Manila and to resettle families affected by government projects. Some relocation of communities into virgin forest took place in the early 1970s. Examples include the San Pedro Resettlement Site (Aguilar, 1982), Pantabangan (Floro, 1980), Angat (Calanog, 1977), and Ambuklao and Binga (Llapitan, 1977) watershed areas.

Table 1 provides a summary of public land applications and patents from 1903 until 1978 which covered over 5.3 million hectares of public forest lands. In table 2, government-sponsored settlement efforts are outlined, showing that about 57,000 families were supported through direct aid in transfer from 1913 to 1971. The BFD, however, reported in 1972 that 300,000 to 600,000 families had been resettled (BFD, 1980).

PRELIMINARY ESTIMATE OF UPLAND POPULATION AND MIGRATION

To focus on upland population and migration, the ideal procedure would be to identify all existing upland barrios first, using a combination of a slope and area coverage in delineating upland communities.

Table 1. Public land applications and patents since passage of the public land law of the Philippines.

Years	Applications Approved	Area (Hectares)	Patents Issued	Area (Hectares)
1903 - 1909	–	–	–	–
1910 - 1920	–	–	10,577	59,992
1921 - 1930	–	–	18,065	184,400
1931 - 1940	–	–	64,066	684,805
1941 - 1950[a]	38,354	300,495	16,323	193,308
1951 - 1960	296,593	1,646,142	241,156	1,410,074
1961 - 1970	375,182	1,223,098	374,264	1,189,731
1971 - 1978[b]	266,572	590,775	649,127	1,543,757
Total[c]	–	–	1,393,912	5,342,686

[a]During World War II all prewar records of the Bureau of Lands were destroyed.

[b]This includes only the first quarter of 1978. In 1972, the President allowed provincial offices of the Bureau of Lands to issue patents for less than 3 hectares.

[c]It is estimated that applications for 2.5 million hectares are currently pending approval or await patenting.

Source: Records Division, Bureau of Lands, Central Office, Manila, September 1978. In: James, W.E. 1979. An Economic Analysis of Public Land Settlement Alternatives in the Philippines. Ph.D. Dissertation, University of Hawaii, Honolulu, Hawaii, U.S.A

This could be done at the provincial level using overlays of topographic maps and aerial photographs. Because this procedure is time-consuming and costly, an alternative approximation of upland population is used by taking a sample of the total number of upland communities (map 4).

The sample upland barrios are identified by using a listing of 1983 BFD social forestry project communities. Table 3 provides a list of the 1,179 social forestry communities or barrios and the corresponding total and migrant population as of 1970. Two reasons motivated the

Table 2. Government sponsored settlement efforts in the Philippines

Date	Agency	Families Settled	Total Expenditure[a]
1913-1917	Insular Government	1,500	–
1918-1939	Bureau of Labor and Migration	9,172	–
1939-1949	National Land Settlement Administration (NLSA)	8,300	11.0
1949-1950	Rice and Corn Production Administration (RCPA)	–	–
1950-1954	Land Settlement and Development Corporation (LASEDECO) and Economic Development Corporation (EDCOR)	1,503	3.5
1954-1963	National Resettlement and Rehabilitation Administration (NARRA)	30,646	449.0
1964-1971	Land Authority (LA)	2,855	55.5
1971-	Bureau of Resettlement, Department of Agrarian Reform (DAR-BURE)	3,315	483.1

[a]All expenditure figures are in millions of current, unadjusted pesos.

Sources: Pelzer (1945), Starner (1961), Golay (1961), Hule (1963), IARST (1984), Center for Urban Studies (1978), and Budget Commission Annual Report (1963 to 1977). In: James, W.E. An Economic Analysis of Public Land Settlement Alternatives in the Philippines, Ph. D. Dissertation. University of Hawaii, Honolulu, Hawaii, U.S.A.

adoption of this list. In the absence of a delineation of upland communities at the national level, the BFD listing provides a convenient sample from which a partial estimate of the upland population can be derived. While it cannot be assumed that the communities under the BFD social forestry program cover the entire area listed by NCSO, it is expected that the sample coverage will at least give a more realistic overview of upland population movements than what is currently provided by the BFD and others. The second reason has to do with the usefulness of the partial estimate to BFD since it provides an analysis of community populations in areas under BFD jurisdiction. It is estimated that the sample municipalities represent about 34% of the total number of communities in the country. The sample estimate also represents 29.3% of the total population and over one-half (55.5%) of the total migrant population for the period 1960 to 1970 (Table 3).

Census Estimates: Advantages and Disadvantages

The major advantages of using census data, compared to data derived from smaller surveys and case studies, are that the information provided by the census covers the entire country, and the census data are gathered at the barrio level, the smallest administrative unit of the government (Flieger, 1977).

On the other hand, there are five major limitations to the census estimate. First, the national coverage of the census precludes intensive data gathering. Explanatory variables, such as motives for moving and other determinants of migration movements, are not included.

Second, in reference to the nature of the census data itself, information on population and migration movements between census years are not captured. The data is static. Flieger (1977) points out that the exclusion of this information underestimates the proportion of migrant population by about half the total amount registered for the intercensal period.

Third, information regarding migrant characteristics, for example, average family income and land availability, need to be inferred from other sources of information.

Fourth is the general difficulty of defining what constitutes the uplands, and, especially for population counts, the occupations that can be included within the domain of upland population. Are uplanders teaching or doing office work in the lowlands to be counted? How are lowland residents working in timber concessions or plantations in the total labor force for forestry to be counted? Since the major concern is to estimate the total upland population, all persons residing in the uplands, regardless of occupation are counted in the population estimate. Workers in timber concessions, because they are nonresidents, are not counted.

Lastly, the sample social forestry communities do not necessarily constitute the entire upland land area and the population of the barrio units defined by the census. It is possible, for example, to have only one-half of a barrio's land area in the uplands. This problem is partially resolved by grouping the sample communities and taking only those

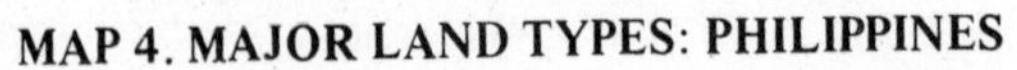

MAP 4. MAJOR LAND TYPES: PHILIPPINES

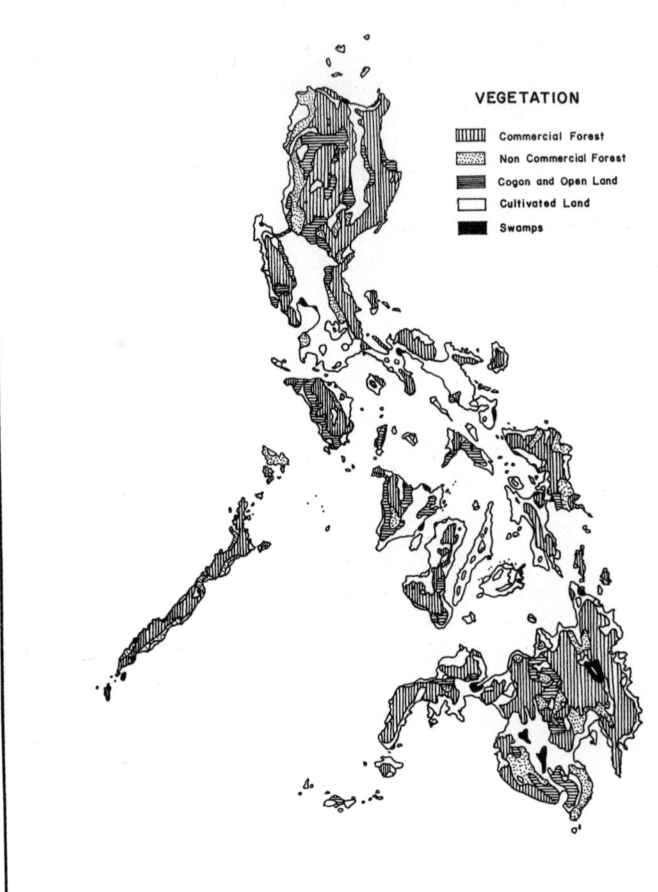

MAP 4. MAJOR LAND TYPES: PHILIPPINES

Table 3. Summary of integrated social forestry communities included in the study.

Region/Province	Number of Municipa- lities	Barrios	Total Population (1970)	Total Migrant Population (1970)
All Regions	486	1,179	10,725,064	2,797,983
By Region.				
1. Ilocos Region	96	348	1,188,934	156,374
La Union	9	15	149,047	15,374
Pangasinan	15	55	300,133	37,888
Ilocos Norte	17	76	266,276	26,000
Abra	24	94	107,953	9,441
Ilocos Sur	14	33	137,112	16,488
Mt. Province	6	20	71,737	7,315
Benguet	11	55	156,676	43,663
2. Cagayan Valley	30	43	511,753	97,161
Cagayan	10	11	148,151	23,324
Isabela	7	14	185,558	35,152
Nueva Viscaya	8	9	124,502	27,042
K. Apayao	1	2	10,470	2,758
Ifugao	4	7	43,072	8,885
3. Central Luzon	22	46	599,262	126,759
Zambales	6	11	185,602	63,777
Tarlac	2	2	19,631	3,141
Nueva Ecija	5	14	147,726	14,141
Pampanga	3	9	139,969	21,775
Bulacan	1	4	N.A.	N.A.
Bataan	5	6	108,334	24,125
4. Southern Tagalog:	50	118	793,377	235,159
Rizal	3	9	44,004	16,485
Cavite	1	1	12,743	124
Occidental Mindoro	9	19	109,898	49,616
Laguna	3	3	47,177	16,863
Marinduque	2	3	63,280	4,150
Quezon	8	19	197,113	46,690
Oriental Mindoro	5	16	119,540	33,639
Romblon	11	26	41,314	2,153
Palawan	8	22	158,308	65,439

Region/Province	Number of Municipalities	Barrios	Total Population (1970)	Total Migrant Population (1970)
5. Bicol Region:	36	57	951,274	196,318
Camarines Sur	13	19	403,535	75,421
Camarines Norte	5	9	121,084	42,767
Catanduanes	6	11	107,423	11,477
Albay	3	4	91,943	22,957
Masbate	5	5	96,585	19,797
Sorsogon	4	9	130,704	23,899
6. Western Visayas:	26	60	1,021,760	173,424
Antique	9	12	123,384	9,351
Aklan	3	4	31,460	2,399
Iloilo	11	10	240,285	27,666
Negros Occidental	11	31	571,364	125,981
Capiz	3	3	55,267	8,027
7 Central Visayas:	62	131	1,236,072	259,715
Cebu	14	35	691,343	159,100
Bohol	17	30	174,261	27,584
Negros Oriental	18	23	370,468	73,031
8. Eastern Visayas:	25	27	456,403	80,884
Northern Samar	3	2	27,671	4,885
Western Samar	7	8	115,640	18,755
Eastern Samar	4	4	39,486	4,035
Southern Leyte	6	7	128,725	16,367
Leyte	5	6	144,881	36,842
9. Western Mindanao:	26	57	595,423	180,502
Zamboanga del Sur	14	28	473,836	165,881
Zamboanga del Norte	4	12	65,558	12,635
Tawi-Tawi	1	2	N.A.	N.A.
Sulu	3	5	56,029	1,986
Basilan	4	10	N.A.	N.A.

Region/Province	Number of Municipalities	Barrios	Total Population (1970)	Total Migrant Population (1970)
10. Northern Mindanao:	44	69	866,831	327,534
Surigao del Norte	3	8	28,419	4,425
Agusan del Norte	9	31	277,303	89,744
Misamis Oriental	17	11	341,442	103,546
Misamis Occidental	4	2	72,073	9,684
Camiguin Province	2	—	27,672	4,045
Bukidnon	4	9	108,276	54,894
Agusan del Sur	5	8	111,646	60,796
11. Southern Mindanao	49	198	1,779,839	757,270
Surigao del Sur	16	73	237,153	63,889
Davao Oriental	10	33	233,129	84,833
South Cotabato	13	37	382,208	198,277
Davao del Sur	10	55	656,090	267,802
Davao del Norte	13	43	271,259	142,469
12. Eastern Mindanao:	20	25	624,136	207,088
Lanao del Norte	1	9	104,493	43,345
Lanao del Sur	3	1	38,683	1,634
Cotabato	16	15	480,960	162,109
SOCIAL FORESTRY COMMUNITIES (ALL REGIONS)	486	1,179	10,725,064	2,797,983
TOTAL NUMBER OF ENTIRE COUNTRY[a]	1,440	34,000	36,642,666	5,045,450
PERCENT OF SOCIAL FORESTRY SAMPLE TO ENTIRE COUNTRY	34	3.5	29.3	55.5

[a]Perez, Aurora. 1978 Internal Migration, Table 20: Economic and Social Commission for Asia and the Pacific Country Monograph Series No. 5. Population in the Philippines, p. 45.

barrios with 75% or more of their land area in the uplands. This procedure is expected to bias the population estimate downwards.

The Census Estimate

National census data are classified by administrative boundaries — province, municipality, and barrio — and by total population broken down into male and female. Place-of-birth in 1970, place-of-residence in 1960 and 1965 by male, female, and age cohorts provide the major sources of estimating net migration rates and indices of migrant selectivity.

Perez defined migrants as ". . . persons who moved across migration defining areas and incurred a change of residence during a specific time interval" (1978: 4). Using this definition, migrants are then classified in terms of the volume and type of migration movement. Three general groupings include:1) lifetime migrants whose place-of-birth differs from place-of-residence at the time of enumeration, 2) short-distance or intra-regional and inter-provincial migrants whose place-of-birth and place-of-residence varied within the region, and 3) long-distance migrants whose place-of-birth and place-of-residence varied interregionally (Perez, 1978). For the last groupings, migration-defining areas follow the twelve regions of the Philippines.

The 1970 Census does not provide information on migration streams except for general patterns of movements described in the place-of-birth and place-of-residence enumeration. The earlier discussion of case studies of migration in Mindanao and parts of Luzon provide some directions of migratory movements.

Summary: Population and Migration Estimates for the Sample Sites

In contrast to the 800,000 population figure cited by the BFD and other sources placing the upland population between 400,000 to 7.5 million, the sample estimate (using 486 municipalities and 1,179 barrios) shows that as of 1970 about 11 million (10,725,064) people reside in the uplands (refer to table 3). This figure represents 29.2% of the total Philippine population of 36.7 million in 1970 and slightly over one-half (55.5%) of the total migrant population for the entire country. Of the 11 million upland residents enumerated in 1970, 21.2% (2,797,983) are classified as lifetime migrants. The figures (including table 4) indicate that the uplands comprise a large percentage of the national population and a significant proportion of the country's overall migration movement.

Upland migrants are less mobile. Only 18.3% changed their region of residence, compared to the national average where 76.4% (Smith,

Table 4. Volume of life-time migration in integrated social forestry communities by region in 1970.

Region of Residence in 1970	Same Municipality	Same Province	Same Region
	Percent of Enumerated Population Born in:[a]		
All Regions	71.5	78.8	81.7
By Region:			
1. Ilocos Region	87.1	92.1	92.6
2. Cagayan Valley	81.1	87.6	88.6
3. Central Luzon	76.3	80.6	82.6
4. Southern Tagalog	68.8	76.9	78.4
5. Bicol Region	72.0	80.7	90.0
6. Western Visayas	81.2	91.1	92.9
7. Central Visayas	79.1	83.2	91.2
8. Eastern Visayas	68.0	75.5	93.2
9. Western Mindanao	69.5	75.2	76.0
10. Northern Mindanao	62.3	68.1	71.9
11. Southern Mindanao	57.2	62.2	63.0
12. Eastern Mindanao	60.0	66.3	73.2

[a]Each column was divided by the total enumerated population, except for residents born abroad or birthplace not stated.

Source: National Economic and Development Authority. 1970. Integrated Census of the Population and Its Economic Activities, Manila: NEDA-NCSO.

1976) or 51.2% (Flieger, 1977; Perez, 1978) are said to have changed residence across regional boundaries. In terms of inter-provincial population movements, however, an opposite trend is observed. In 1960, 88.1 percent of upland migrants changed residence, in contrast to the national average of 11.3% (Perez, 1978). These trends are consistent with Flieger's (1977) analysis of 1960-1970 migration patterns indicating a positive correlation between interregional in-migration and intraprovincial population movement in the agricultural frontier areas.

The net lifetime migration rate for all regions is 28.5%. This rate means that about a third of total population in the uplands was born in a place other than their current place of residence. The highest in-migration rates are found in the Mindanao provinces — Northern Mindanao

Table 5. Migration distance streams in integrated social forestry communities by region of residence in 1970.

Region of Residence in 1970	% of Total Migrant Population Migrated Within		% of Total Migrant Remaining within Province in 1960
	Same Province	Same Region	
All Regions	49.7	52.6	21.2
By Region:			
1. Ilocos Region	43.0	56.5	7.4
2. Cagayan Valley	34.0	65.0	13.4
3. Central Luzon	43.1	45.1	19.4
4. Southern Tagalog	54.3	55.8	23.1
5. Bicol Region	47.3	56.6	19.3
6. Western Visayas	27.7	29.5	8.9
7. Central Visayas	29.7	29.9	8.8
8. Eastern Visayas	56.5	74.2	24.5
9. Western Mindanao	18.7	55.5	24.8
10. Northern Mindanao	17.0	63.4	31.9
11. Southern Mindanao	11.8	50.9	37.8
12. Eastern Mindanao	17.0	40.6	33.7

[a]Divided by total enumerated population with birthplace different from municipality of residence at time of enumeration.

[b]Divided by total enumerated population with birthplace different from province of residence at time of enumeration.

Source: National Economic and Development Authority. 1970. Integrated Census of the Population and Its Economic Activities. Manila: NEDA-NCSO.

(Region 10) with a lifetime migration rate of 37.7%, Southern Mindanao (Region 11) with 42.8%, and Eastern Mindanao (Region 12) with 40%. In Luzon, the region with the largest proportion of lifetime migrants is Southern Tagalog (Region 4). This finding is partly explained by the heavy influx of populations into the upland provinces of Palawan and Mindoro (see table 5).

With respect to migration distance, place-of-birth enumeration was compared to place-of-residence information at the provincial and regional levels. Of the total number of lifetime migrants, one-half were interprovincial migrants and 47.4% migrated interregionally. Mindanao had the largest percentage of interprovincial and interregional population movements.

Intercensal migration can be observed by comparing place-of-residence in 1960 with place-of-birth data for 1970. For the entire sample of BFD social forestry communities, a significantly large percentage (78.8%) of migrants moved residence in the 10 year period. There is less movement in Mindanao intraregionally compared to the two Visayas regions which have been traditional out-migration areas.

A larger proportion (62.1%) of males is found among the lifetime, long-distance migrants (table 6). This finding is supported by the Wernstedt and Simkins (1965) study of the settlement of Mindanao where the male-female ratio was 108.6 in 1960, with the frontier upland municipalities having a larger (109.9) ratio than the settled areas (106.1).

Table 7 contains an estimate of population density for the BFD upland communities for 1960 and 1970. Interregionally, the Southern Mindanao provinces experienced the highest rate of increase in population density of 71% over the 10-year period. Within Luzon, the largest increase (35%) was found in the Sierra Madre, Zambales, and Caraballo highland areas along Zambales, Nueva Ecija, and Pangasinan provinces.

The table also shows that Western Mindanao (Region 9) had an unusually high population density of 388 persons per square kilometer. There are, however, no critically overpopulated upland regions as of 1970 if Conklin's (1961) 200 persons per square kilometer maximum carrying capacity estimate for tropical forests is adopted. Of course, this observation must be taken with caution since the density of some areas approaches this ceiling. Eastern Mindanao (Region 12) had 198 persons per square kilometer; Eastern Visayas (Region 8) had 184 persons per square kilometer. In addition, this observation does not allow for an evaluation of density relative to existing soil (and other important environmental) conditions significant in the evaluation of an area's carrying capacity.

When compared to the national average population density of 140 persons per square kilometer (Perez, 1978) and the 370 persons per square kilometer maximum density for lowland rice-growing areas (Hanks, 1972) the average population density in the uplands of 103 persons per square kilometer appears low. However, since land quality in the uplands has a greater variation cross-sectionally, due to differences in gradient, soil type, and other agroclimatic conditions, the present density figure may in fact be close to or beyond the system's carrying capacity.

Table 6. Estimates of net migration for males and females using the census survival ratio method[a] in integrated social forestry communities by region and sex distribution in 1970.

Region	Both Sexes	Male	Female
All Regions	2,169,631[a]	1,539,188	1,330,443
By Region:			
1. Ilocos Region	141,543	67,344	74,199
2. Cagayan Valley	80,454	37,853	42,601
3. Central Luzon	128,018	59,415	68,603
4. Southern Tagalog	215,489	112,991	102,498
5. Bicol Region	177,824	86,328	91,496
6. Western Visayas	162,840	83,392	79,448
7. Central Visayas	301,969	175,617	126,352
8. Eastern Visayas	75,706	36,393	39,313
9. Western Mindanao	171,030	85,028	86,002
10. Northern Mindanao	330,612	168,697	161,915
11. Southern Mindanao	771,172	415,715	355,457
12. Eastern Mindanao	312,974	210,415	102,559

[a]The difference in the estimate is due to data in consistencies for total migrant population and male-female population figures.

The census survival ratio is estimated from the following formula:

$$tM_o = Pt - S(P_o)$$

where:
M = net migration
P_t = population at time t
P_o = population at base period
S = survival ratio

Reference: Yun Kim. 1972. Net Migration in the Philippines, 1960-1970 Journal of Philippine Statistics, 23(2): 10. Data taken from NCSO (1970).

Basic Source of Data: Republic of the Philippines, National Economic and Development Authority. 1970. Integrated Census of the Population and Its Economic Activities, Manila: NEDA-NCSO.

Table 7. Estimates of population density by region for integrated social forestry communities, 1960 and 1970.

Region	Total Land Area (Sq. km.)	Total Population 1960	1970	Population Density (Persons/ Sq. km.) 1960	1970
All Regions	100,783	8,243,557	10,725,064	82	103
By Region					
1. Ilocos Region	7,433	858,682	1,188,934	155	160
2. Cagayan Valley	5,846	379,311	511,753	65	88
3. Central Luzon	4,688	383,302	599,262	82	128
4. Southern Tagalog	20,096	508,771	793,377	25	39
5. Bicol Region	7,336	836,692	951,274	114	130
6. Western Visayas	14,456	908,494	1,021,760	63	71
7. Central Visayas	9,410	999,899	1,236,072	106	131
8. Eastern Visayas	3,078	390,495	456,403	127	148
9. Western Mindanao	1,534	846,182[a]	595,423	551[a]	388
10. Northern Mindanao	8,815	672,599	966,831	76	110
11. Southern Mindanao	14,932	1,048,770	1,779,839	70	119
12. Eastern Mindanao	3,159	410,360	624,136	130	193

[a]A possible error in census enumeration may account for the wide discrepancy in the estimates.

Souce: For land area, BFD listing for 1983; census data taken from NCSO (1960 and 1970).

SUMMARY AND POLICY IMPLICATIONS

This chapter started by showing the major policy significance of realistic estimates of population growth and its characteristics in the uplands. The second section presented the conceptual foundation for a study of the relationship between demographic pressure and environmental constraints, including implications for technical and institutional change. Such a framework is critical for an understanding of the problems of population growth and migration arising not merely from short-term policy concerns, but more importantly, from fundamental processes in upland development.

The third section presented studies of the determinants of migration and the historical and social circumstances of movement. In general, the inducements to migrate were of two types. One concerned those factors which operated more or less constantly for all migratory movements, such as those associated with the push-pull model (Lee, 1966; Levy and Wadycki, 1974). A majority of the case studies observed the "pull" (the prospect of owning land) rather than "push" (hardship or poverty) factors motivated movement to the uplands. In studies of pioneer settlement in Mindanao, cultural and linguistic variations among migrant groups increased proportionately with distance between areas of origin and destination.

The other set of factors that affected migratory movements occurred on a seasonal or intermittent basis at different levels of intensity (Vayda, 1980). The historical accounts of Christian missionaries n the uplands, and much later, the national land settlement scheme in Mindanao and Luzon provided examples of such external inducements to upland migration. Also, the case studies show that certain ethnolinguistic groups, such as the Visayans and Ilocanos behave in widely divergent patterns of movement and settlement. Central Visayan migrants settled in Davao and Zamboanga del Sur, while migrants from Negros Occidental, Antique, and Aklan preferred the Cotabato highland areas. Migrants from Leyte occupied Surigao, as indicated in the major migration streams shown earlier in map 2.

More comparative studies of migration on the micro-level are needed. There is insufficient information on the characteristics of chronic migrants (or movers) and stable migrants (or stayers). Such information on migrant behavior and characteristics will be important in designing long-term strategies for social forestry projects.

Because of the absence of any scientific basis for current data offered by official and unofficial sources, population levels and migration rates using the 1970 census were estimated. Admittedly, the data source was not ideal, but the fact that the census is the only valid available data on the national level indicates that the estimates are of greater reliability than previous ones. It is on the basis of these estimates that the problem of population pressure in the uplands appears to be much greater than current literature leads one to expect. Thus, the official upland population figure of 800,000 reported by the BFD substantially underestimates the extent of the problem of population pressure in the uplands. That the sample BFD municipalities represent about 29% of the total population and 55% of the migrant population for the country indicates that the official estimate of BFD is in error by a factor of at least 10.

Finally, when more hard data become available, research into the causal relationships between changes in demographic stress (e.g., from migration) and other environmental and socio-economic changes should be made. As Netting argues ". . . population pressure appears to be a critical variable, the engine which sets in motion adaptive changes in a set of related technological and social variables among subsistence cultivators" (1974:54). Further research on such issues, however, will be feasible only when the basic numbers have come in.

REFERENCES CITED

Aguilar, Filomeno V., Jr. 1982. Social Forestry for Upland Development: Lessons From Four Case Studies. Quezon City: Institute of Philippine Culture, Ateneo de Manila University.

Allan, W. 1965. The African Husbandman. Edinburgh: Oliver and Boyd.

Anderson, James N. 1974. Social Strategies in Population Change: Village Data Economic Central Luzon. Berkeley: University of California. processed.

Baker, P.T. and W.T. Sanders. 1972. Demographic studies in anthropology, Annual Review of Anthropology 1:151-178.

Blair, Emma H. and James A. Robertson, eds. 1903-1909. The Philippine Islands, 1493-1898. 55 Vols. Cleveland: A.H. Clark Company.

Boserup, Ester. 1965. The Conditions of Agricultural Growth: The Economics of Agrarian Change Under Population Pressure. London: Allen and Unwin, Ltd.

Brookfield, Harold C. 1968. New directions in the study of agricultural systems in tropical areas, In: E.T. Drake, (ed.), Evolution and Environment. New Haven: Yale University Press.

Cadelina, Rowe V. 1983. Lowland Migrant Upland Swiddeners Around Lake Balinsasayan Area, Negros Oriental: A Unique Case of Upland Poverty. Manila: De La Salle University, Research Center.

Carr-Saunders, A.M. 1922. The Population Problem: A Study in Human Evolution. Oxford: Clarendon Press.

Chang, Tuck Hoong Paul. 1981. A review of micro migration research in the third world context. In: G.F. de Jong and R.W. Gardner, (eds.) Migration Decision-Making: Multidisciplinary Approaches to Microlevel Studies in Developed and Developing Countries. New York: Pergamon Press.

Clarke, William C. 1976. Maintenance of agriculture and human habitats within the tropical forest ecosystem. Human Ecology 4(3): 247-257.

Cohen, Mark. 1977. The Food Crisis in Prehistory: Over-population and the Origins of Agriculture. New Haven: Yale University Press.

Conklin, Harold C. 1961. The study of shifting cultivation. Current Anthropology 2(1): 27-61.

Cruz, Wilfrido D. 1982. Technical and institutional change in renewable resource development with application to traditional fisheries. Ph.D. Dissertation. Madison: University of Wisconsin.

Darity, William A. 1980. The Boserup theory of agricultural growth: A model for anthropological economics. Journal of Development Economics 7: 137-157.

De Jong, Gordon. 1968. Demography and research with high altitude populations. In: Paul Baker et. al., (eds.), Adaptation in a Peruvian Community. University Park: Pennsylvania State University Occasional Papers in Anthropology No. 1.

Doeppers, Daniel Frederick. 1968. Hispanic influences on demographic patterns in the Central Plain of Luzon, 1565-1780. University of Manila Journal of East Asiatic Studies 12: 13-96,

Duldulao, A.C. et.al. 1981. An integrated project for kaingin control in the Philippines, Phase I to Phase III Reports.

Dumond, D. E. 1961. Swidden agriculture and the Rise of Maya civilization. Southwestern Journal of Anthropology 17: 301-316.

Eder, James. 1977. Agricultural intensification and returns to labour in the Philippine swidden systems. Pacific Viewpoint 19: 1-21.

Flieger, Wilhelm, S.V.D. 1977. Internal migration in the Philippines during the 1960s. Philippine Quarterly of Culture and Society 199-231.

Freedman, Ronald, ed. 1964. Population: The Vital Revolution. Garden City, New York: Doubleday.

Gourou, P. 1966. The Tropical World: Its Social and Economic Conditions and Its Future Status, Translated by S.H. Beaver and E.D. Laborde. Longmans.

Grigg, David. 1977. Ester Boserup's theory of agrarian change: A critical review. Progress in Geography 64-84.

Grigg, David. 1976. Population pressure and agricultural change. Progress in Geography 8: 135-176.

Hawley, Amos. 1973. Ecology and population. Science 179: 1196-1201.

Helleiner, G.K. 1966. Typology in development theory: the land surplus economy (Nigeria), Food Research Institute Studies 6(2).

Herrin, Alejandro N. 1982. Population and development research in the Philippines: A survey, In: Survey of the Philippine Development Research II. Makati: Philippine Institute of Development Studies 288-343.

James, William E. 1979. An economic analysis of public land settlement alternatives in the Philippines. Ph.D. Dissertation. Honolulu: University of Hawaii.

James, William E. 1983. Settler selection and land settlement alternatives: New evidence from the Philippines. Economic Development and Cultural Change 31(3): 576-586.

Keesing, Felix M. 1962. The Ethnohistory of Northern Luzon. Stanford, California: Stanford University Press.

Keyfitz, Nathan and W. Fleiger. 1971. Population: Facts and Methods of Demography. San Francisco: W.H. Freeman and Co.

Kim, Yun, 1972. Net migration in the Philippines, 1960-1970. Journal of Philippine Statistics 23(2): 9-27

Lee, Everett S. (1966). A theory of migration. Demography 3:47-58.

Levi, John F. S. 1976. Population pressure and agricultural change in the land-intensive economy. Journal of Development Studies 13: 61-78.

Levy, Mildred B. and Walter J. Wadycki. 1974. What is the opportunity cost of moving? Reconsideration of the effects of distance in migration. Economic Development and Cultural Change 22: 198-214 (January).

Llapitan, E.A. 1977. The impact of shifting cultivation in hilly country, Proceedings Int. Workshop on Hilly Land Development. Legaspi City, Philippines 235-240.

Lopez-Gonzaga, Violeta 1983. Peasants in the Hills. Quezon City: University of the Philippines Press.

Lynch, Owen, Jr. 1984a. Natural resource disposition in the Philippine uplands: A search for equitable policies and practices. Paper presented at the Workshop on Policy Oriented Research on Equity in Natural Resource Development: Issues and Data Needs. Tagaytay City: Development Academy of the Philippines.

(1984b) The invisible Filipinos: Indigenous and migrant citizens within the 'public domain'. Philippine Law Register 18-22.

Madigan, Francis C., S.J., Marilou Palabrica-Costello, and Lita C. Paloma 1982. Population and development research in the North Mindanao region. In: Survey of Philippine Development Research II. Makati: Philippine Institute of Development Studies 350-441.

McCay, Bonnie J. 1978. Systems ecology, people ecology, and the anthropology of fishing communities. Human Ecology 16(4): 397-422.

McLennan, Marshal Seaton. 1973. Peasant and hacendero in Nueva Ecija: The socio-economic origins of the Philippine commercial rice-growing region. Ph.D. Dissertation. University of California, Berkeley.

Netting, Robert McC. 1974. Agrarian ecology. In: B.J. Siegel, A.R. Beals, and S.A. Tyler (eds.), Annual Review of Anthropology 3: 21-56.

Olofson, Harold. 1982. Social forestry from one anthropologist's point of view. Proceedings of the Workshop sponsored by De La Salle University and University of the Philippines at Los Baños. College of Forestry.

Pascual, Elvira M. 1965. Internal migration in the Philippines. First Conference on Population. Quezon City: University of the Philippines Press.

1966. Population Redistribution in the Philippines. University the Philippines Population Institute.

Perez, Aurora E. 1978. Internal migration. In: Population of the Philippines. Bangkok, Thailand: UN-ESCAP pp. 44-78.

Prill-Brett, June. 1982. The Bontok model of village participatory decision-making. Participatory Approaches to Development Seminar Series. Integrated Research Center, De La Salle University.

Revilla, A.V., Jr. 1984. Forest development by public corporations. Paper presented at the Workshop on Policy-Oriented Research on Equity in Natural Resource Development: Issues and Data Needs. Tagaytay City: Development Academy of the Philippines.

Schlegel, Stuart A. 1981. Tiruray traditional and peasant subsistence: A Comparison. In: Harold Olofson, (ed.), Adaptive Strategies and Change in Philippine Swidden-Based Societies. Los Baños: Forest Research Institute. 105-116.

Schwartz, Aba. 1973. Interpreting the effect of distance on migration. Journal of Political Economy 81: 1153-1169 (Sept-Oct.).

Scott, Geoffrey A.J. 1981. Soil erosion and the kainginero: The evolution of the socio-economic approach to soil conservation. Philippine Geographical Journal 25(2): 57-63.

Shyrock, Henry. 1969. Survey statistics on reasons for moving. Paper presented to International Union for the Scientific Study of Population, London Conference (September).

Slobodkin, L.B. and A. Rapport. 1974. An optimal strategy of evolution, Quarterly Review of Biology 49: 181-200.

Smith, Peter, C. 1976. The changing character of interregional migration in the Philippines. Philippine Geographical Journal 20(4): 146-162.

Spencer, J.E. 1966. Shifting Cultivation in Southeastern Asia. Berkeley: University of California Press.

Stouffer, Samuel. 1940. Intervening opportunities: A theory relating mobility and distance. American Sociological Review 5(6): 845-867.

Pernia, E.M 1978. Individual and household migration decisions. Philippine Economic Journal 17:259-284.

Philippines (Republic) Bureau of Forest Development. 1981. Philippine Forestry Statistics. Quezon City: Planning and Evaluation Division, Bureau of Forest Development.

Street, J.M. 1969. An evaluation of the concept of carrying capacity. Prof. Geography 21: 104-107.

Tapawan, Zenaida G. 1981. Economics of farming systems in the upland areas of Hamtic Antique, Philippines. Unpublished M.S. thesis. University of the Philippines at Los Baños.

Townsend, Patricia K. 1980. New Guinea sago gatherers: A study of demography in relation to subsistence. In: J.R.K. Robson, (ed.), Food, Ecology, and Culture: Readings in the Anthropology of Dietary Practices. New York: Gordon and Breach Science Publishers. pp. 21-26.

UNFPA-NCSO J (1976). Geographical Patterns of Internal Migration in the Philippines. 1960-1970. Manila: National Census and Statistics Office.

United Nations. n.d. Methods of measuring internal migration. Manual IV. Sales No. E. 70. XIII.3.

Vayda, Andrew P. 1980. Buginese colonization of Sumatra's coastal swamplands and its significance for development planning. In: Eric C.F. Bird and Apulani Soegiarto, (eds.), Proceedings of the Jakarta Workshop in Coastal Resources Management.

Ware, Helen. 1977. Dissertification and population: Sub-Saharan Africa, In: Michael Glantz, (ed.), Dissertification: Environmental Degradation in and Around Arid Lands. Boulder: Westview Press. pp. 165-202.

Wernstedt, Frederick L. and Paul D. Simkins. 1965. Migration and the settlement of Mindanao. Journal of Asian Studies 25: 83-103.

Zachariah, K.C. and Ernesto M. Pernia. 1975. Migration in the Philippines with particular reference to less developed regions of the country. processed.

Zosa, Imelda A. 1973. An explanatory survey on the determinants of inter-provincial migration. Unpublished M.A. thesis. University of the Philippines Population Institute.

5
UPLAND DEVELOPMENT TECHNOLOGIES

Benjamin K. Samson

The productivity and sustainability of agro-ecosystems are dependent on the conservation of the resource base. A reason for low production and for production instability in the uplands is the fragility of that environment. Steep slopes, high rainfall, and unstable soils contribute to soil erosion. Removal of the native vegetation cover (usually forests) aggravates the problem by exposing the soil to the elements and disrupting the regulative mechanisms controlling nutrient, moisture, energy, and information flow (Sajise, 1977a). Government agencies, research institutions, universities, and private organizations now recognize the plight of people subsisting on the fragile upland resources and are trying to develop and introduce technologies to assist upland farmers to better meet their needs. Such technology development and extension address problems associated with the production of food, fuelwood, and cash crops. Productivity and stability are critical to upland farming. Soil erosion, fallow periods, pests, and diseases constitute important research topics. Marginalized upland areas, especially watersheds, present rehabilitation and protection problems. This chapter examines some of the technologies being applied.

The chapter is divided into three sections. The first describes technologies that address specific problems. A second section examines integrated schemes in upland development. These are discussed in terms of specific projects. Indigenous technologies are compared to introduced technologies. A third and concluding section evaluates current directions. Gaps and new directions for research are identified.

Major concerns of upland development include increasing, stabilizing, and sustaining the productivity of upland farms, and the rehabilitation of marginal upland areas. Soil erosion control, improvement of soil nutrient status, and intensification of land use are important to productivity. Reforestation and grassland use are important elements of rehabilitation.

PROBLEM ORIENTED TECHNOLOGIES

Introduced technologies for the upland farmer address production system problems. High soil erosion rates cause rapid soil nutrient losses and declining crop yields. Adverse environmental conditions include high wind speeds, high insulation and high seasonal variability in the availability of moisture (Sajise, 1977a).

Soil erosion has been recognized as perhaps the most critical factor responsible for the instability and non-sustainability of upland production systems. Soil loss from cultivated upland areas, with slopes ranging from 36% to 70%, range from 10 tons to 11 tons/ha/yr (Pacardo and Samson, 1979). This is high compared to soil erosion in secondary forest (157.83 kg/ha/yr), an ipil-ipil plantations (753.87 kg/ha/yr), and a protected cogon grassland (173.13 kg/ha/yr.) Hillside farms and kaingins in the Magat watershed were estimated to lose 100 tons of soil/ha/yr (David, 1984). All other land uses, except for overgrazed pastures, had lower soil-loss rates. Bare plots in Negros lose as much as 290 tons of soil/ha annually, and approximately 90 tons of soil/ha erode from cropped land per year (Sajise, 1982). In Benguet Province, Colting (1981) reported a soil erosion rate of 62.3 tons/ha/yr in a bare plot of land with 29% slope. An erosion rate of 23 tons/ha/yr was recorded for bare plots with 27% slope in Bicol (Bocato, 1981). These erosion rates represent losses of mineral nutrients and organic matter required for crop production. Such losses necessitate erosion control technologies.

Soil erosion and sedimentation by water result from detachment, transportation, and deposition of soil particles (Foster and Meyer, 1977). Detachment is the dislodging of soil particles from the soil mass by the erosion agent. Once dislodged, soil particles are moved by transportation. Sedimentation is the deposition of soil particles carried by the run-off. The major erosive agents in water erosion are impacting raindrops and run-off water flowing over the soil.

Impacting raindrops strike the soil surface at up to 9 meters per second (Foster and Meyer, 1977) and can concentrate intense energy at impact. These forces disrupt the soil structure, causing the detachment of soil particles from the soil mass. Transportation by impacting raindrops follows the transfer of kinetic energy from the raindrop to the detached soil particles. Overland flow detaches and entrains soil particles as erosive forces exceed soil resistance. Soil particle sedimentation from run-off occurs when the total sediment load exceeds a flow's total transport capacity.

Types of erosion include splash, sheet, and channel. Splash erosion results from the soil splash caused by the impact of raindrops. Sheet erosion is the removal of a thin, relatively uniform layer of soil particles by rainfall and run-off. Channel erosion occurs where soil removal by water causes cuts or channels on the soil surface.

Soil erosion control is based on three approaches: 1) reducing the velocity of run-off water, 2) increasing the infiltration rate of the soil, 3) and attenuating or dissipating the kinetic energy of raindrops before they hit the soil surface (PCARRD, 1984). Reducing the velocity of run-off reduces scouring and promotes the sedimentation of suspended solids. Scouring is the dispersing or rubbing action of rocks and soil particles on the soil surface over which the run-off is flowing. The amount and size of eroded material and the magnitude of scouring is proportional to the run-off velocity. Aside from decreasing the magnitude of scouring, slowing down run-off velocity promotes the deposition of sediments or eroded materials. Erosion losses can also be decreased by increasing the infiltration rate of the soil. This decreases run-off by promoting the downward movement of water through the soil. The third approach considers the high kinetic energy of falling rain. When rain hits a soil clod, kinetic energy is transferred to the clod, causing dispersion. Dispersed soil is more easily picked up and transported. Erosion control includes engineering and vegetative techniques.

Engineering Methods

Engineering techniques change land slope characteristics to reduce the amount and velocity of surface run-off (Foster, 1973) and include terraces, dams, dikes, and channels.

Terrace Construction. Terraces are earthen embankments or ridges constructed across a slope to control run-off and soil loss. They effectively decrease the length of the hillside slope, decrease run-off velocity, retain run-off in areas of inadequate precipitation, and are needed where slopes are greater than 2% and slope length is longer than 91 meters (PCARRD, 1984). Bench terraces are a series of earthen embankments with level to nearly-level tops and sloping (to vertical) down-hill faces. These terraces convert a slope (20% to 30%) into a series of level or nearly level benches (Martin, 1977). Ridge type terraces are long, low ridges of earth with gently sloping sides, and a shallow channel along the upper side. Soil erosion is controlled by diverting run-off across, rather than down, slope. The broadbase terrace is a ridge terrace typically 10 inches to 20 inches high and 15 feet to 30 feet wide, with gently sloping sides, a rounded crown, and a dish shaped channel along the upper side. The broadbase terrace controls erosion by diverting run-off along the contour at a non-scouring velocity.

Terrace construction requires a survey to formulate spacing between terraces, grade, terrace length, the location of the top terrace, and lay-out. Position of the field relative to other fields, trails, watercourses, and water outlets must be considered. If the field to be terraced adjoins another field, drainage must be established to channel water coming from the field. Terraces are laid out using a transit, Abney, dumpy, Wye levels or a carpenter's level mounted on a A-frame. A weight suspended on a line tied to an A-frame's apex can also be used.

In locating and staking terrace lines, the vertical interval between terraces must be determined. PCARRD (1984) recommends the following to determine the spacing:

$$VI = S/2 + 2$$

where VI is the vertical interval (in feet) between terraces and S is the average slope expressed in percentage. Accordingly, a field with an average slope of 8% should have terraces with a vertical interval of 6 feet. This formulation is apparently borrowed from the rule of thumb employed in determining vertical spacing of terraces in the Upper Mississippi States (Foster, 1973).

The upper terrace is laid out first, and can be done by two persons with a simple A-frame and bundle of stakes. The first contour is located at the upper edge of the field. The A-frame is set-up, the plumb-line or carpenter's level is centered, and a stake is placed at each point where the frame's legs touch the earth. The next point on the contour is determined by pivoting the A-frame on either leg and moving the free leg of the frame up or down slope until the plumb line or the carpenter's level is again centered. This is repeated until the field has been traversed. A stake is driven at each point of the contour as determined by the A-frame. Each new point is the next pivot point of the A-frame.

The next line is established after determining the vertical interval (according to the formula). An alternative is to approximate the position of the next contour by moving with one arm extended perpendicular to the body, down the slope from one of the stakes on the first contour line until the fingertips point at the stake. The next contour is located where the base of the stake is level with the tip of the extended arm. Subsequent contour lines are determined similarly. The area above the top terrace should not exceed 2 ha; and the maximum length of the terrace should be 800 meters (PCARRD, 1984). Saplaco (1983) notes that inclusion of natural waterways into the terrace system limits the length of terrace drainage from 80 m to 100 m.

The top terrace is built first and must be well constructed since the lower terraces depend on whether the top terrace can resist the run-off and sediment from the watershed. If the top terrace gives way to run-off, lower terraces will also probably fail. A grassed diversion canal can be established above the first contour as a safety device. This canal safely channels run-off from above the terraced field.

The terrace can be constructed manually by using a "cut and fill" method (Saplaco, 1983). The earth is cut and moved above the center line (an imaginary line in between contour lines) to fill the portion below the center line. The upper portion is the "cut" and the lower portion is the "fill." The wall of the "cut" is shaped to form a 1:1 riser (cut riser). A corresponding 1:1 riser on the "fill" completes the contour riser. Soil should be moved from the cut until the bottom of the cut riser and the top of the fill riser are level. About a 4% backslope from the top of the fill riser to the bottom of the cut riser should be provided to conduct run-off away from the fill riser. A lip, a mound of earth about 10 cm in height, should be established at the top of each riser to prevent the spilling over of excess run-off. A pick-up ditch, 10 cm wide by 10 cm deep, can be dug at the botton of each riser. The pick-up ditch is constructed level throughout to slow the movement of water.

Pick-up ditches and diversion canals are linked to waterways on both sides of the terraces. Width varies from 0.5 meter to 1.0 meter and can be increased if run-off volume is high. The waterways are best kept grassed to prevent excessive erosion.

Terracing has been introduced in the Philippine uplands by both government and private entities (Bernales and de la Vega, 1982a and b). Published data as to effectiveness and acceptability to upland farmers is limited, however. Tapawan (1981) gathered crop yield data for terraced and unterraced lands in the Antique Upland Development Project area at Hamtic, Antique, during the 1980 cropping season (tables 1 and 2). Terracing on residual soils had little effect on yields of corn and peanut, but gave significantly higher rice yields. Terraced alluvial soils, on the other hand, had significantly higher rice and corn yields than terraced and unterraced residual soil.

Table 2 shows the influence of terracing in combination with fertilizer use on rice yields. Rice yield in the terraced-unfertilized plot (89 kg/ 1000 sq m) is slightly higher than that in the unterraced-unfertilized plot (58 kg/1000 sq m). The difference between terraced and unterraced

Table 1. Mean yield (kg/1000 sq m) of major crops in terraced and unterraced plots, AUDP Areas, Hamtic, Antique, 1980. (From Tapawan, 1981).

	Unterraced Plots	Terraced Plots	
Crop	Residual Soils	Residual Soils	Alluvial Soils
Rice	119.96 c	178.77 b	237.12 a
Corn	27.98 b	36.30 b	64.50 a
Peanut	58.00 a	111.20 a	96.52 a

Means with the same letter are not significantly different from each other at 5% level of significance.

Table 2. Mean yield of rice in terraced and unterraced areas with and without fertilization, AUDP Area, Hamtic, Antique, 1980. (From Tapawan, 1981).

	Rice Yield (kg/100 sq m)
Unterraced areas, not fertilized	84.59 b
Terraced areas, not fertilized	88.76 b
Unterraced areas, fertilized	132.26 b
Terraced areas, fertilized	210.24 a

Means with the same letter are not significantly different from each other at 5% level of significance.

plots is heightened in the fertilized treatment. The terraced plot had a rice yield of 210 kg/1000 sq m, compared to 132 kg/1000 sq m in the unterraced plot. This implies that terracing is effective in preventing nutrient loss, which in the uplands usually takes place via run-off and soil erosion. Nutrient conservation is accomplished by preventing high run-off and promoting infiltration into the soil.

Indigenous uplanders have employed terracing as an erosion control technique. The Bontocs of the northern Philippines have terraced for hundreds of years in order to use mountain slopes for paddy rice and vegetable production (Omengan and Sajise, 1981). The high rice production in these fields was attributed to the high soil and water conservation properties of the terraced paddy field and the appropriate management of water, soil, nutrients, crop, and animals. The dry season rice yield was 124 cavans/ha, approximately 6 tons per ha (Omengan, 1981).

The *gen-gen* of the Ikalahans is a variant of the bench terrace: crop waste is piled in contour strips within the cropping area and, after several croppings, mounds of crop waste accumulate and serve to trap soil and water (Barker, 1984). Some uplanders lay log barriers to catch soil carried by run-off water (UHP, 1979a).

Adoption of terracing has been problematical. Land tenure uncertainty hindered the adoption of terracing as an agroforestry project in Buhi, Camarines Sur, (UHP, 1979a). Terracing was undertaken using funds provided by both the project and owners of titled upland areas. Tenant farmers on titled lands were hesitant to plant permanent crops on terraces because of the uncertainty of sharing in the benefits. Landowners refused to grant permits for terracing out of the fear that tenants would claim the land they terraced. Bernales and de la Vega (1982a, b) reported that the Antique Upland Development Project and the Farmer Occupancy Program in Doña Remedios, Bulacan, encountered problems in convincing farmers to adopt bench terracing. Antique farmers considered the construction of bench terraces as a back-breaking activity, requiring collective effort of 5 to 10 farmers to terrace a one hectare lot. The Bulacan farmers, on the other hand, contend that they cannot bench terrace because construction requires a carabao, which most do not have. The objections in Buhi revolved around equity and distribution of benefits of soil conservation improvements, while the Antique and Bulacan cases had more to do with labor requirements.

Canals and Dams. Canals and checkdams used together with terraces protect the system from heavy run-off loads. These structures have also been employed separately. Diversions are channels across slope designed to intercept and conduct surface run-off to a safe outlet (Agpaoa et al., 1973; Saplaco, 1983). Checkdams are small, low dams constructed in a gully or water course to reduce gradient, lessen flow velocity, minimize channel scour, and promote water percolation into the soil and deposition of eroded materials (Agpaoa et al., 1973; Foster, 1973). These structures usually require some capital outlay, labor, and engineering know-how, especially if they are to be permanent structures. Hence, these are usually government public works projects.

Vegetative Methods

Vegetative soil erosion control methods are less expensive alternative to engineering methods. Vegetative techniques seek to maintain a living or dead vegetation cover on the soil (Highfill and Kimberlin, 1977). Vegetation cover attenuates the force of falling raindrops and increases the resistance of the soil to the erosive force of run-off. The vegetation used determines the degree of erosion control. For Mt. Makiling, a cropped area had the highest erosion rate, followed in decreasing order by a cogon grassland, an ipil-ipil plantation, and a secondary forest (Pacardo and Samson, 1979). David (1984) estimated that overgrazed grassland had the highest, and primary forest areas the lowest, erosion rates in the Magat river watershed. David rated the relative influence of the different vegetation covers in terms of indices. These cover indices, used in an estimation of sheet and rill erosion from the Magat watershed, range from zero to 100. The zero value corresponds to maximum cover and the greatest degree of protection. The value of 100 corresponds to absolutely no ground cover and the soil subject to greatest erosion. Among the cover types used in the study, primary forest had the least cover index, while overgrazed areas, old kaingins, and areas partly disturbed by earth moving operations had the highest cover index. Others have demonstrated similar effects for single crops and combinations of crops on soil loss (Palis, 1977; Colting, 1981; Bocato, 1981; Serrano, 1982; Sajise, 1983; Palis et al., 1983). Vegetative erosion control methods include soil-conserving tillage techniques, contour farming, contour strip cropping, and soil conserving cropping systems.

Tillage Techniques. Soil-conserving tillage technologies use tillage methods that do not require soil inversion or disturbance and that retain protective plant residues on the surface (Highfill and Kimberlin, 1977). Methods include minimum tillage, zero tillage, and stubble mulching.

Minimum tillage has been traditionally practiced in the uplands. Farmers usually do not extensively cultivate their farm lots. Planting takes place once they clear the field and rains start. They usually use a dibble stick to make holes in the ground for seeds (UHP, 1979a). The soil around the plants is not disturbed as it would be with plowing and harrowing. The technique maintains the natural structure of the soil which promotes resistance to dispersion by raindrops.

Sajise (1982) studied the effect of tillage intensity on soil erosion and other variables. Using an area with 21% slope, the experiment compared the effects of conventional tillage (plowing until the soil is pulverized) and furrow tillage (minimal plowing) on soil erosion, run-off, and nutrient loss. Four crop treatments were used: fallow (which corresponds to the bare control treatment), sugarcane, corn and cassava. The experiment showed that conventional tillage leads to a significantly higher sediment yield, run-off volume, nitrogen, phosphorus and potassium loss over furrow tillage (table 3).

Yield, however, was not affected by the different tillage treatments. This confirms the work of other researchers that reduced tillage does

Table 3 Mean calorie yield, sediment yield, run-off volume, nitrogen, phosphorus and potassium loss in conventionally — and furrow — tilled plots. (Adapted from Sajise, 1983).

Parameters	Tillage Practice	
	Conventional Tillage	Furrow Tillage
Calorie yield (Mcal/ha)	19.40	19.98
Sediment load (t/ha)	105.88	69.29
Run-off volume (mm)	410.94	430.84
Nitrogen loss (kg/ha)	93.01	65.45
Phosphorus loss (kg/ha)	2.58	1.50
Potassium loss (kg/ha)	10.16	7.19

not result in reduced yield (Free et al., 1963; French and Blake, 1965; Mabbayad et al., 1968; Gines et al., 1976). Benefits include soil and water conservation, minimized nutrient losses, and relatively lower energy and labor inputs. Zero tillage also has these potential benefits.

Zero tillage has been extensively studied in the United States because of its value as a soil-conserving and labor-saving technology. A long-term field test was conducted at the North Appalachian Watershed Research Station of the Department of Agriculture in Ohio. The soil was well-drained Mangum silt loam with 9.4% slope. Zero and conventional tillage were compared in terms of corn yield, soil erosion, and surface water run-off. Approximately seven tillage operations prior to planting corn, including the incorporation of manure, were used in the conventional tillage system. The no-tillage system, on the other hand, was made up of only two operations: spray application of herbicides (Atrazine, Simazine, and 2, 4-D) to control the growth of undesired vegetation and the corn planting itself (Harrold et al., 1967; Rask et al., 1976). Manure and corn mulch (from previous crops) were left on the surface in the zero tillage system. This mulch covered 75% of the no-tillage surface and reached an estimated weight of 15.7 t/ha by 1968, five years after the start of the experiment. Corn yield in the no-tillage treatment was generally higher or equal to corn yield from the conventionally-tilled land (table 4).

The Philippine version of zero and minimum tillage is more labor-intensive. Uplanders usually cannot afford the cost of fertilizers and herbicides. Human labor and organic wastes are substituted for chemicals. In spite of differences, benefits in soil erosion control would probably be similar given minimum soil disturbance. Benefits of zero and minimum tillage are greater when combined with mulching. Harrold et al. (1967) attributed the difference in soil erosion from the conventional vs. zero tillage experiment to the vegetative residue and manure mulch.

Table 4. Corn yield, run-off and erosion from no-tillage and conventional watersheds: May-Sept, 1964-1968, Coshocton, Ohio (Harrold et al., 1970).

Year	Treatment	Corn Yield (kg/ha)	Run-off (mm)	Soil Erosion (kg/ha)
1964	Conventional	6390	15	6380
	No-tillage	9150	5	132
1965	Conventional	7130	4	146
	No-tillage	7130	0	0
1966	Conventional	6520	0	0
	No-tillage	7870	0	0
1967	Conventional	7870	28	2170
	No-tillage	7330	0	0
1968	Conventional	6990	9	0
	No-tillage	7600	0	0

Mulching. Mulching is the application of such materials as plant residues, stones, and paper on the soil. The technique is applicable where there is a lot of crop waste. A three-year experiment in Rhodesia tested the effect of wire gauze as soil mulch. Soil erosion in the wire gauze covered soil ranged from 0 to 5.02 tons/ha/years, compared to bare control plots in which soil eroded at the rate of 154.4 to 567 tons/ha/year (UNESCO/UNEP/FAO, 1979). An experiment on the effect of slope and soil management on soil loss at the International Institute for Tropical Agriculture (IITA) in Ibadan, Nigeria, showed that soil loss was lowest for all slope classes in the experiment in mulched soil which had been subjected to two consecutive croppings of maize. This result was in spite of the observation that the unmulched maize-maize cropping pattern was the most erosion-prone among the cropping systems tested (Lal, 1976). This experiment was followed by a study on the effect of mulching on run-off and soil loss. Tables 5 and 6 show that run-off and soil loss decreased with increasing amounts of mulch applied to the soil. Run-off decreased from 286 mm for zero mulch, to 7.8 mm and 6 tons mulch/ha. Similarly, soil loss decreased from 13.0t/ha (zero mulch) to 0.07 t/ha (6 tons mulch/ha). Soil loss in Carcar, Cebu, decreased when corn stubble from the previous cropping was retained (Pacardo et al., 1983). Cuevas (1983) used cogon as a mulch for rice on the slopes of Mt. Makiling. Soil erosion was lower in mulched plots compared to unmulched plots.

Mulches reduce the force of impact of raindrops on the soil and increase soil infiltration. Results include decreased surface sealing, increased surface storage, decreased run-off velocity, improved soil

Table 5. Effect of mulch rate on run-off (mm) (From IITA, 1973).

Slope (%)	Mulch Rate (t/ha)				
	0	2	4	6	No-tillage
1	283.2	6.0	4.2	0.0	6.4
5	345.9	61.2	10.1	6.7	8.9
10	218.5	45.5	20.9	12.3	15.2
15	294.3	46.9	20.0	12.1	14.1
Mean	285.5	39.9	13.8	7.8	11.2

Table 6. Effect of mulch rate on soil loss (9t/ha). Total rainfall = 64.01 mm (From IITA, 1973).

Slope (%)	Mulch Rate (t/ha)			
	0	2	4	6
1	0.48	0.01	0.0	0.0
5	12.19	3.49	0.67	0.16
10	27.06	0.82	0.11	0.03
15	12.25	0.64	0.31	0.08
Mean	13.0	1.24	0.27	0.07

structure and porosity, and improved biological activities related to soil cover (Lal, 1976), Mulches shield soil from rain. A favorable side effect is the minimization of the sealing of soil pores and the unhindered infiltration of water into the soil body. The rate of water infiltration into the soil can decrease if the soil is directly exposed to raindrop compaction or splash erosion. Decline in infiltration could also be brought about by the clogging and/or sealing of soil pores with sediments suspended in run-off water (Sopper and Lull, 1967).

Soil moisture storage is increased by the absorptive colloidal products of decomposition of organic mulches (Brady, 1974; Wade and Sanchez, 1975). Mulch also bars the movement of run-off (Foster and Meyer, 1977). Run-off velocity and surface scouring are reduced and soil loss is reduced. The organic products of decomposition and the waste products of organisms which feed on the mulch serve to bind soil particles together (UNESCO, 1969), improving soil structure and porosity. The soil is then more resistant to erosion. The organic matter also stimulates the activities of soil flora and fauna. Earthworms in particular increase their burrowing and soil porosity is again improved (Etherington, 1975).

Contour Farming. Contour farming is another vegetative erosion control method involving cultivating, planting and harvesting along the contour at right angles to the natural direction of slope (Foster, 1973). Contour farming creates barriers and slows the flow of water downslope. Run-off water can be absorbed by the soil. With less water flowing, less soil is carried down-slope. Contours act as micro terraces. In contrast, furrows created by cultivation with the slope act as canals for water and sediment and promote gully erosion (Martin, 1979).

A three-year study in Bugcaon, Bukidnon, monitored surface run-off in plots subjected to contour tillage and up-down tillage. The plots were on 10% and 20% slope. Surface run-off was lower in plots where contour tillage had been practiced compared to those plots which had been tilled along the slope (Palis et al., 1983) (table 7).

Table 7. Run-off (mm) in contour tilled and up-down tilled soil, Bugcaon, Lantapan, Bukidnon. (Adapted from Palis et al., 1983).

| | | Tillage Treatment | |
	Slope (%)	Contour Tillage	Up-down Tillage
1979	10	26.75	28.90
	20	20.51	21.19
1980	10	137.28	141.57
	20	103.93	126.35
1981	10	65.15	76.07
	20	77.42	82.55

Contour farming is most applicable where slope is fairly uniform and least practicable with irregular topography and wide slope variation (PCARRD ibid.) After a site has been chosen, at least one contour on the slope (using any of the methods discussed above) is determined. Tillage and planting are conducted on the contour. In highly erosion-prone areas, PCARRD (1984) recommends contoured rows combined with terracing, strip cropping, and grassed waterways.

Contour strip cropping reduces soil erosion, especially where terraces are not practical because of uneven slope or unsuitable topography (Foster, 1977). Contour strip cropping is the planting of alternating bands or strips of close-growing crops with clean-tilled or fallow crops (Highfill and Kimberlin, 1977). Crops can be planted in relatively narrow strips along the contour (across the slope), such that strips of erosion-prone crops like corn, rice, and peanut are separated by dense, erosion-preventing crops such as mungo, soybean, and other legumes (PCARRD, 1984). Erosion is minimized by the impedance of surface run-off by the close growing crops. Extension soil cover filters sediment carried by run-off, and water absorption is also aided by roots permeating the strips in-between.

The contour of a field to be contour strip cropped can be determined as discussed previously. There has been so far no study done in the Philippines on the most effective strip width, which will vary according to the degree and length of slope, soil permeability, susceptibility to erosion, quantity and intensity of rainfall, and kinds and arrangement of crops (Foster, 1973).

The UPLB-Program on Environmental Science and Management (PESAM) studied the effect of contour strips of ipil-ipil *(Leucaena leucocephala)* on soil erosion and productivity of upland rice (PESAM, 1982). The hypothesis was that the tree component (ipil-ipil) would be able to produce effective root systems to hold the soil together and vegetation to cover the soil surface. The need for soil cover was thought to be most critical during the early growth stages of the crop when the soil is most bare.

The study was conducted on a south-facing, 25° to 45° Mt. Makiling slope. For the stripped treatment, vegetation strips consisting of a double row of ipil-ipil and *kakawate (Gliricidia sepium)* were established along the contours. Interplanting distance within the strip was 0.5 meters, and the rows were 1.0 meter apart to ensure a dense stand. Strips were located along the contour using an A-frame. The area between strips, 3 m to 4 m wide, was reserved for the crop. Ipil-ipil and kakawate trees were trimmed to 1.0 m every 45 days, and trimmings were placed between strips. No burning was done. All trees were cut to ground level in the control plots. The cut vegetation was dried and then burned near the end of the dry season. Zero tillage was used in both the treatment (stripped) and the control (no-strips) plots. Planting was by dibbling (traditionally grown) rice seed. No inorganic fertilizers, insecticides, or weedicides were applied. Weeding was done four times during the growing season.

Contour stripped plots had a much lower mean grain yield (9.8 g/0.25m^2) than the non-stripped control plots (47.0 g/0.25m^2). Crop biomass and height were also significantly higher in the non-stripped plots. However, total sediment load in the stripped plots was lower (0.41 t/ha) than the non-stripped plots (1.65 t/ha). Analysis of the individual rain events showed that sediment load in the non-stripped plots was significantly higher than the stripped plots only during the early part of the rainy season. This showed the usefulness of incorporating ipil-ipil and kakawate strips in the farming systems aimed at sustained yield cropping in sloping upland areas. The low productivity of the stripped plots was attributed to limited light availability between the strips, germination, and growth-suppressing substances produced during the decomposition of ipil-ipil leaf litter and weed competition.

In the following year, an effort was made to improve yield while maintaining the soil-conserving ability of the stripped treatment (PESAM, 1983). To remedy shading, the strip was cut to about half a meter and the cutting cycle was shortened to 30 days to 35 days. To determine the effect of the presence or absence of strips and the cutting height of the strips on weed growth, weeding treatments were also introduced. Weeded plots were kept weed-free throughout the duration of the study, while non-weeded plots were left alone. The results showed a markedly higher grain yield in the partially stripped treatment (0.5 m cutting height) over the full-stripped treatment (1.0 m cutting height). Non-stripped treatments still had the highest grain yield. However, soil erosion was highest in the non-stripped and lowest in full-stripped treatment (table 8). Weed biomass was highest in the full stripped treatment. The partial stripped and non-stripped treatments had approximately the same mean weed biomass.

Table 8. **Mean grain yield, soil loss, run-off, and weed biomass in ipil-ipil and kakawate stripped, partial-stripped, and non-stripped plots on Mt. Makiling, Laguna (From UPLB-PESAM Annual Report, 1983).**

| | | Treatments | |
Parameters	Stripped	Partial-strip	Non-stripped
Grain yield (kg/ha)	10.5	16.91	26.79
Soil loss (kg/ha)	6.64	17.94	107.05
Run-off (li/ha)	3.18	4.01	14.15
Weed biomass	53.62	29.73	23.37
Rice biomass	35.96	50.02	56.05

A similar experiment in Carcar and Barili, Cebu, assessed the productivity and ecological stability of a corn/ipil-ipil cropping system (Pacardo et al., 1983). The study addressed the low corn yields attained by hilly land farmers and the harm that the cropping system is doing to local soil resources. Soil loss in the experiment follows the order: bare soil > corn alone > corn/ipil-ipil with stubble removed > corn/ipil-ipil with stubble retained. Corn yield was observed to be 58% to 75% greater when planted between the ipil-ipil rows than when planted alone. This result may be due to the "pumping-up" of nutrients from the lower soil horizons by the deep rooted ipil-ipil. Observed were higher concentrations of phosphorus and potassium in the run-off from plots planted to ipil-ipil. A similar observation was made by the UPLB-Upland Hydro-ecology Program concerning erosion and run-off from an ipil-ipil stand (UHP, 1979a). Such "pumping-up" may be of benefit since additional nutrients for plant growth are brought to the surface. However, where run-off and soil erosion are serious problems, ipil-ipil may aggravate the problem by bringing up nutrients that will then be carried away by run-off.

Improvement of Soil Nutrient Status

Another major problem is that associated with the rapid decline in soil fertility. The rapid decline in soil fertility with cropping is partially responsible for low productivity and the lack of sustainability of upland farming systems. Soil fertility is the presence of the necessary elements, in sufficient amounts and in the proper proportion, available for the growth of specified plants (when other factors such as light, temperature, moisture, and the physical conditions of the soil are suitable (Foster, 1973; Brady, 1974). Traditional shifting cultivation does not greatly damage the soil if there is a long fallow period for soil regeneration (Sanchez, 1972). Given present population pressures, however, fallows have been declining, and soils have not been able to regenerate.

Soil-improving technologies include the production and use of organic fertilizer, composting and mulching, use of nitrogen fixing plants and of inoculant microorganisms, green manuring, and the utilization of mycorrhizal relationships to increase the availability of phosphorus. Conventional approaches such as the application of inorganic fertilizers and liming are also practiced for some high-value cash crops.

Technologies which make use of organic and inorganic materials to add nutrient elements to the soil and to improve its physical and chemical properties include mulching, composting, green manuring and biological nitrogen fixation, and the use of inorganic fertilizers. These technologies, practiced in the uplands to varying degrees, have not been well-documented. The situation is changing as there are now several studies on minimum input farming systems (Gomez et al., 1984). Studies indicate the mechanisms underlying these technologies and their performance under field conditions. Technologies being developed include the screening of microorganisms which can speed decomposition of organic matter and the use of the mycrorrhizal association.

The use of organic farm wastes is not new. Uplanders cultivating rice terraces of the mountain provinces have relied on composted pig manure and plant materials for fertilizer (Omehgan, 1981).

Compost application in this system, estimated at 247 kg/paddy, is approximately equal to an application of 250 kg of nitrogen and 120 kg of phosphorus per hectare. The use of organic matter results in the release of nutrient elements for plant growth, soil improvement and possible disease and pest controlling effects. Humus formation is the ultimate end of composting and the decomposition of organic mulches.

The Use of Humus and Related Organic Materials. Humus, the relatively stable fraction of soil organic matter remaining after most of the added plant and animal residues have decomposed (Brady, 1974), is highly colloidal and amorphous in nature. It is composed of hard-to-decompose organic substances such as lignin, fats, and waxes (Ghildyal and Gupta, 1959) and has a high surface area and adsorptive capacity. Like clays, humus carries a net negative charge. Humus, however, carries a larger negative charge than clay, and hence, is able to attract and absorb more positively charged elements (e.g., the cations calcium, magnesium and potassium) This ability is measured in terms of the "cation exchange capacity" (CEC). CEC is the sum total of exchangeable cations that the soil or substance can adsorb. The CEC of well developed humus ranges from 150 meq/100 gms to 300 meq/100 gms, while the CEC of siliate clays ranges from 8 to 150 meq/100 gms. Because of this high adsorptive capacity, humus prevents the deep percolation of nutrient elements into the soil profile (Etherington, 1975). Humus also favorably affects the availability of nutrients essential for plant growth. Hydrogen ion-saturated humus colloids react with soil minerals and extract their bases (calcium, magnesium, potassium, etc) Once extracted, these nutient elements are loosely adscrbed on the surface of the colloid and are easily available for adsorption by plant roots (Brady, 1974).

Humus also acts as a chelating agent for metal ions. A humus-metal ion complex keeps the metal in solution but prevents it from causing any toxic effect (Odum, 1971).

Organic materials also reduce and regulate plant pathogen and pest populations. Organic matter in the upper soil level controls peanut rot and pod rot caused by *Sclerotium rolfsii* Sacc. and *Rhizoctonia* sp (Papavizas, 1970). Volatile compounds from decomposing alfalfa stimulate germination of sclerotia of *S. rolfsii* (Linderman and Gilbert, 1968) which then die for lack of a host plant. A similar stimulatory effect by decomposition products was observed for *Fusarium oxysporium* (Linderman, 1970). Tomato root-knot nematode infestation can be controlled by plant residues (Sayre, 1971) because of the reduction of pathogen population during anaerobic decomposition and the denial of a host for the pathogen upon germination (Huber and Watson, 1970). Three mechanisms control the organic residue on nematode populations. The production of alkaloids and volatile fatty acids may be harmful to the nematode. Organic residues may also stimulate the activity of nematode predators and may alter the physiology of the host plant, that is, make it more resistant to nematode infestation (Sayre, 1971).

Mulching. Mulching, described above in terms of erosion control, is also beneficial for soil maintenance and improvement, and crop production. Table 9 shows the positive effects of mulching on maize and cassava yields in Nsukka and Ibadan, Nigeria. The significantly

Table 9. Effect of mulching on the yield of corn and cassava.

| Crop | Component | Year | Yield (kg/ha) | | Source |
			No mulch	With mulch	
Corn	Stover	1971	4893	5877	Ayanaba and
		1972	2659	3300	Okigbo, 1974
	Grain	1971	1387	1850	Ayanaba and
		1972	1542	2621	Okigbo, 1974
		1971	5480	6170	IITA, 1973
		1972	6680	7320	
		1973	5710	8350	
Cassava	Fresh	1971-72	2733	3931	Ayanaba and
	tubers	1972-73	2016	2489	Okigbo, 1974

higher yields can be attributed to the moisture-conserving effect of mulches and a lower soil temperature that favors seed germination (IITA, 1973), reduced soil erosion, fewer weeds during early crop growth (Okigbo, 1965; Agboola and Udom, 1967), and improved soil structure (Jacks et al., 1955; UNESCO, 1969). Mulching also increases humus content (Alexander, 1961; Russel, 1973), increases cation exchange capacity of the soil (Ayanaba and Okigbo, 1974) and contributes soil nutrients. Table 10 shows the effect of residues on organic

Table 10. Residue management and its effect on organic carbon content and ccation exchange capacity (CEC) of the soil. (From IITA, 1973).

Residue Treatment	Amount Returned (t/ha)	Organic Carbon (%)	CEC (meg/100 g)
Retained	16.4	1.63	6.82
Removed	—	1.04	4.64

carbon and cation exchange capacity for the research in Ibadan, Nigeria. Plant residues were returned to the soil at a rate of 6.4 t/ha. Soil organic carbon content was increased by 0.59% and the CEC was

2.17 meq/100 gms. of soil greater than for the treatment in which plant residues were not returned (IITA, 1973). Yield of *Eleusine* is higher when mulched than when fertilized with inorganic nitrogen (Griffith, 1951). A similar result was obtained for cotton seed yield (table 11).

Table 11. **Yield of eleusine and cotton in Uganda as influenced by fertilizer, manure, and mulch. (From Griffith, 1951).**

Crop	Control	(NH) SO	Na NO	Kraal manure	Mulch
Eleusine (t/ha)	18.0	19.0	18.0	42.0	24.0
Cotton seed (kg/ha)	258.0	370.0	347.0	448.0	1053.0

In mulching, crop residues and animal manure are left on the ground until the next cropping. The practice among uplanders of harvesting only the usable or edible portions of crops with hand-held cutters, picking only the rice panicles, and hand-picking corn and leaving the residue to decompose are, thus, forms of mulching.

Composting transforms organic wastes into partly decomposed material rich in humus and mineralized inorganic nutrients. The mean percentage composition of some common animal and plant manures and residues is given in table 12. Poultry manure has the highest, and cassava refuse has the lowest nitrogen and phosphorus content among all plant and animal residues and manures. Decomposition is fostered by conditions favorable to the growth and activity of decomposer organisms. Decomposition hastens the release of inorganic nutrients from their bound organic forms. Composting is being improved through selection of fungal species that rapidly decompose lignin and cellulose containing materials. In an experiment by Cuevas (1982), two species of *Trichoderma (T. aureoviride* and *T. pseudokoningii* Rifai aggr.) and a species of *Verticillium (V. cellulosae* Daszewska) were isolated from decomposing materials such as paper, kitchen refuse, wood, leaf litter, and soil from compost pits and forest floors. Identification and use of the fungal species that rapidly decompose cellulose will make possible the use of variety of crop and wood waste for organic fertilizer.

Nitrogen Fixation. The nutrient cycle is opened when a forest is cleared and cropped. Soil nutrients can be eroded away. Crop uptake reduces soil nutrients and the quantity of nitrogen is diminished (Manguiat, 1984). Nitrogen fixation, therefore, plays an important role in soil replenishment.

Table 12. **Mean nitrogen, phosphorus and potassium content of some common plant and animal residues in the Philippines (from Misra and Hesse, 1984).**

Material	C/N	Percentage (over-dry weight basis)		
		Nitrogen	Phosphorus	Potassium
cattle dung		1.50	1.0	0.94
sheep dung		2.02	1.75	1.94
goat and sheep dung		1.42	1.02	0.71
horse dung		1.59	1.65	0.65
pig dung		2.81	1.61	1.52
poultry manure		4.0	1.98	2.32
rural composts		0.58	0.17	0.59
maize straw		0.46	0.11	0.97
rice straw	105	0.58	0.10	1.38
soybean straw	32	1.30	–	–
bean straw		1.57	0.32	1.34
cowpea stems		1.07	1.14	2.54
sugarcane trash	112-120	0.35	0.04	0.50
green weeds	13	2.45	–	–
cassava refuse		0.10	0.04	–

Biological nitrogen fixation is the converting of elemental nitrogen into organic forms readily utilizable in biological processes (Brady, 1974). The process is usually accomplished through symbiotic associations of bacteria, algae, legumes, and nonlegumes. Free-living algae and several genera of bacteria and fungi are also capable of nitrogen fixation (Epstein, 1972). Much attention has been paid to the legume *Rhizobium* bacteria association (Dinglasan, 1936; Aquino and Madamba, 1939; Paterno, 1983; Paterno et al., 1983; and Tilo, 1983).

Legumes effectively add nitrogen to the soil. Subba Rao (1975) determined the difference in the amount of soil nitrogen for Rhizobium inoculated and for uninoculated cowpea, pea, and gram. The uninoculated legumes added 8 kg to 36 kg of nitrogen per ha. Rhizobium inoculated plants added 40 kg to 134 kg of nitrogen per hectare. The uninoculated treatment added nitrogen because of the Rhizobium bacteria in the soil. Rhizobium was not, however, sufficiently numerous near the legume roots to effect high levels of infection, nodulation, and, hence, nitrogen fixation. Maguiat (1984) concluded that legumes are necessary to build up the nitrogen fertility of degraded *kaingins*. He monitored changes in the amount of residual nitrogen (organic nitrogen + clay-fixed ammonium ions) in plots planted to two successive crops of cowpea (legume) and corn (non-legume). A fallowed area was used as a control. The legume was not inoculated with Rhizobium. Residual nitrogen in the corn plot decreased by 25 kg nitrogen per hectare, while the cowpea plot gained 77 kg nitrogen per hectare (table 13). Legumes may not always leave the soil with a higher nitrogen content, but since they fix part of the nitrogen that they use for growth, extraction of soil nitrogen is usually much less than for non-leguminous crops. In this sense, Brady (1974) refers to legumes as "nitrogen savers."

Table 13. Changes in residual nitrogen (N) in legume and non-legume (corn) planted plots after two croppings. (From Manguiat, 1984).

Plot	Residual Nitrogen (kg/ha)	N
Continuously fallowed	2781.0	—
Non-legume (corn)	2756.2	(-) 24.8
Legume	2858.4	(+) 77.4

Research in the Philippines has examined rhizobia and environmental stresses such as acid soils and low soil moisture content, the mass production of inocula, and inoculation techniques to enhance biological nitrogen fixation. Acid soil-tolerant strains of rhizobia for mungbean, peanut, and soybean were also found to be tolerant of the low phosphorus and high aluminum levels characteristics of cropped upland soils (Paterno, 1983). Inocula production has been aided by the finding that rhizobia can be grown in coconut water (Paterno et al., 1983; Mamaril et al., 1984). Inocula application is now easier with a granular inoculant which can be used with insecticide and/or fungicide treated seeds (Paterno et al., 1983). Inoculation itself is being studied. Where moisture is limiting, high inoculation rates compensate for rhizobia mortality (Paterno and Tilo, 1980). Pelleting phosphorus with the inocula on the seed also promises increasing survival and growth of some plant species (Manguiat et al., 1983).

Green Manuring. Green manuring is an extension of biological nitrogen fixation technologies. It consists of growing either legumes (which have been inoculated with a corresponding bacterial species and strain) or a high biomass producing species, and then incorporating the grown crop into the soil (Foresca and Lanuza, 1963). The organic matter is decomposed. While similar to composting and mulching, the crop is grown solely for providing organic matter. According to Floresca and Lanuza (1963), the green manure crop must produce a high amount of nitrogen-rich dry matter, and must have a high nitrogen content so that plant materials will decompose quickly and not cause nitrogen starvation of the succeeding crop. The crop must also grow quickly, especially initially, so that weed growth is suppressed; it must have more leafy than woody growth to facilitate rapid decomposition; it should have a deep and fibrous root system to penetrate lower soil horizons and to bring to the surface nutrients which would otherwise be unavailable. Finally, the green manure crop must be able to produce abundant easily harvestable seed to facilitate the establishment of succeeding crops. Some promising legumes are *Cajanus cajan* (pigeon pea) *Canavalia ensiformis* (jack bean), *Crotalaria* sps., *Glycine max* (soybean), *Phaseolus* sp., *Sesbania* sp., and *Vigna* sp.

In the Philippines, corn yield was significantly higher than a control with green manuring of mungbean (Barros, 1940). Cowpea and tapilan as green manure crops perform as well as mungbean in improving corn yield. Among the three, however, only cowpea green manure was able to maintain high yields up to the second corn crop (Eusebio and Umali, 1952). Hybrid corn yield was improved by green manuring (Aala and Gonzales, 1964). These studies took place in rainfed flatlands. In the uplands, green manuring is practiced differently. Crops generally cannot be grown exclusively for green manuring, and the biomass is not incorporated into the soil. Rather, legumes are grown to maturity and are either cut close to the ground and laid on the soil or are simply allowed to dry. Litterfall from the drying plants is decomposed and contributes to soil nutrients. Other schemes involve growing leguminous tree species, such as ipil-ipil *(Leucaena leucocephala)* and kakawate *(Gliricidia sepium),* along slope contours (PESAM, 1982) or on the farm borders. These trees are trimmed periodically to prevent crop shading. Cut branches and leaves are placed on the ground as green manure.

Mycorrhizal Association. The use of mycorrhizal association should have a significant impact on upland farming and land rehabilitation. Mycrorrhizae are the mycelia of fungi living in a mutualistic association with plant roots (Odum, 1971). The fungi interact with root tissues to form structures that increase the ability of the roots to extract minerals. The plant in return supplies the fungi with photosynthates. Mosse (1981), reviewing research on vesicular-arbuscular (VA) mycorrhiza for tropical agriculture, noted that the VA mycorrhiza is a physiologically economical way for the plant to increase its absorbing surface in the soil. Mycorrhiza is important in the absorption of the phosphate ion which is easily absorbed in clay complexes and diffuses slowly. Because of the absorption, a depletion zone of phosphate ions builds up rapidly around an actively absorbing root hair. Hyphal strands of the fungi go beyond the depletion zone and explore a greater volume of soil for mineral elements. Once the phosphate ion has been absorbed by the hyphae, it is transported back to the roots to be utilized for further root and shoot growth.

Gerdemann (1964) experimented with corn grown with 30 lbs of acid-soluble phosphorus. The mycorrhizal corn had a dry weight of 13.3 gms per plant, or about 300% greater than the dry weight of the non-mycorrhizal corn (3.7 gms per plant). Islam (1977) and Nyabyenda (1977) conducted similar studies with cassava *(Manihot* sp.) and sorghum *(Sorghum bicolor)*. Dry matter yield of mycorrhizal cassava was twelve times greater than the non-mycorrhizal treatment. Mycorrhizal sorghum, on the other hand, had double the dry matter production of non-mycorrhizal sorghum. Percent phosphorus in the tissues of the test plants was found to be higher in the mycorrhiza treatment in both cases. The uptake of other elements such as zinc, copper and sulfur are also improved by mycorrhiza. Mycorrhizal fungi have also been noted to have outstanding effects on soil aggregation. The mycorrhiza may, thus, possibly improve crop production in marginalized upland areas. The production of the mycorrhiza inoculant in large quantities is still needed, however.

Organic materials are not always a good thing. Decomposition can produce materials injurious to plants. The decomposition of barley, cowpea, soybean and cotton mulches produced acids (benzoic, phenylacetic, 3-phenylpropionic, and 4-phenylbutyric acid) toxic to tobacco (Linderman, 1970). Phenolic compounds such as vanillin, ferulic acid, and protocatechuic acid may also be harmful. Organics do not always decrease disease incidence or numbers of pathogens. *Rhizoctonia* sp. infected peanut in the presence of extracts of organic materials (Papavizas, 1970). Black rot of tobacco and bean, caused by *Thielaviopsis basicola,* was increased by residue extracts (Linderman, 1970). The potential positive and negative impact of using organic residues for soil improvement must be carefully assessed. The combined usage of organic and inorganic fertilizers may be a "best bet."

INTEGRATED TECHNOLOGIES

Increased production normally follows agricultural intensification, which, in turn, is achieved by: 1) using more land for agricultural production, 2) using land more often, or 3) increasing the use of technological inputs (Trenbath, 1976). Increased production has conventionally been achieved by increasing area cultivated. This is no longer feasible in most developing countries. Similarly, the third option of increasing inputs such as fertilizers is no longer feasible because of high cost.

Thus, increased production is now being achieved by using lands more often. Multiple cropping uses land more often and has been introduced in the uplands.

Multiple Cropping

Multiple cropping is the planting of more than one crop at the same time or in close sequence to one another on the same land. Three forms include sequential cropping, intercropping, and relay cropping. Sequential cropping is the continuous use of the land for a crop or a series of crops planted one after the other. Sequential cropping allows at most a very short fallow period. The most common intensive cropping sequence in the Philippines is irrigated continuous rice. Intercropping is the planting of two or more crops of different growth characteristics and requirements that are met by their simultaneous cultivation. A fast growing, tall crop requiring a lot of sunshine can be combined with a slower growing, shorter crop requiring less sunlight. Examples include rice-corn and corn-legume (soybean) intercrops. A variation of multiple cropping is relay cropping, the planting of a second succeeding crop before crop harvest. The second crop makes use of residual resources in the soil such as water and nutrients.

Multiple cropping is not new to Southeast Asia, and is a well developed cropping practice among traditional uplanders. Conklin (1957) reported as many as 100 different species in a single Mindoro Mangyan swidden out of a total of 4300 cultigents regularly grown by those swidden cultivators. The Lua of northern Thailand raise rice (a number of varieties), sorghum, chili, pepper, cotton, maize, root and tuber crops, viney plants, herbs, and seasonings on their swidden fields (Kunstadter et al., 1978). Christanty (n.d.) described an established home-garden in West Java :

> The lowest layer is dense with tuberous and shade tolerant plants such as taro, sweet potato, ginger, arrow root, chili pepper, etc. The second layer is formed by tall shrubs and fast growing plants such as salacca, papaya, and banana, usually occupying the height stratum from 2-5 meters. The greatest number of tree species occurs in the 5-10 meter height stratum, which is well utilized by foliar biomass. The space above 10 m height is occupied by coconut and tall trees such as *Albizzia* and *Parkia.*

Multiple cropping is desirable because of potentially higher yields, lower variability of yields from season to season, a better spreading of production over the growth period, less susceptibility to diseases or lodging, and improved crop quality (Trenbath, 1974). Other benefits are high efficiency in the use of soil and sunlight, better resistance to pests, diseases, and weeds, production of more varied and nutritious food, and better use of local nonhybrid, open-pollinated, locally adapted seeds (Altieri et al., 1983). These systems serve to minimize the risk of total crop failure, even the demand for household labor over time, and decrease soil erosion (Capistrano, 1983).

Polycultural systems result in higher yields per unit area than a similar sized area divided into monocultural components of the same polyculture (Jensen, 1952; Simmonds, 1962). The polycultural system also "overyields," or yields more than an equivalent area of a mono-culture of the highest yielding component (Trenbath, 1974). Sajise (1983) intercropped mungbean with sugarcane, corn, and cassava. He compared the performance of the intercropped treatment with the monocropped treatment using the land equivalent ratio (LER) which is the sum of the quotients of the yield of components of the intercrop from a unit area (yi) and the yield of these components grown as sole crops over the same area (yii).

$$LER = SUM \ yi / yii$$

If LER is equal to 1, then the various yields from the intercrop could have been obtained from the same unit areas planted solely to the monocrops of the components, each occupying the appropriate fraction of the total land area. If LER is greater than 1, the intercrop overyields the sole crops (Trenbath, 1976). Intercropping mungbean with sugarcane, corn, and cassava overyielded the single cropping of these species by 36% to 65% (table 14). Total calorie yield for intercropping and mono-cropping were not, however significantly different.

Table 14. Mean land equivalent ratio (LER) of monocropped and mungbean – intercropped sugarcane, corn, and cassava. (Adapted from Sajise, 1983).

Treatment	Sugarcane	Corn Crop	Cassava
Monocropped	1.0	1.0	1.0
Intercropped	1.65	1.40	1.57

Benitez and Naderman (1979) tested two cropping patterns in the uplands of Peru. System 1 was corn and rice sown at the same time and cassava relay planted into the corn rows 54 days after initial planting.

Peanuts were interplanted between the cassava after the corn and rice were harvested. System 2 consisted of corn and peanuts sown at the same time, with cassava relay planted 54 days after the initial planting. After the corn and peanuts were harvested, rice was planted in between the cassava. Nitrogen was supplied at 0, 80, 160 and 240 kg/ha. Phosphorus at the rate of 160 kg P_2O5/ha and 2 tons of lime/ha was broadcast into each plot and incorporated at 0 to 10 cm. depth. Sulfur, copper, iron, boron, potassium, and magnesium were applied. Phosphorus, potasium, and magnesium were applied after each harvest.

Table 15 shows the yields, relative yields (RY = yield as an intercrop component/yield as a sole crop), land equivalent ratio (LER = summation of relative yields) obtained by corn, rice, peanut and cassava in two multiple cropping systems and four levels of nitrogen. Yields of individual components were below those that would have been attained if they were monocultured . This is reflected in the RY values of less than one. However, the LER values showed that both intercropping systems doubled yields per unit of land. Overyielding has also been observed in mixtures of rice, barley, grass, flax and linseed, and legume-nonlegume mixtures.

Trenbath (1974, 1975, 1976) reviewed how yields of mixtures compare to yields of pure cultures. Most research has used mixtures of similar components, such as varieties of grain crops or species of grasses. Sixty-four% of a series of 139 experiments on 50:50 two-component mixtures yielded more grain than the average yield of their components' pure cultures (Clay, 1967; cited by Trenbath). About 60% of 300 mixtures of grasses and cereals had greater dry matter yields than the average dry matter yield of the pure cultures of the components (Trenbath, 1974). The margin by which the mixtures "overyielded," however, was not large and may be due to experimental error. Multiple cropping for increased yields may be confined to several "lucky" crop combinations. Research is determining why some systems and trials overyield and others do not.

A concern in upland farming is yield stability, the extent to which the aggregate yield of a mixed crop remains constant over time in the face of climatic and other fluctuations (Trenbath, 1982). Mixed cropping allows for a more consistent yield from season to season (Aiyer, 1949; Trenbath, 1974; Kass, 1978; Willey, 1979). Harvesting can be done throughout the year. A concept of production stability in mixtures has emerged because natural species population stability is greater in complex communities than in simple ones (Elton, 1958; Pimentel, 1961). Complex communities like the tropical rain forests are more stable in their composition from year to year compared to simpler ecosystems.

Table 15. Relative yields and land equivalent ratios (LER) obtained by corn, rice, peanut, and cassava in two multiple cropping systems and four levels of nitrogen, Yurimaguas, 1977. (From Benitez and Naderman, 1979).

Cropping System	Nitrogen Level		Yield (t/ha)				LER
			Corn	Rice	Peanut	Cassava	
1	0	Intercrop	0.48	2.04	1.06	14.65	
		Monocrop	1.57	4.24	2.66	18.37	
		RY (%)	0.31	0.48	0.40	0.80	1.79
	80	Intercrop	2.26	1.79	1.0	13.33	
		Monocrop	3.14	3.95	2.39	24.08	
		RY (%)	0.72	0.45	0.42	0.55	2.14
	160	Intercop	2.58	1.49	0.99	13.0	
		Monocrop	3.66	3.42	2.43	25.60	
		RY (%)	0.70	0.44	0.41	0.51	2.06
	240	Intercrop	2.79	1.34	0.87	12.99	
		Monocrop	3.51	3.0	2.68	26.70	
		RY (%)	0.79	0.45	0.32	0.49	2.05
2	0	Intercrop	1.33	0.50	1.35	1.88	
		Monocrop	1.57	3.09	2.93	18.37	
		RY (%)	0.85	0.16	0.46	0.65	2.16
	80	Intercrop	2.59	1.10	1.12	13.15	
		Monocrop	3.14	2.91	2.93	24.08	
		RY (%)	0.82	0.38	0.35	0.55	2.13
	160	Intercrop	2.64	0.92	1.03	12.70	
		Monocrop	3.66	2.26	2.93	25.60	
		RY (%)	0.72	0.41	0.35	0.50	1.97
	240	Intercrop	3.15	0.63	0.95	13.55	
		Monocrop	3.51	1.45	2.93	26.70	
		RY (%)	0.90	0.43	0.32	0.51	2.16

Hutchinson noted that population oscillations observed in arctic and boreal fauna may be partially due to those communities not being complex enough to damp out oscillations (1954). The idea that increased complexity causes increased stability has been taught to several generations of students. Mathematical modelling and experiments, however, suggests that greater diversity and complexity may lead to less stability. May, evaluating and comparing simple mathematical models of complex systems with many species and simple two-species models, concluded that models with many species are generally less stable than models of few species (1974). Turnbull and Chant (1961) argued that simpler predator-prey systems are often more stable than more complex systems. Observed stability in nature may be due to the stability of the environment, or, in the case of unstable systems, the lack of time for the species making up the system to coevolve necessary stabilizing factors (Trenbath, 1975). The attainment of stability is not as simple as the creation of diverse-species systems.

Also supporting yield stability of mixtures over time are the differing responses of species to environmental conditions (Trenbath, 1975). A mix of crops that interact with the environment in different ways ensures that there will always be something to harvest. For example, a corn-sweet potato mix can be hit by typhoon causing the decimation of the corn. The sweet potato will survive because it grows close to the ground and is not as affected. The uplander's preference for multiple cropping reflects a working knowledge of the phenomenon.

A diverse cropping system lessens the extent of insect damage. Corn borer damage is minimized by intercropping peanut with corn. Diamond-back moth damage on cabbage is minimized by intercropping with tomato, and root knot nematode *(Meloidogyne incognita)* populations are suppressed by intercropping *Tagetes* or *Crotolaria juncea* with tomato (Carandang, 1978). Effects to crop mixtures that confer protection from insect or pest damage to one or all of the components of the mixture include: a) the fly-paper effect, b) the compensation effect and, c) the microenvironmental effects (Trenbath, 1975). The fly-paper effect is based on the high specificity of the host-pest relationships. The presence of non-host plants in the mixture dissipates the effect of the pest. The fly-paper effect is most effective during the insect's early life -stages when it cannot steer towards its host plants. This effect is also most effective during the spore dispersion stage of fungi diseases. Spores landing on non-host plants cannot take off and are effectively removed from the population. Diminution of the population of the insects or fungi depends on the proportion of non-host plants in the mixture. By contrast, the compensation effect is based on the decline in competition for resources by the infected or infested plants, thereby allowing neighboring non-host plants to draw more of the available resources and compensate for the loss in production from the affected plants. Micro-environmental effects result from modification of the micro-environment of the susceptible crop. The changed microenvironment may act to decrease the susceptibility of the host crop, may create conditions wherein the pest or disease cannot thrive, or may encourage the growth and reproduction of natural enemies of the pest.

Multiple cropping is suited to situations of labor surplus given the labor required for planting and harvesting. These activities are spread throughout the cropping cycle given different planting times and selective harvesting of crops with different times of maturity (Nguu and Corpus, 1979). Soundly practiced mixed cropping can require less pesticide, herbicide and fertilizer, and can thus be a less polluting and less expensive farming method (Trenbath, 1975).

Table 16 summarizes multiple cropping introduced in upland communities on Nueva Ecija, Batangas, Laguna, Cavite, Antique, Capiz, Bukidnon and Agusan. These cropping patterns usually included a staple crop, such as rice or corn, and vegetables, legumes, or cash crops. The non-staple crops allow the farm family with a balanced diet, help renew the soil, and provide cash. Cropping system content also considered product demand, ease of establishment, capital requirements, waiting time for plants to bear, farm to market distance, and infrastructure.

Table 16. Introduced and traditional agricultural production systems in some upland areas of the Philippines.

Site	Production System/Cropping Pattern		

Villarica, Pantabangan, Nueva Ecija

| (Intercrop of beans, tomato, bitter gourd, squash, and sweet potato) | (Intercrop of rice and sweet potato) | (Annual raising) | |
| (Intercrop of okra, pigeon pea, and other legumes) | (Intercrop of rice and stylosanthes, cassava, sweet-potato, ube) | (Intercrop: beans, egg-plant, and tomato) | |

Mt. Makiling, Laguna

Corn - ube - gabi
Ube - rice - garlic - mustard
Sweet potato (camote)
Ginger, gabi, kalamismis
Rice - ube - hot pepper - gabi
Ginger - peanut - sweet potato
Corn - ginger - papaya - peanut - sweet potato - kadios

Ginger - corn - winged bean - bataw
Rice - peanut - mustard - corn - camote
Garlic - gabi

Antique Upland Development Program, Hamtic, Antique

Rice - fallow
Rice - rice - rice - upland crop
Rice - upland crop or (rice + corn) -
    fallow or (rice + corn) - upland
    crop
Upland crop - fallow
Upland crop - upland crop
Upland crop, peanut, mungbean

Capiz

Corn - peanut - corn/peanut
Upland rice - corn - corn/mungbean
(Corn + ipil-ipil) - corn/mungbean
(Corn + ipil-ipil) - (upland rice +
    ipil-ipil) - (corn + ipil-ipil)
Upland rice - (peanut + corn) - corn/
    cowpea
Cowpea or mungbean - (upland rice +
    corn) - corn - mungbean or cowpea

Bukidnon

Introduced
    a. Plateau
        Upland rice - corn - mungbean
        Corn - corn - mungbean
        (Corn + peanut) - (corn + mungbean)
    b. sideslope
        Coffee + corn
        Corn - corn - mungbean
        Rice - corn - mungbean
        (Rice + corn) - (corn + peanut)
        (Rubber + corn) - corn - mungbean

Traditional
    Upland rice - corn
    Corn - corn
    Corn

Agusan

Introduced
    a. Plateau
        (Corn + peanut) - (corn + mungbean)
          - upland rice
        (Corn + mungbean) - corn

b. Sideslopes

(Corn + peanut) - (corn + mungbean)
- upland rice
(Corn + mungbean) - corn

Traditional

Transplanted rice - transplanted rice
Corn - corn
Mungbean - corn

Most introduced cropping systems include one or two legumes. These are either intercropped, relayed or sequence cropped with the non-leguminous components. Legumes are included because they tolerate drought, fix atmospheric nitrogen, and have a short life cycle, good nutritive value, multiple uses and good market value (Gomez and Zandstra, 1976). Mungbean, pigeon pea, and soybeans, legumes tolerant of low soil moisture, can be grown late in the wet season, and can survive, grow and bear fruit at about the middle of the dry season when soil moisture is low. Legumes can add enough nitrogen to the soil to benefit subsequent or non-leguminous intercrops. The short life cycle of legumes (e.g., mungbean matures in 65 days and cowpea in 75 days) makes it possible to use land intensively and to free land for other crops. High protein content makes legumes a good food. Legumes can also be used for livestock feed and firewood and can be converted to a number of processed products. Most grain legumes require a minimum of postharvest treatment, allowing farmers to sell when prices are higher.

An economic analysis of cropping systems in Bukidnon showed that introduced or experimental cropping systems had higher yields than traditional systems. High yield, however, was attained by the application of costly fertilizers and pesticides. In Bukidnon, raising upland rice cost P304 using the existing cropping system, but cost P1,111 using the experimental cropping system. There were some savings on labor costs, however. Returns to labor under the experimental system was higher, but returns to material costs were lower. The opposite was observed for the traditional system. The experimental cropping systems ignored that required costs were excessive for most upland farmers. The experimental system economized on labor by substituting more technological inputs (Capistrano, 1982). These factors undoubtedly contribute to yield gaps between experimental plantings and farmers' fields and to the low adoption of introduced cropping systems.

Agroforestry. Agroforestry is a multiple cropping technology, or an intercropping of woody plants with food and/or forage crops. A more comprehensive definition: Agroforestry "is any sustainable land-use system that maintains or increases total yields by combining food (annual) crops with tree (perennial) crops and/or livestock on the same unit of land, either alternatively or at the same time, using management practices that suit the social and cultural characteristics of the local people and the economic or cultural conditions of the area" (Bene et al. cited by Vergara 1982). Pollisco defined agroforestry as "the intensive development of the land by devoting that portion which is suitable to agriculture for the production of farm crops and livestock, and the remaining area which is marginal to sub-marginal to tree farming" 1975).

Agroforestry is attractive because of the similarity of its structure to the tropical rain forest ecosystem and its multilayered canopy. The canopy increases the efficiency of light capture and utilization. Light transmitted or reflected by one layer can be utilized by a succeeding layer of leaves. Moisture loss from the soil by evaporation is minimized by a complete canopy. The canopy also protects the soil from rain, and decreases soil erosion. High species diversity is characteristic of the ecosystem. High diversity and the multiple layered canopy allow an essentially closed nutrient cycle. Agroforestry mimics the structure of the forest; thus, corresponding functional attributes are hopefully replicated as well.

Agroforestry can be cyclical, *taungya,* or integral (Vergara, 1981). Cyclical agroforestry includes traditional kaingin farming in which food cropping and forest fallow alternate. Taungya agroforestry is the early simultaneous planting food crops and trees, followed by natural evolution towards pure forest. Integral agroforestry consists of the simultaneous but continuous planting of food crops and trees. Vergara says that integral agroforestry is best for stabilizing hillside farming. Integral agroforestry has been characterized as acceptable to target beneficiaries, inexpensive to implement, more productive and sustainable than other systems, efficient in utilizing available land resources, and ecologically sound.

Many upland areas now depauperate of soil nutrients and topsoil and subject to drought have been taken over by grasses capable of tolerating such environmental conditions. Grasslands in the Philippines include cogon *(Imperata cylindrica),* samsamon *(Themada triandra),* Misamis grass *(Capillipedium parviflorum)* and amorseko *(Chrysopogon aciculatus:* Sajise et al., 1976).

Marginalization of the uplands is a result of resource mismanagement. Logging and kaingin farming have been blamed. Seventy % to eighty % of the nutrients in a tropical rainforest is in the tree biomass. Removal of this biomass opens the closed nutrient cycle and allows nutrients losses, leaving a limited nutrients in the soil which, in turn, are soon removed from the ecosystem. Open sites lead to soil erosion and surface water run-off. Erosion contributes to soil marginalization. If the land is cropped, nutrient uptake constitutes a further loss from the soil. The amount of nutrients lost increases with agricultural intensification. The loss of exchangeable bases leads to a preponderance of hydrogen ions and a decline in soil pH. The erosion of top soil rich in organic matter destroys soil structure and its water holding and strong capacity.

The dominance of grasses in such depauperate areas can be either beneficial or harmful. Grass forms a protective cover against rain, and the root system minimizes soil and nutrient loss by water erosion. Litterfall from grasses accumulate and aid in restoring soil structure, water holding capacity, and fertility. These physical changes then allow the establishment of other plant species that further modify the environment to allow even more species to germinate and grow, until eventually a forest community is attained. Fire and human exploitation, however, are the grassland's worst enemies. Fire, in particular, prevents natural forest regeneration. Sajise (1977b) called it the "vicious cycle of grass-fire-grass."

Grasslands are burned during the dry season for a variety of reasons (UHP, 1979b): Pasture managers burn to encourage the production of new shoots, which make better forage. Hunters burn since the better forage acts as a lure to game animals. Some fires start from swidden fires or are intentionally set by farmers wishing to use grasslands for crop production. Firing saves labor in cutting and clearing and produces ash to the soil. Pest and disease control are also reasons for putting grasslands on fire.

Burning adds to the marginalization of the ecosystem. The UPLB-Upland Hydroecology Program (1979b) studied the effect of grassland-burning at various times during the dry season. Burning volatilized nitrogen locked up in the plant biomass. Additional nitrogen losses took place via the burning of soil organic matter. Other nutrients such as calcium, magnesium, potassium, and magnesium were unaffected by burning. The impact of burning on soil erosion and water run-off was related to the time of burning. Burning leaves the soil devoid of cover, contributing to soil and nutrient losses and making the area less suitable for plants other than grasses. Allelopathic chemicals exuded by grasses further insures their dominance in marginal sites. Overgrazing and intensive cropping also contribute to the decline of these upland ecosystems, as indicated by sparse savanna vegetation cover such as *Piloistigma Themeda and Antidesma Themeda* or *Piliotigma/Imperata* and *Antidesma Imperata.* Marginal conditions are also indicated by a dominance of *Chrysopogon aciculatus, Hyptis spp., Sida acuta, Pseudoelephantopus spicatus, Panicum walense, Stachytarpheta jamaicensis, Brachiaria repens,* and *Shizanchyrium fragile* (Sajise, 1977a).

Regeneration technologies for such denuded and marginal uplands range from those requiring high inputs of fertilizers, soil amendments, pesticides, herbicides, labor and management to low-input requiring technologies dependent on natural biological processes.

Reforestation

Reforestation has been a major technology employed. Reforestation regenerates a woody vegetation stand either by active or passive means. Research and development efforts have been concerned with different kinds and treatments of fertilizers, seed treatments and germination, and the effects of potting media, moisture, and light on the growth, survival, and performance of reforestation species. High input and capital intensive reforestation has been attempted to attain a quick solution to upland degradation. The establishment of fast growing tree species in denuded areas has been emphasized and is a short-circuiting of natural succession. While natural succession may require several years for trees to come into the grassland, reforestation seeks to telescope time. While theoretically possible, the strategy requires large amounts of resources.

Government reforestation includes the introduction of exotic trees, establishment of nurseries, inputs, land preparation, planting, and periodic weeding. Exotic trees species are imported because of their ability to thrive under local environmental stresses and because of their potential economic value. Giant ipil-ipil *(Leucaena leucocephala)*, developed in Hawaii and South America, was imported because of its reported ability to grow under almost any conditions, and because of its potential to fix nitrogen from the atmosphere, and produce biomass for fuel, fertilizer and animal feed. Similarly, alder (Alnus sp.) species have been introduced because of their ability to withstand moisture and low mineral nutrient stresses. Yemane *(Gmelina arborea)*, albizia *(Albizia falcataria)*, *Eucalyptus*, *Anthoecephalus*, *Pinus caribaea*, and *Cecropia* are being planted because of their fast-growing abilities. These species are successional pioneers with adaptive mechanisms for colonizing open lands, have rapid growth rates and low wood densities, are self-pruning, and possess large, thin leaves, often riddled with insect damage (Ewel, 1979). The ability to continuously flush new leaves probably enables these species to cope with insect attack. They are light-wooded, thin-barked, and lack the saps, resins, and silica inclusions that characterize mature-forest species. These species have been monocultured in large areas of the Philippine uplands.

Time has shown that these exotic species are not "miracle trees." Giant ipil-ipil is unable to attain "giant" proportions in areas where soil fertility is low and soils are acidic. Albizia is unsuited to the seasonally dry and storm crossed areas of Luzon and Visayas because it requires even moisture throughout the year and its soft wood is prone to wind breakage. There is now a trend towards utilization of native pioneer species for reforestation. Lanite *(Wrightia laniti)*, kakawate *(Gliricidia sepium)*, tibig *(Ficus nota)*, kamachili *(Pitchecelobium dulce)* and *Macaranga* are being examined. Use of these species may reduce the cost of reforestation by bringing down expenses of nursery management and increasing the rate of seedling survival.

Nurseries are basic to reforestation. Nurseries grow seedlings to a size, age, and vigor that will enable them to compete with other plants. The seedlings are watered, fertilized, and treated with chemicals to dampen insect, fungal, and bacterial infections and infestations (Agpaoa et al., 1975). From the nursery, seedlings are transported to the site and planted manually, usually during the late dry season or the onset of the wet season. This gives the seedlings a chance to establish their root systems and lessens seedling mortality due to moisture stress. Seedlings are, however, more prone to pest and disease attack at this time. Grass growth is also at a peak, increasing the chances of seedlings being shaded-out.

Seeds can also be broadcast. Experience with ipil-ipil, however, has shown that this is not very effective because of competition by grasses. The tree seeds and seedlings are either shaded-out or affected by allelochemics produced by the grasses. Allelopathy is one of the bases of the competitive edge of cogon over other species (Sajise, 1980).

The best way to decrease the competitive ability of native grasses in order to increase the survival of the reforestation species has been studied, as has land preparation prior to seedling planting. Land preparation ranges from lodging the grass with a log to the application of herbicides. Spraying cogon with glyphosate, N-phosphonomethyl glycine ("Round Up"), is more effective than lodging with a log prior to the planting of ipil-ipil (del Rosario, 1982). However, chemicals cost much more. Sajise (1983) recommended that lodging (done 2 to 3 times during the first year) instead of herbicide application be used for site preparation in the reforestation of *Imperata* dominated areas.

Reforestation using minimal inputs has also been studied. Such a regeneration technique is the utilization of the "life-form-effect" of savanna tree clumps. Plant growth conditions in stands of *Antidesma frutescens* and *Piliostigma malabaricum* are better than in open grassland (Tupas and Sajise, 1977). Savanna tree clumps reduced the incident light intensity from 3% to 65%. Air and soil temperatures under the tree clumps were 4.0° and 2.0° Celsius lower than corresponding open grassland measurements. The tree clumps also maintained a higher relative humidity (approximately 2% higher than the open grassland). These findings led to the idea of using the savanna tree clumps as starting points for grassland regeneration activities. The favorable microenvironmental conditions are used to give reforestation species established within the clumps a better chance of surviving. These plants, in turn, will later contribute to the further modification of the microenvironment to allow other less adaptable species to survive and grow. Hence, reforestation is envisioned to grow from the clump outwards in a widening, ripple-like manner.

Critical to this method is the selection of a pioneer species that will survive, grow, and reproduce given the environmental constraints. Glori (1977) reported that yamane *(Gmelina arborea)* is more resistant to drought than Kaatoan bangkal *(Anthocephalus chinensis)*. Tanguilig (1979) found that *Leucaena leucocephala* had the least transpiration rate compared to four other reforestation species under water stress, a competitive advantage where moisture is limiting. Samson (1981) showed that the drought avoidance mechanisms of Leucaena is apparently due to its capacity to develop high osmotic pressure which enables the plant to continue extracting water from the soil during drought, and to slow its transpiration rate.

The soil fertility has conventionally been modified by fertilization and liming. This approach is, however, very expensive and may not be feasible for the large area and sloping topography of the uplands. Hence, use of legumes and non-legumes capable of fixing nitrogen from the air is being studied. Ipil-ipil inoculated with Rhizobium bacteria has been a main technology of many reforestation projects. Although nitrogen requirements can be met by nitrogen fixation, growth of legumes and nitrogen-fixing non-legumes is ultimately limited by other nutrients, notably phosphorus. Techapinyawat (1982) inoculated ipil-ipil with both *Rhizobium* sp., strain L-15, and vesicular-arbuscular mycorrhiza from the rhizosphere of ipil-ipil. Ipil-ipil inoculated with both rhizobium and mycorrhiza had a higher growth rate and nutrient absorption than the non-inoculated, rhizobium inoculation alone, and mycorrhiza inoculation alone treatments. This points to the future use of both rhizobium and mycorrhiza for increasing the survival and competitive edge of reforestation species over grasses.

Another strategy with a ripple-like effect is the selection of sites in which a high survival of the reforestation species can be expected. The Upland Hydroecology Program (1977) reported that dry season seedling survival in north-facing slopes and gullies is better than in south-facing slopes and ridges. A 47% difference in survival of *Leucaena* was observed between plantings in northeast-and southeast-facing slopes. This result was attributed to difference in soil moisture. Ground water draining into depressions and gullies enable these places to have higher soil moisture than ridges. The difference in moisture with slop direction is attributed to the more southerly position of the sun during the dry season, hence, higher insolation and greater moisture losses through evapotranspiration. Reforestation can be started from the more favorable sites and allowed to expand outward as microenvironmental conditions are modified.

Fire protection of grasslands to allow for natural regeneration is also a relatively low cost reforestation technique. Sajise (1972) observed that an *Imperata*-dominated grassland in Siniloan, Laguna, was succeeded by a shrub community of *Melastoma-Nephrolepis* and *Solanum-Mikania* after fire was controlled for two years. Cogon was gradually shaded out by the taller canopy of *Salanum* and *Mikania*. Shading, by as much as 50%, decreased the cogon's net photosynthesis and resulted in a significant decrease in rhizome production. Since 60% of the total cogon biomass is stored underground, reduction in rhizome production severely limited ability to regenerate. After three years the shrub-vine community was replaced by secondary forest pioneer species such as *Ficus* spp., *Mallotus* spp., *Homolanthus populneus,* and *Trema orientaus.* These then produce changes in the soil, micro-climate, and energy balance, promoting natural plant succession.

Reforestation is for protection or production, depending on species and management system used. Grasslands, used for livestock, can also be productive. Sajise (1982) proposed the improvement of cogon grasslands by introducing stylo *(Stylosanthes guyanensis)* coupled with the raising of goats. Stylo competes with cogon and is able to tolerate soil moisture stress. A well-established stylo-native grass mixture produces an average of 18.3 tons of biomass per hectare per year when 50 percent of the standing biomass is allowed to remain for subsequent regeneration (Vidal, 1979). This amount could support about 115 penned goats fed on a cut-and-carry system. Ramirez (1980) looked at free-grazing goats in a stylo-cogon pasture. Soil bulk density increased from 1.01 gm/cc to 1.14 gm/cc within a period of six months in the grazed area. Total soil nitrogen and organic matter were significantly higher in the ungrazed than in the grazed area. The goats damaged ipil-ipil saplings within the area. Thus, improved pastures and goat production may be feasible, but problems exist. Goats, while tolerant to low supplies, still require some water.

CURRENT DIRECTIONS AND RESEARCH GAPS

This chapter has dealt with problem areas in upland development. These problems, addressed as if separate from one another, are closely interlinked. Soil erosion, agricultural intensification, and the degradation of upland resources are meshed together in a cause-and-effect relationship such that a fragmented problem solving approach cannot succeed. Unfortunately, fragmented approaches are currently being used by those entities responsible for the uplands. Table 17 shows a sampling of some upland development projects and their respective technologies. Success in implementation, however, depends on the perception of how a particular set of problems are interrelated and how a set of solutions can be coordinated.

Table 17. Production technology components of some upland development projects and programs in the Philippines. (Adapted from Capistrano and Fujisaka, 1984).

Project/Program	Technology Components
Forest Occupancy Management	Agroforestry
Communal Tree Farming	Agroforestry
Social Forestry Program	Agroforestry
PANAMIN Projects	Reforestation, goat dispersal, pig dispersal
Antique Upland Development Program	Reforestation, pasture improvement, orchard/field crops production, vegetable/cash crop production, bench terracing, strip cropping, contour farming, cattle fattening, goat raising/dispersal program
Hanunoo Mangyans (Peace Corps)	Reforestation, introduction of new cultigens, water and weed control, organic fertilizers
Kalahan Educational Foundation	Reforestation, orchard development, tree nursery establishment, fire control, research on highland agricultural systems
Bicol River Basin Development Project – Buhi	Bench terracing, nursery/orchard/firewood lot development, vegetative terracing, contour ditching, cover cropping
Baptist Out-of-School Training Program – Sloping Area Land Technology	Ipil-ipil contour strip planting, multiple cropping
PICOP	Intercropping of annual food crops and fast-growing tree species
World Neighbors	Contour ditching, Napier grass on ditch bunds for erosion control, composting, organic fertilizers, water and soil traps, crop diversification

CONCLUSIONS

Technologies are tools developed to achieve given goals. Use of such tools presupposes that a problem has been evaluated and understood to the extent that it is possible to advance a technological solution. A technology will address a problem only in so far as the problem has been understood. It is thus disturbing to hear upland.development implementors pushing "technology packages" for broad application in upland areas. Packages are of very general application and often do not answer specific needs nor address unique problems to specific target areas. They do not work in all cases, especially when the more unique features are of importance. This is not to denigrate technology packages; however, a solid understanding of the unique needs and characteristics of the upland is necessary to provide solutions to the gamut of ecological and agricultural problems. Just as understanding cannot be achieved by generalizing and treating the uplands as homogenous, neither can it be achieved by focusing on specific problem areas alone. Because upland problems are closely interlinked, focusing on a limited set of problem areas would only further the use of fragmented solutions. Evaluation of technologies and technology use must be undertaken from a variety of perspectives and should be placed within a coherent framework of needs and problems.

Low success rates of upland development projects are also due to the inappropriate use of technologies. Technologies include assumptions, usually implicit, that affect usefulness. For example, a project that includes ipil-ipil growing may assume that conditions such as moisture, soil nutrients, and soil chemical properties are adequate to support growth. Where insufficient, it is often assumed that these inputs will somehow become available. Another assumption may be that ipil-ipil will produce relatively higher or at least equal returns compared to agricultural crops, so that the planting of ipil-ipil is economically justifiable. Technology proponents and target end-users need to be aware that such assumptions are not always warranted.

Technologies can also be overvalued. Studies tend to extrapolate from ideal experimental conditions directly to the farmer's field. Technologies need to be tested under field conditions. While optimum controllable conditions are necessary at the experimental stage, the performance of the technology under ideal experimental conditions should not be taken as the only measure of expected performance in the field. Overvaluation also takes place because drawbacks and failings of technologies are overlooked or ignored. The problems associated with technological solutions should be known to all participants in the development effort. Knowledge of potential problems would make it possible to minimize or cushion the adverse impacts of technologies on end-users and the environment.

Many technologies and research findings show promise for alleviating problems in the uplands. Research is needed, however, on various areas and on ways to translate the resulting information into the development of workable technologies. Research needs include basic scientific studies that may not have immediate field applications. Some areas in need of further research and development work are listed below:

— research on low-cost and less labor intensive engineering structures, alternative ways of building terraces, structures for erosion control, and vegetative erosion and run-off control

— studies on farmers' practices and types of Philippine agricultural production systems and their seral stages, if any, in order to identify ecologically critical or conservative systems of production

— research on use of mycorrhiza with agricultural crops, production of inoculum and use under field conditions, and the combined use of nitrogen-fixing bacteria and mycorrhiza for tree and crop species

— basic research on decomposition, the allelopathic by-products of decomposition and on optimum amounts of organic material needed to improve fertility

— research to determine how organic matter can be managed to best benefit soil and crops and the use of (combinations of) organic and inorganic fertilizers in the uplands

— research on multiple cropping systems designed for the uplands, multiple cropping overyields, ways to favor overyielding, and the benefits of crop mixes over monocultures

— studies of a) both agroforestry schemes and of the natural forest in order to develop a better mimicking of the forest structure, functions, and homeostatic mechanisms, and b) the appropriate mixes of trees and crops that will be both protective and productive

— research on the ecology and physiology of pioneer tree species of the Philippines for eventual use in reforestation programs

— studies of fire protection strategies that favor natural succession

— the development of productive and protective pasture areas and animal production systems, that is, development that considers the unique constraints of the uplands

REFERENCES CITED

Aala, F.T. and T. Gonzales. 1964. The effects of green manuring and fertilizers on the production of hybrid corn seeds. Philippine Journal of Plant Industry 29:65-75.

Agboola, A. and G.E. Udom. 1967. Effects of weeding and mulching on the response of late maize to fertilizer treatments. Nigerian Agricultural Journal 4:69-72.

Agpaoa, A., et al. 1975. Manual of Reforestation and Erosion Control. W. Germany: German Agency for Technical Cooperation Ltd. (GTZ).

Aiyer, A.K.Y.N. 1949. Mixed cropping in India. Indian Journal of Agricultural Science 19: 439-543.

Alexander, M. 1961. Introduction to Soil Microbiology. John Wiley and Sons, Inc.: New York.

Altieri, M., D. Latournes, and J. Davis. 1983. Developing sustainable agroecosystems. BioScience 33(1): 45-49.

Aquino, D.I. and A.L. Madamba. 1939. A study of root nodule bacteria of certain leguminous plants. Philippine Agriculture 28:120-132.

Ayanaba, A. and B.N. Okigbo. 1974. Mulching for improved soil fertility and crop production. Paper presented at the FAO/SIDA Expert Consultation Meeting on Organic Materials as Fertilizers. Rome.

Barker, T.C. 1984. Shifting cultivation among the Ikalahans. Working Paper: Series 1. UPLB-PESAM, College, Laguna, Philippines.

Barros, F. 1940. A comparative study of the effects on yield of corn of some leguminous crops used as green manure. Philippine Agriculture 29(2): 142-147.

Benetes, J. and G.C. Naderman. 1979. Multiple cropping — Nitrogen experiment. In: Agronomic — Economic Research on Soils of the Tropics. Annual Report. North Carolina State University, Raleigh, North Carolina.

Bernales, B.C. and A.P. de la Vega. 1982a. Case Study of Forest Occupancy Management Program in Dona Remedios, Trinidad, Bulacan. Integrated Research Center, De La Salle University, Manila, Philippines.

1982b. Case Study of the Antique Upland Development Program. Integrated Research Center, De La Salle University, Manila, Philippines.

Bocato, F.M. 1981. Effects of different vegetal cover on runoff and soil loss. MS Thesis. University of the Philippines at Los Banos, College, Laguna, Philippines.

Brady, N.C. 1974. The Nature and Properties of Soils. 6th ed. New York: Macmillan Publishing Co., Inc.

Capistrano, A.D.N. 1982. Economic analysis of upland crop production systems. Paper presented at Livelihood Project and Curriculum Development Workshop. University of the Philippines at Los Baños, College, Laguna, Philippines.

1983. Polycultural agricultural ecosystems in Southeast Asia: A preliminary survey. Working Paper. East-West Center, Environment and Policy Institute. Honolulu, Hawaii.

Capistrano, A.D.N. and S. Fujisaka. 1984. Tenure, technology, and productivity of agroforestry schemes. Philippines Institute for Development Studies. Working Paper No. 84-06.

Christanty, L. (n.d.). Traditional agroforestry in West Java, Indonesia. Institute of Ecology, Bandung, Indonesia.

Colting, R.D. 1981. Effect of vegetative cover on runoff and soil loss in Benguet. M.S. Thesis. University of the Philippines at Los Baños, College, Laguna, Philippines.

Conklin, H.C. 1957. Hanunoo agriculture: a report on an integral system of shifting agriculture in the Philippines. Forestry Development Papers No. 12. FAO. Rome.

Cuevas, V.C. 1982. Survey and screening for cellulosic and lignin decomposing fungi. UPLB Basic Research, Annual Report. University of the Philippines at Los Baños, College, Laguna, Philippines.

1983. The effect of cogon mulch on the rate of soil erosion and productivity of a rice-based kaingin system. PESAM Annual Report.

David, W.P. 1984. Environmental effects of watershed modifications. Paper presented at Seminar-Workshop on Economic Policies for Forest Resrouces management sponsored by the Philippine Institute for Development Studies. Calamba, Laguna, Philippines, February 17-18.

del Rosario, E.V. 1982. Effects of mode of site preparations and introduction of inoculated and uninoculated *Leucaena leucocephala* (Lam.) De Wit. on the regrowth of *Imperata cylindrica* (L.) Beauv., B.S. Thesis, Univeristy of the Philippines at Los Baños, College, Laguna, Philippines.

Dinglasan, M.L. 1936. A study on the formation and nitrogen content of root tubercles of cowpea. Philippine Agriculture 25: 168-190.

Elton, C.S. 1958. The Ecology of Invasions by Animals and Plants. London. Methuen and Co.

Epstein, E. 1972. Mineral Nutrition of Plants: Principles and Perspectives. John Wiley and Sons, Inc.

Etherington, J.R. 1975. Environment and Plant Ecology. John Wiley and Sons.

Eusebio, R.V. and D.L. Umali. 1952. A test of four green manure crops for corn. Philippine Agriculture 36: 251-258.

Ewel, J. 1979. Secondary forests: the tropical wood resource of the future. In: Chavarria, M., (ed.), Simposio Internacional Sobre las Ciencias Forestales y su Contribucion al Desarrollo de la America Tropical. 11-17 October, San Jose, Costa Rica.

Floresca, E.T. and A.Q. Lanuza. 1963. Legumes and green manuring. IRRI Saturday Seminar. IRRI, College, Laguna, Philippines. 8 June 1983.

Foster, A.B. 1973. Approved Practices in Soil Conservation. Danville, Illinois. The Interstate Printers and Publishers, Inc. (4th ed.).

Foster, G.R. and L.D. Meyer. 1977. Soil erosion and sedimentation by water — an overview. In: Proceedings of the National Symposium on Soil Erosion and Sedimentation by Water. Chicago, Illinois. 12-13 December, 1977.

Free, G.R., S.N. Fertig and C.T. Bay. 1963. Zero tillage for corn following sod. Agronomy Journal 55: 207-208.

French, G.W. and G.R. Blake. 1965. Primary tillage for potatoes. Trans. ASAE 8(2): 246-248.

Ghildyal, B.P. and U.C. Gupta. 1959. A study of the biochemical and microbial changes during the decomposition of *Crotolaria juncea*

(Sann hemp) at different stages of growth in the soil. Plant and Soil 11: 312-330.

Gines, H.L., L. Lavapiez, R.L. Tinsley, S.N. Lohani, H.G. Zandstra, R. Torralba, and J. Manzon. 1976. Evaluating alternative cropping patterns in Pangasinan. IRRI Saturday Seminar. June 25, 1977. IRRI, College, Laguna, Philippines.

Glori, A.V. 1977. Drought resistence of yemane *(Gmelina arborea Roxb.)* and kaatoan bangkal *(Anthocephalus chinensis* (Lamk.) Rich ex. Walp.). Ph.D. Dissertation. University of the Philippines at Los Baños, College, Laguna, Philippines.

Gomez, A., et al. 1984. Minimum input farming systems. Annual Report. PCARRD, College, Laguna, Philippines.

Gomez, A.A. and H.G. Zandstra. 1976. An analysis of the role of legumes in multiple cropping systems. Paper presented at a workshop on "Exploiting the Legume — Rhizobium Symbiosis in Tropical Agriculture". Hawaii. 23-28 August, 1986.

Griffith, G. 1951. Factors influencing nitrate accumulation in Uganda soil. Experimental Agriculture 19: 1-12.

Harrold, L.L., G.B. Trilett Jr., and R.E. Youker. 1967. Less soil and water loss from no-tillage corn. Ohio Report on Research and Development (52(2): 22-23.

Highfill, R.E. and L.W. Kimberlin. 1977. Current erosion and sediment control technology for rural and urban lands. In: Proceedings of the National Symposium on Soil Erosion.

Huber, A. and B. Watson. 1970. Effect of inorganic enrichment on soil-borne pathogens. Phytopathology 60: 22-26.

Hutchinson, G.E. 1954. Theoretical notes on oscillatory populations. Journal of Wildlife Management 18: 107-109.

International Institute for Tropical Agriculture (IITA). 1973. Annual Report. Nigeria.

Jacks, G.V., W.D. Brind, and R. Smith. 1955. Mulching. Technical Communication No. 49. Community Agriculture Bureaux. Farmingham Royal, Bucks, England.

Jensen, N. F. 1952. Agronomy Journal 44: 30-34.

Kass, D.C. 1978. Polyculture cropping systems: a review and analysis. Cornell International Agriculture Bulletin No. 32. Cornell University, New York.

Kunstadter, P., E.C. Chapman, and S. Sabhasri (eds.) 1978. Farmers in the Forest: Economic Developmental and Marginal Agriculture in Northern Thailand. Honolulu: University press of Hawaii.

Lal, R. 1976. Soil erosion problems on an alfisol in Western Nigeria and their control. IITA Monograph No. 1. Ibadan, Nigeria.

Linderman, R.G. 1970. Plant residue decomposition products and their effects on host roots and fungi pathogenic to roots. Phytopathology 60: 19-22.

Linderman, R.G. and R.G. Gilbert. 1968. Stimulation of *Sclerotium rolfsii* and its antagonists by volatile components of alfalfa hay. Phytopathology 58: 1057-ff.

Mabbayad, B.B., B.N. Emerson, and E.L. Aragon. 1968. Further tests on minimal tillage and rates of nitrogen in transplanted rice. Philippine Agriculture 55: 216-220.

Mamaril, J.C., F.T. Begonia, and R.B. Aspiras. 1984. The utilization of coconut water as a medium for rhizobia. In: G. Veeger and W.E. Newton, (eds.), Advances in Nitrogen Fixation Research. Martinus Rijhoff/Dr. W. Junk Publishers. Pudoc. Wageningen.

Manguiat, I.J., D.M. Mendoza, and V.M. Padilla. 1983. Nitrogen fixation by Rhizobium — legume symbiosis: agro-forest crops. BIOTECH Annual Report. BIOTECH, UPLB, College, Laguna.

Manguiat, I.J. 1984. Biological nitrogen fixation and soil rehabilitation. Paper presented at the conference on "Soil Rehabilitation: The Philippine Experience", Manila, 12-14 April.

Martin, C.R. 1979. Mechanical erosion control aspects of soil conservation. Proceedings of the First Regular Short-Term Course on Soil Erosion Control Management for Field Trainors. Baguio City, 22-29 Oct.

May, R.M. 1974. Stability and Complexity of Model Ecosystems. Princeton University Press.

Mosse, B. 1981. Vesicular-arbuscular mycorrhiza research for tropical agriculture. Research Bulletin 194. Hawaii Institute of Tropical Agriculture and Human Research.

Odum, E.P. 1971. Fundamentals of Ecology. Third edition. W.B. Sanders Co.

Okigbo, B.N. 1965. Effects of mulching and frequency of weeding on the performance and yield of maize. Nigerian Agriculture Journal 2: 7-9.

Omengan, E. 1981. Nitrogen and phosphorus cycles in a Bontoc rice paddy system. M.S. Thesis. UPLB, College, Laguna, Philippines.

Omengan, E. and Sajise, P.E. 1981. Ecological study of the Bontoc rice paddy system: a case of human — environment interaction. Paper presented at the IRRI Thursday Seminar, 26 March.

Pacardo, E.P. and Samson, B.K. 1979. Erosion studies at Mt. Makiling. In: Annual Report. UPLB — Upland Hydroecology Program: Upland Production System and Occupancy: Institutional, Socio-economic and Ecological Analysis of Five Upland Areas in the Philippines. UPLB, College, Laguna, Philippines.

Pacardo, E.P., G.O. San Valentin, R.M. Macandog, and D.L. Comia. 1983. Effect of corn/ipil-ipil cropping system on productivity and stability of upland agroecosystem. Annual Report PCARRD.

Palis, R.G., Monte, R.A., Diaz, C.V. and Naboa, V.F. 1983. Soil erosion control/management studies. Report presented during the Third PCARRD Coordinated Review, 6-9 July.

Palis, R.G. 1977. Soil erosion under different cover crops. MS Thesis. UPLB. College, Laguna.

Papavizas, G.C. 1970. Colonization and growth of *Rhizoctonia solani* in soil. In: *Rhizoctonia solani:* Biology and Pathology. University of California Press. Berkeley.

Paterno, E.S. 1983. Ecological studies of rhizobia introduced as seed inoculants. In: The Nitrogen Fixation and Mycorrhiza Program: A Three-Year Executive Summary. National Institute of Biotechnology and Applied Microbiology (BIOTECH). UPLB. College, Laguna.

Paterno, E.S., F.G. Torres, M.L.Q. Sison, and E.S. Garcia. 1983. Nitrogen fixation by the Rhizobium-legume symbiosis in food crops. BIOTECH Annual Report. UPLB. College, Laguna.

PCARRD. 1984. The Philippine Recommends for Soil Conservation. PCARRD Tech. Bull. Seires. No. 28-A.

Pollisco, F.S. 1975. Reforestation and silviculture techniques for the regeneration of Philippine forests. Canopy. 1(6): 1-3, 6-8.

Pimentel, D. 1961. Species diversity and insect population outbreaks. Annuals of the Entomology Society of America 54.

Program on Environmental Science and Management, UPLB. 1982. The effect of biological contour strips composed of ipil-ipil *(Leucaena leucocephala)* and kakawate *(Gliricidia sepium)* on the productivity and rate of erosion of a rice-based kaingin system on Mt. Makiling, Puting Lupa, Calamba, Laguna. Annual Report. Mt. Makiling Cropping Systems Research Team. UPLB, College, Laguna.

1983. The effect of cutting height on crop production and site degradation on Mt. Makiling, Puting Lupa, Calamba, Laguna. Annual Report. Mt. Makiling Cropping Systems Research Team. UPLB, College, Laguna.

Ramirez, D.M. 1980. Goat grazing effects on a native grassland over-seeded with stylo *(Stylosanthes guyanensis HBK)*. MS. Thesis. EPLB. College, Laguna.

Rask, N., G.B. Triplett, Jr., D.M. van Doren, Jr. 1967. A cost analysis of no tillage corn. Ohio Report on Research and Development 52(1): 14-15.

Russel, E.W. 1973. Soil Conditions and Plant Growth. 10th ed. Longmans, Green and Co. Ltd. London.

Sajise, Jr. G.E. 1983. Effects of intercropping and tillage practices on crop productivity and soil conservation. MS Thesis. UPLB. College, Laguna.

Sajise, P E. 1972. Evaluation of cogon *(Imperata cylindrica (L.) Beauv.)* as a seral stage in the Philippine vegetational succession. I. The cogon seral stage and plant succession. II. Autecological studies on cogon. Ph.D. Thesis. Cornell University N.Y.

1977a. Regeneration of critical upland areas: An ecological imperative. Professorial chair lecture. UPLB. College, Laguna. 5 Jan.

1977b. Some factors to consider in hilly land development. Paper presented during the International Workshop on Hilly Land Development sponsored by PCARRD, FFTC, SEARCA, DA, DNR, FORI, NFAC and FSDC. Legaspi City, Albay, Phil. 3-6 Aug.

1980. Alang-alang *(Imperata cylindrica* (L.) Beauv.) and upland agriculture. In: Proceedings of the BIOTROP Workshop on Alang-alang. BIOTROP Special Publication No. 5. BIOTROP. Bogor, Indonesia.

1982. Ecological approaches to managing degraded uplands in the Philippines. Paper presented at the Joint Chinese Environmental Protection Office/East-West Environment and Policy Workshop on Ecosystem Models for Development. People's Republic of China. 25 Sept. – 9 Oct.

Sajise, P.E., N.M. Orlido, J.S. Lales, L.C. Castillo, and R. Atabay. 1976. The ecology of Philippine grasslands: Floristic composition and community dynamics. Philippine Agriculture 58(9-10): 317-334.

Samson, B.K. 1981. Effect on N-P-K fertilization and soil moisture stress on the water relations of some reforestation species. M.S. Thesis, UPLB, College, Laguna.

Sanchez, P.A. 1972. Soil management under shifting cultivation. In: A review of soils research in Tropical Latin America. Soil Science Dept., North Carolina State University at Raleigh, North Carolina.

Saplaco, S.R. 1983. Terracing. In: Manual on Upland Management, Research and Extension. UPLB-PESAM, UPLB, College, Laguna.

Sayre, R.M. 1971. Biotic influences in soil environment. In: Plant Parasitic Nematodes. Academic Press: New York.

Serrano, R.C. 1982. Hydrology of different coconut *(cocos nucifera* L.) based agroecosystems. M.S. Thesis. UPLB, College, Laguna.

Simmonds, N.W. 1962. Biology Review 37: 442-465.

Sopper, W.E. and H.W. Lull. (eds.) 1967. Erodibility and erosion potential of forest watersheds. In: Proceedings of the International Symposium of Forest Hydrology. Pennsylvania State University.

Subba Rao, S.N. 1975. Nitrogen gains by legumes and residual nitrogen left behind in the soil through *Rhizobium* application. Indian Journal Genetics and Plant Breeding 35. 236-38.

Tanguilig, V.C. 1979. Comparative response of some reforestation species to soil related constraints. B.S. Thesis. UPLB, College, Laguna

Tapawan, Z.N. 1981. Economics of farming systems in the uplands of Antique province, Philippines. Paper presented during the Workshop on Human-Agroecosystems Interaction sponsored by the Program on Environmental Science and management (PESAM) and the Environment and Policy Institute (EAPI). 24 Nov.-Dec. 1981. UPLB, College, Laguna.

Techapinyawat, S. 1982. Studies on inoculation of ipil-ipil *(Leucaena leucocephala.* (Lam.) de Wit.) with rhizobia and mycorrhizal fungi. Ph.D. Thesis. UPLB, College, Laguna.

Tilo, S.N. 1983. Nitrogen fixation in pasture and forage legumes. In: The Nitrogen Fixation and Mycorrhiza Program: A Three-Year Executive Summary. BIOTECH. UPLB. College, Laguna.

Trenbath, B.R. 1974. Biomass productivity of mixtures. Advanced Agronomy 26: 177-210.

1975. Diversify or be damned? Ecologist 5: 76-83.

1976. Plant interactions in mixed crop communities. In: Multiple Cropping. Papendick, R.I. (ed.). American Society of Agronomy Special Publication No. 27. Madison, Wisconsin.

1982. The dynamic properties of mixed crops. In: Proceedings International Conference on Frontiers of Research in Agriculture. Indian Statistical Institute. Calcutta, India.

Tupas, G.L. and P.E. Sajise. 1977. The role of trees in plant succession. In. Ecological conditions associated with tree clumps. Kalikasan, Philippine Journal of Biology 6(3):229-244.

Turnbull, A.L. and D.A. Chant. 1961. The practice and theory of biological control of insects in Canada. Canadian Journal of Zoology 39: 697-753.

UNESCO. 1969. Soil Biology: Reviews of Research. Paris: UNESCO.

UNESCO/UNEP/FAO. 1979. Tropical Grazing Land Ecosystems. UNESCO.

Upland Hydroecology Program (UHP). 1977. Annual Report. UPLB. College, Laguna.

1979a. Upland Production Systems and Occupancy: Institutional, Socio-economic, and Ecological Analysis of Five Upland Areas in the Philippines. Annual Report. UPLB. College, Laguna.

1979b. Fire Ecology. Annual Report. UPLB. College, Laguna.

Vergara, N.T. (ed.) 1982. New Directions in Agroforestry: The Potential of Tropical Legume Trees. Environment and Policy Institute, East-West Center, Hawaii.

Vidal, E.T. 1979. Imperata grassland regeneration: Nutrient cycling patterns in a legume — native grassland — goat production system. M.S. Thesis. UPLB. College, Laguna.

Wade, M.K. and P.A. Sanchez. 1975. Mulching and green manuring studies. In. Agronomic — Economic Research on Tropical Soils. Annual Report. Soil Science Dept., NCSU, Raleigh, North Carolina.

Willey, R.W. 1979. Intercropping — its importance and its research needs. Part I. Competition and yield advantage. Field Crops Abstracts 32: 1-10.

6
UPLAND ECONOMICS AND ECONOMIC IMPACT ANALYSIS

Marian Segura-de los Angeles

.This chapter reviews and synthesizes various studies of Philippine upland development and examines the economic effects of specific strategies for improving the uplands and the communities therein and the respective research methodologies used in assessing such effects. To provide a fuller understanding of upland development, the behavior of upland communities is analyzed from the economist's viewpoint. Upland community reaction to a changing environment, whether formally induced or not, is then analyzed based on various research findings. Evaluation of impacts of planned intervention in the uplands are presented. Corresponding data and research gaps are identified and the implications for prioritization of future upland economics research are suggested.

THE ECONOMICS OF UPLAND USE BY FOREST OCCUPANTS

The following factors influence upland use by forest occupants: 1) open access to the uplands and upland resources, 2) differential effects of upland cultivation on the uplands and lowlands and between present and future users, 3) lack of more permanent alternative sources of livelihood for uplanders, and 4) a high preference for present consumption over future consumption.

Contrary to provisions of Philippine forest laws, effective access to the uplands has been virtually "free," resulting in minimal cost of migrating into forest land and near-zero cost of using forest land for the uplander. Migrants pay only transportation and information costs since neither rent nor other fees are collected from them by the government. Among the several conditions that have caused this situation are inadequate protection of forests by government and loggers, increased accessibility of forest lands due to roads built by timber concessionaires, and ease in the clearing of inadequately stocked, logged-over forests.

Once in the uplands, a cultivator is faced with a "shortage" of time for agricultural production. The farmer relies mainly on family labor and is faced with seasonal factors and usually with limited capital (and cash). Thus, the upland cultivator maximizes returns to the limited factors of production, labor, and capital through the adoption of slash-and-burn farming, which hastens the conversion of biomass into more direct inputs (temporarily fertile soil cover).

The cultivator discovers that soil productivity soon decreases, and demand for weeding labor increases after one to a few years of cultivation. The cultivator then finds it more rewarding to shift to another field. Field rotation consisting of cultivation and fallow periods is eventually established. Such a system of land use allows the cultivator to avoid permanent decreases in soil fertility. Communities experiencing "first-hand" agricultural use of forest land, thus, realize unusually high economic rents during the early phases of upland production. Eventually, however, the possibility of continuously using new tracts of land for field rotations diminishes due to increased competition from other forest users. Once cultivators find that they can shift fields only within a given area, the following options or combinations thereof can be tried: 1) development of land-saving farming technologies, 2) diversification of income sources, 3) alteration of expectation from upland production, or 4) establishment of a system of rights to the land to preserve harmonious relations among fellow upland farmers from whom a supply of exchange labor is drawn (for the theoretical treatment of technological and institutional changes under demographic pressure, (see Boserup, 1965; Darity, 1980; and W.D. Cruz, 1982).

The development of soil-conserving technologies, diversification of upland activities, lowering of demand for land through population control measures, and establishment of property rights governing "common property ownership" have indeed been observed for numerous upland groups who have long occupied forest lands. Stable production systems characterized by controlled burning and well-timed field rotations have been documented for cultural minorities with low population densities, such as the Tagbanua of Palawan (Warner, 1981) and the Hanunoo of Mindoro (Conklin, 1957). Fertility control has also been noted among cultural minorities (Cadeliña, 1981). The Ifugaos developed an important land-saving and soil-conserving technology — terracing — and were observed to possess "the most complex and elegant indigenous property laws in the nation" (Lynch, 1983).

Meanwhile, however, perceived gains from upland cultivation, displacement of lowland farmers, a "land for the landless" policy in the early fifties, better access to the uplands in the late sixties, and lack of protection of logged-over areas may all have contributed to increased lowland to upland migration. Migration eventually led to overcrowding manifested in shorter rotation periods. Shorter fallow periods, in turn, partly caused the earlier high economic rent from upland cultivation to be dissipated to zero, and, hence, the onset of subsistence farming. Moreover, the ill effects of decreased fallow periods have been compounded by the inappropriate tilling technologies used by inexperienced upland farmers.

For the more recent migrants into the uplands, the option of field rotation available to the earlier upland settler may not be as feasible due to rising man-land ratios and increases in other competing forest land uses. At the same time, a lack of awareness of the long-term effects of decreasing upland soil productivity and a preference for present consumption (eagerness to consume now versus the future) have possibly inhibited the development of ecologically sound cultivation by upland farmers.

Even among the earlier settlers of the forest, long fallow periods (and ecologically stable field rotations) may likewise be disrupted. The consumption patterns of such groups may be influenced by their new neighbors and by more interaction with low-landers. Gains from improved health and nutrition may occur without corresponding increases in work opportunities for the female family members, possibly resulting in larger families.

The above imply that changes in upland cultivation patterns are needed and that such changes must occur faster than the development by the earlier upland users of land-saving technologies. Such changes are needed on a wide scale since recent estimates on upland population show the predominance of new migrants over pioneer migrants in the Philippine uplands. Well-planned interventions are needed. Externally initiated upland development projects may need to predominate in the very near future. The subsequent sections discuss attempts at changing forest resources use.

Subsistence Farming in the Uplands

Anthropological and sociological studies analyze how the economic life of the uplander is interrelated with his social, environmental, religious and other activities. Lopez-Gonzaga discusses the Mangyan swidden cycle and mentions the following (translated) innovation chanted during the field burning, "Oh field burn, burn; be a well-burnt field, I've no axe, and am feeling lazy, so lazy" (1983:109). This coincides with my observation that burning maximizes returns to scarce labor (time) and capital. The study yields detailed information on household income, land holding, taxes paid, household labor force, and value of produce for some 50 members of the community. The data analysis was mainly by correlational methods (Lopez-Gonzaga, 1983).

Cadeliña (1981) constructed time graphs for a one year period of food gathering from various sources and the corresponding allocation of labor for specific activities. The research was conducted in Negros Island and demonstrates the upland farmers' diversification of food sources in consonance with seasonal variations.

Estioko-Griffin and Griffin (1981) documented the Agta's seasonal and subsistence cycle and highlighted effects of population growth and the influx of migrants on forest destruction.

The few studies by economists on Philippine upland cultivation are diagnostic. Floro (1980), Capistrano (1982), and Nguu and Corpuz (1979) explored the economics of upland farming systems in Central Luzon and Southern Tagalog. Capistrano and Floro studied uplanders in Pantabangan, Nueva Ecija, while Nguu and Corpuz investigated shifting cultivators in Mt. Makiling, Laguna. These studies yielded information on cash income, labor, and capital inputs at the farm level. The Nguu and Corpuz study focused on only two Mt. Makiling shifting cultivators but yielded detailed information on time allocation of activities. In Villarica, Pantabangan, Capistrano used Hayami's (1978) framework for analyzing a village economy and drew from the works of Samonte (1980) and Floro. The study used correlation analyses of labor utilization, income, and asset ownership.

An early comprehensive attempt at estimating upland farms' gross income as composed of both cash income (income from sales) and imputed income (value of output consumed in the farm) was conducted by Floro (1980) at Villarica, Pantabangan. Floro computed corresponding expenditures of the two households and concluded that farm production was at the margin or at the subsistence level. This was compared to the households' pre-resettlement economic status (since the households were among those displaced by the construction of the Pantabangan Dam).

Ellevera-Lamberte examined poverty in the uplands (1983). Average annual income of upland provinces was significantly lower than income of lowland provinces. In terms of land utilization, most upland provinces contained a higher number of idle farms. These findings imply the prevalence of fallowing in the uplands and poverty as measured in terms of cash income.

Importantly, however, non-estimation of income in kind for the uplanders leads to an exaggeration of poverty in the uplands. Subsis-

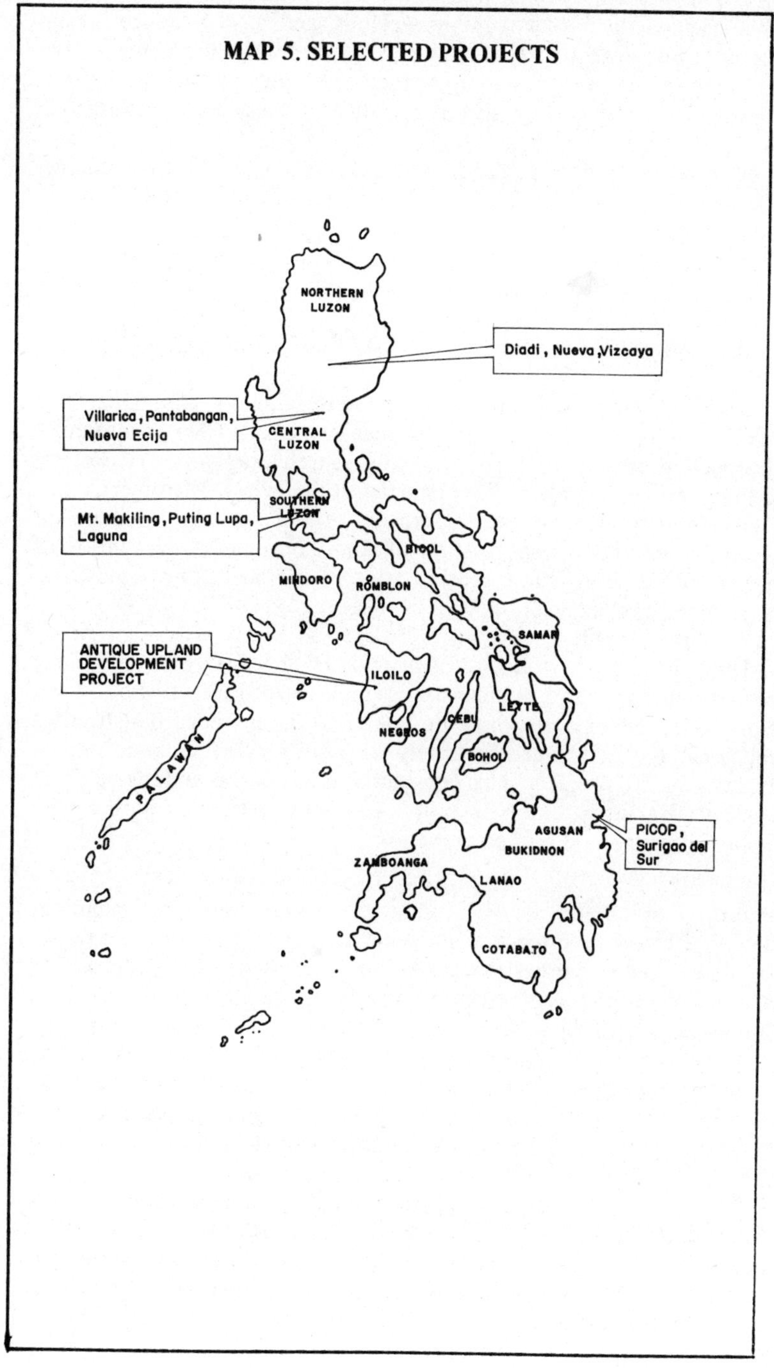

MAP 5. SELECTED PROJECTS
NORTHERN LUZON
Diadi , Nueva Vizcaya
Villarica , Pantabangan, Nueva Ecija
CENTRAL LUZON
SOUTHERN LUZON
Mt. Makiling , Puting Lupa, Laguna
BICOL
MINDORO
ROMBLON
SAMAR
ANTIQUE UPLAND DEVELOPMENT PROJECT
ILOILO
LEYTE
CEBU
NEGROS
BOHOL
PALAWAN
AGUSAN
PICOP , Surigao del Sur
BUKIDNON
ZAMBOANGA
LANAO
COTABATO

tence farming is prevalent in the uplands, and most produce is consumed at home. Ono (1982) studied underreporting of income in kind and showed that the highest incidence of underreporting occurred in the forestry sector among all types of Bulacan households investigated.

A notable gap in the research on upland farming is thus the construction of a uniform method of estimating income and poverty threshold levels. Also, estimates of the magnitude of forest occupancy have been inadequate. Hence, a realistic assessment of the contribution of upland communities to the national economy has yet to be made.

ECONOMIC IMPACT ANALYSIS AND PROJECT EVALUATION

Social forestry projects cover the government's communal tree farming (CTF), forest occupancy management (FOM), and family approach reforestation (FAR) programs as well as agroforestry projects initiated by the private sector (Bernales et al., 1982). The objectives of social forestry include the upliftment of the standard of living of upland communities, stabilization of forest occupancy, prevention of further forest destruction, rehabilitation of degraded forest lands, and minimization of soil erosion.

Upliftment of the standard of living is to be achieved by improving land productivity and by increasing employment-generating activities. While this first objective may be linked to the intended beneficiary's concerns, the last four are largely societal goals. Economic impact analysis of SF projects thus necessarily needs to be conducted from both private and public viewpoints. Analysis is needed of both on-site and off-site effects of soil conservation.

Project evaluation may be conducted during at least two points in time: 1) before a project is implemented, during which its **ex-ante** feasibility and hypothesized effects are projected, and 2) **expost**, or after it has begun (upon completion or later). In general, the term "project evaluation" refers to assessing projects in terms of monetary measures of costs and benefits. On the other hand, impact analysis has been taken to mean assessment of project effects after a project has been completed or has reached an advanced stage of implementation. Economic impact analyses are usually ex-post evaluation studies.

The analysis of potential project impacts, even hypothetically, is included among the following steps of project evaluation: 1) identification of costs and benefits (negative and positive impacts, respectively), 2) transforming costs and benefits into comparable units, such as money terms, through a system of prices, and 3) using a criterion to measure the project's desirability.

Ex-post impact assessment may thus be used to verify ex-ante project evaluation, especially when the impacts are valued in monetary terms. Alternatively, when benefits are expressed in units other than money values (e.g., decreases in soil erosion rates), impact assessment may also be extended into evaluating the cost-effectiveness of various attempts/alternatives of rehabilitating the uplands.

The project evaluator may proceed from the viewpoint of intended beneficiaries, from project management's financial standpoint, or from societal concerns. Selection of a particular viewpoint affects the choice of indicators of project effects as well as economic valuation of project impacts.

EX-ANTE EVALUATION OF UPLAND DEVELOPMENT

Ex-ante evaluation of upland development may be conducted to form the basis of project feasibility studies or to examine policy options open to decision makers. The analytical tools used for feasibility studies include optimization models (which guide project investors with respect to the level of inputs and scale of project), after which cost-effectiveness, cost benefit, or social cost-benefit evaluation are conducted. The NIA and MADECOR (1979) ex-ante evaluation of watershed conservation effects used projections without the project versus project effects with the project. Baseline economic surveys and computer-simulated land optimization strategies were employed. No attempts at valuing ecological effects were made because the experts who conducted an earlier study on the same area concluded that sedimentation was not expected to significantly affect the watershed's yield (NIA and ECI, 1978). Rather, the feasibility study focused on the economic gains from crop production. Attempts at evaluating the project from society's viewpoint were made by varying the costing of capital (interest rate) and labor (wages). No special consideration was made for income redistributive effects, such as the assigning of heavier weights to changes in income of the poorer project beneficiaries.

A similar methodology was employed for evaluating upland rice cropping systems (Paris, 1978). The effects of environmental and economic influences and alternative cropping patterns on the economic performance of cropping systems were analyzed through a computer simulation model. Important policy variables, such as fertilizer levels, weed control, pest and disease control, crop variety, area and planting date were examined in terms of net returns, labor utilization, and yields. The study incorporates biophysical and meteorological factors. The information was obtained from weekly economic surveys conducted over a two-year period by the International Rice Research Institute.

The NIA-MADECOR (1979) study differs from the Paris study. The former was conducted largely from the viewpoint of the project manager (NIA). The Paris study represents normative analysis of how an upland farmer would optimally behave given decision–making based on the information available.

Both studies do not simulate the off-site, environmental effects of altering upland cropping systems. This lack may be due to a dearth of information on the widescale effects of altering ecological systems. Most of the available studies conducted on factors such as sedimentation due to various cropping patterns have been quite experimental (e.g., R.V. Cruz, 1982 and UHP, various reports). Thus, none of the analytical tools which have been used elsewhere for ex-ante evaluation of alternative cropping systems have been applied in the Philippine setting. Such tools include a linear programming model which either incorporates restrictions in upland use (Taylor and Frohberg, 1977) or includes sedimentation effects arising from cost-minimization decisions for cropping patterns by farmers (Wade and O'Heady, 1979). Another tool is the use of optimal control theory for examining rotation choices and the effects of those choices on soil fertility and erosion (Burt, 1981).

Additionally, none of the feasibility studies attempt to incorporate varying rates of adoption of suggested cropping patterns or the various degrees of participation in upland development efforts. Indeed, most feasibility studies are quite optimistic in this respect. The assumption of full cooperation by intended beneficiaries is often made. This optimism is true not only for Philippine research, but also for many project feasibility case studies conducted elsewhere (e.g., UN-FAO, 1979).

At the macro-level, no attempt has been made to normatively investigate the appropriate mix of upland projects, including those implemented by non-forestry agencies. A national forest land use planning model has yet to be devised for the country. Such a model would need to incorporate the production, distributive and ecological effects of various ways of using the uplands. Perhaps for this reason, forest policy-making has largely been reaction-based, rather than anticipatory in nature.

Two areas of research need to be explored to take into account relevant macro considerations. One concerns the effects of activities in the uplands by various entities (including government) on national revenue, investment, socio-political measures of welfare, and other economic variables. For example, in Indonesia, the interaction between commercial timber users and local communities was assessed in terms of taxes, labor, and infrastructure (Sumitro, 1984a, b). Second is the determination of the impacts of macro-variables on upland development. Nelson

(1984) developed a conceptual framework for investigating the effects of government policies on timber utilization. The framework could be adapted to examine the use of forest lands and resources by uplanders. Operationalization, however, would require detailed information on production function — unfortunately a notable research gap in Philippine forestry. While both research areas need quantitative studies for sound policy formulation, other approaches have been utilized. Kummer (1984) argued for a critical analysis of social forestry, while Revilla (1983) explored strategies for cost-effective forest renewal and for funding such strategies (1984).

EX-POST ANALYSIS OF UPLAND DEVELOPMENT

Studies of social forestry projects conducted after project commencement may be classified into those which look into project participation, management concerns, changes in the beneficiaries' welfare, or societal concerns.

Economic Analysis of Participation in Upland Development Projects. Several investigations by researchers with economics training yielded insights on participation in upland development in the Philippines. Mindajao (1978) interviewed 100 Development Bank of the Philippines (DBP)-financed farmers and 100 non-DBP financed tree farmers from the PICOP tree-farming project in Mindanao. The study looked into farmers initial participation, financial impact on the farmers, and society's cost/benefits. No significant difference was found between the two types of project cooperators in terms of the time it took them to decide to participate in the project (four years on the average). The factors which lead to participation varied, however.

Among the non-DBP assisted farmers, participation was positively influenced by income, but negatively influenced by education and the degree of risk aversion. In the case of the DBP-assisted farmers, only land-use at the time of joining the project appeared to be a significant explanatory variable for farmer participation. Farmers with larger proportions of their land cultivated with agricultural crops were more likely to join the projects. Such cooperators already had adequate income from agricultural production, and land clearing would not be problematic. On the other hand, the non-participants interviewed were hampered by such constraints as lack of suitable land, insufficient labor time, and impatience over the long gestation period (Mindajao, 1978).

The mean annual income of the DBP-financed farmer was P10,000. That of the non-financed farmer was P6,000. At the time of the study, earnings from a few of the tree farms came only in the form of thinnings sold to PICOP. Most of the proceeds were spent on food and clothing. Mindajao's financial analysis based on projected net earnings from future harvest (and yield assumptions) estimates 50% financial rate of return for the DBP-financed farmer and 53% for the non-DBP financed farmer. The societal analysis used a labor opportunity cost of P4.00 per day and yielded an economic rate of return of 23%. No attempt was made to quantify the environmental benefits of the tree-farming project.

A later study examined the maintenance activities and harvesting. Only a few of the DBP-financed farmers had harvested by this time since the rotation period for *falcataria* is eight years. The study found that the agriculture and livestock component had not been pursued by the farmers. The loan size did not realistically account for inflation and harvesting costs not included in the original project feasibility studies were more significant than expected (Hyman, 1981). With the corresponding adjustment in the economic analysis, internal rates of return of 21% to 31% (range: 1-31%) to the farmer and 14% (range: 1-38%) to society were calculated for the typical case. The economic evaluation was found to be very sensitive to wage rate, with labor being a major project cost component. A study team recommended the integration of the original agricultural component and the granting of loans for such. Problems stemming from PICOP's buying price of pulplog and suggestions on improving administrative aspects were also discussed.

The Mindajao study looked into participation, which is usually assumed to be a high percentage, if not 100%, by most project feasibility evaluators. This aspect of his PICOP study, however, was not incorporated into his projections of financial and economic returns to the project.

Tapawan (1981) investigated the adoption of cropping systems by upland cultivators in Antique. The study was conducted in an area where direct intervention into upland communities was initiated by local government. It relied on a survey of a randomly selected sample of farmers and secondary data. The study covered economic physico-biological and meteorological aspects, incorporated market linkages effects, and employed several analytical tools including factors analysis and multiple regression. It focused on the farm business and the production functions. In addition to demonstrating the operationalization of these relatively new research methods, the study yielded findings about input productivity, the need for complementary inputs to terracing, labor under-utilization, the farmers' strong market orientation, the subsistence level of production, and the need for supplementary income sources.

Tapawan's analyses resulted in several findings. The primary, although not sufficient determinants of cropping patterns and practices in the area are rainfall pattern, terracing, and access to supplemental irrigation. Farm households, although market-oriented, were subsistence producers. Non-farm employment provided significant livelihood sources. Other results of Tapawan's study are discussed below (see analysis of impacts of upland development).

Tapawan noted some limitations. Measurement problems were associated with recall data and lack of farm accounting records, homogeneity in cropping intensity, and the short time available for the study. In reference to homogeneity, various upland cropping systems and/or various types of upland development need to be examined in a comparable (if not uniform matter) in order to simulate decision variables such as those pertaining to credit. The other shortcomings suggest that such a study should be conducted for at least a year and during two time points in order to allow more for the observation of the dynamics of change to be observed.

Factors affecting farmer participation in upland development projects were investigated by this author. Nineteen cooperators of a pilot upland development project in Pantabangan, Nueva Ecija, were studied. The study examined attempts at different forms of terracing. Ordinary least squares analysis of factors leading to soil conservation showed the following relationships (Segura de los Angeles, 1984):

1) Initial higher organic matter imply lower rate of adoption.
2) Larger farm areas imply lower adoption rate.
3) The higher the initial income of the farmer, the less likely he adopts.
4) More knowledge on conservation results in better changes of adoption.
5) The larger the pool of household labor the more likely the farmer practices terracing.

Farmers, therefore, appear to be responsive to changes in the environment and become even more responsive with knowledge gained through project implementation. Those with smaller farms and lower incomes, however, tended to adopt soil conservation practices more easily. The effect of income may be explained further by the fact that those with higher incomes tended to be those with other income sources besides the farm. Naturally, their time was allocated more towards activities with perceived higher and quicker returns. The effect of available labor may also explain this; however, the variable used may also be interpreted to reflect family size, and hence, consumption needs.

A recent semi-ex-post study on choice of cropping system used economic data previously gathered on two shifting cultivators in Mt.

Makiling (Corpuz, 1984). Data also included potential yields from modified cropping patterns based on experiments conducted by natural scientists in the same area and tree-farming information obtained under similar growing conditions. Employing economic measures such as net present worth and benefit-cost ratio, Corpuz concluded that, from the viewpoint of the farmer, traditional *kaingin* is as advantageous as the modified cropping systems, while tree farming was the least desirable.

The analyses used a discount rate of 18%, which naturally imposed a bias against a longer pay-off alternative such as tree farming. Corpuz, moreover, assumed non-decreasing productivity for traditional *kaingin*. This is realistic only in so far as shifting to another field is feasible and long fallow periods could be maintained.

These studies yield insights into site-specific factors affecting adoption of soil-conserving technologies. They do not, however, provide firm conclusions about the effects of the policy variables of land tenure status and credit. Lack of variation in these variables among the cooperators examined and the one-shot nature of the studies conducted made the investigation of such policy variables unfeasible. Indeed, this is a prominent research gap in upland development (For studies of adoption of soil conservation measures, see Lee, 1980; Seitz and Swanson, 1980; and Lee and Stewart, 1983).

Economic Impact Analysis. Impact analysis may be conducted through the application of the experimental design, use of quasi-experimental designs, or conduct of the non-experimental design (Weiss, 1972). The ideal experimental design involves observations of the variable in question, with observation conducted on project-influenced and control groups for two time periods of observation. A hypothesized impact is attributed to the project if the change in the variable for the influenced group is significantly larger than the change observed for the control group.

Often, however, the experimental design is difficult to implement because of problems in finding an appropriate (randomly selected) control group with corresponding before-project data. In addition, political problems may disallow research in communities where the project is absent. In such cases, quasi-experimental designs must be employed. Such approximations of the experimental design include cross-sectional studies on influenced communities upon which multivariate analysis may be conducted, use of non-randomly selected control groups, before and after studies or similar time series analysis and case studies.

Multivariate analysis conducted without a control group faces the problem of factors competing with project-related variables. A similar confounding problem may likewise be experienced with time series studies if non-project events are also important determinants of economic impacts. For case studies, the problem of generalization of findings is often encountered. In practice, these quasi-experimental designs are generally used simultaneously to minimize their individual weaknesses.

I measured economic and social impacts of agro-forestation. (Segura-de los Angeles, 1983). Comparison of project cooperators' incomes across time was feasible for Diadi, Nueva Vizcaya, since income data were consistently investigated by the project implementors. Caveats with respect to attributing increases in income over the duration of the project are, however, offered because the composition of the surveyed sample changed, alternative sources of income such as logging emerged, and inflationary effects needed to be accounted for. The observed increase in real income of project cooperators was not wholly attributable to the project but was instead traced to the conduct of destructive forest activities. That is, small-scale logging by shifting cultivators (in forests where timber licenses had been cancelled as punishment for destructive logging practices) was cited during later surveys as an income source.

The same research looked into a pilot agroforestry project in Pantabangan, Nueva Ecija. While apparent gains from a demonstration farm was noted, no conclusion was made about similar experiences for the other participating farmers. A range of adoption of recommended agroforestry practices prevailed in the farms. Their corresponding effects on farm yield had not been measured at the time of the study. Detailed economic studies were not feasible because of cooperators resistance to be queried repeatedly about the extent of their poverty. The study then examined interaction among various forestry activities and development projects in the area, their effects on farmer knowledge about soil conservation and the environment, and attitudes towards various government agencies in the area. The need for coordination among project implementors and among various development projects was noted. The study also collated information which could be used as benchmark figures for future studies.

The most-important economic impact documented for other upland development projects has been employment generation (Aguilar, 1982). For 36 Ikalahans of Nueva Vizcaya, 42% of total income consisted of wage labor. Higher percentages were observed for 36 San Pedro, Laguna, CTF cooperators (71% of total income) and for 22 Lake Buhi farmers (58% of total income from wages).

Impact of changes in upland cropping patterns on soil productivity was measured by Tapawan (1981) on farms in Antique. Terraced plots were found to have higher pH and organic matter content as well as higher rice and corn yields than unterraced plots. This result highlights the positive effects of on-site productivity and off-site conservation in terms of the lower erosion rates implied. The same study, however, cautions that the effect on cooperators' income was yet to be experienced. Diversification of income sources was strongly suggested.

None of the studies which looked into economic impacts (in terms of direct effects such as cooperators' income, labor use, and productivity) and ecological effects (as measured by biophysical indicators) have yet attempted to value, in economic terms, the environmental costs and benefits of upland cropping systems and interventions. Thus, a wide gap exists in research on economic impacts on society of social forestry.

SUMMARY, CONCLUSIONS, AND RECOMMENDATIONS

Research on upland development by economists are still few, although important issues on the economics of upland development may be observed. In general, case studies have prevailed, and there is a need to rationalize their conduct. Comparisons need to be made across various types of uplanders in terms of a minimum set of economic variables including gross income (including imputed), allocation of time among various activities, and productivity of labor and other factors of production. Changes in the uplanders' economy in terms of rational measures of development also need to be assessed. Likewise, the (cost) effectiveness of upland development projects need to be evaluated. The stated objectives of upland development should be measured consistently in terms of common measures of income, conservation, and ecological stability. Glaring research gaps are, thus, the measurement of the ecological role of upland development in terms of off-site soil erosion rates, on-site organic matter content, sedimentation in affected lowlands, distribution of income, and production of wood and food. Data on costs of upland development projects were notably absent in a recent list of social forestry projects (Bernales et al., 1982). Project costs could be used simultaneously with indicators of outputs of upland development to come up with measures of the efficacy of upland projects.

Externally-induced social forestry projects have generally been conducted on a pilot or experimental basis. Such projects normally involve relatively long gestation periods and are continuously modified to suit local environmental and social needs. Most are in the early stages of development. Thus, along with the upland development activities are attempts to evaluate the social, cultural, economic, and environmental feasibility of the projects by the implementors themselves. Studies have captured early changes and need to be followed up during later stages of such projects.

There is a need to continuously assess the factors which lead to participation in various upland development projects. Specific areas for investigation are tenure, credit availability, knowledge (or education), labor availability, and population pressure. The latter variable is composite. Demand for products from the uplands should be measured hand-in-hand with population increases.

Realistic assessment of adoption of 'prescribed' technologies could then be employed by project feasibility evaluators, and some allowance (and counteracting measures) can be made for less than 100% project participation. The following table summarizes research priorities for economic evaluation of the uplands and their current state of research in the country.

Table 1. Research gaps in a minimum set of economic concerns in upland development.

Areas of Concern/ Suggested Variables	Existing Studies	Remarks
A. Economic welfare of Upland Cultivator		
1. Gross Income (cash, in kind)	Floro	Existing studies usually contain project value only; inconsistent methods of measuring income prevail.
2. Allocation of Labor	Nguu and Corpuz	Notably focused on by non-economist researchers
B. On-site product		
1. Yield per unit	—	Inadequately evaluated due to inconsistent measurement of input and output.
2. Soil fertility	Tapawan	Evaluated only for pilot/ experimental projects.
3. Extent of soil conservation	Tapawan, Segura-de los Angeles	Adoption factors investigated in very limited manner.

Areas of Concern/ Suggested Variables	Existing Studies	Remarks
C. Off-site effects		Not being measured at all.
1. Migration into forest lands and foregone value of timber	—	
2. Sedimentation of water systems and valuation of decreased productivity of lowland farms/ hydroenergy/etc.		
D. Macro-concerns		Assessments so far conducted have only been feasibility studies. Serious evaluation of existing components of social forestry and forest land management is needed in these terms.
1. Cost effectiveness of alternative upland projects	—	
2. Contribution to National Production		
3. Employment	—	
4. Foreign Exchange generated or saved	—	

REFERENCES CITED

Aguilar, F., Jr. 1982. Social forestry for upland development . Lessons from four case studies. Final Report submitted to the Bureau of Forest Development by the Institute of Philippine Culture, Ateneo de Manila University, Quezon City.

Bernales, et al. 1982. Social Forestry Projects in the Philippines: An Inventory and a Listing of Communal Forests and Pastures. Integrated Research Center, De La Salle University, 1982.

Boserup, E. 1975. The impact of pupulation growth on agricultural output. Quarterly Journal of Economics, Vol. 89, No. 2, pp. 257-270, 1975.

Burt, O.R. 1981. Farm level economics of soil conservation in the palouse area of the Northwest. American Journal of Agricultural Economics. 83-92.

Cadelina, R.V. 1981. Food management under scarce resources by Philippine marginal agriculturists . Technological pluralism towards national building. Philippine Economic Journal, XX:1.

Capistrano, A.D. 1982. Study of a village economy. The case of Villarica, Pantabangan, Nueva Ecija (Philippines). Unpublished Master's Thesis, University of the Philippines, at Los Banos.

Corpuz, E. 1984. A comparative economic study of traditional kaingin, modified cropping patterns and tree farming in Mt. Makiling. Unpublished M.S. Thesis, University of the Philippines at Los Baños, College, Laguna.

Cruz, R.V.O. "Hydrometeorological characterization of selected upland cropping systems in Mt. Makiling. Unpublished M.S. in Forestry Thesis, University of the Philippines at Los Baños, College, Laguna.

Cruz, W. 1982. Technical and institutional change in renewable resource development (with application for traditional Fisheries). Unpublished Ph.D. Dissertation, University of Wisconsin-Madison.

Darity, W.A., Jr. 1980. The Boserup theory of agricultural growth. Journal of Development Economics, 7.

Ellevera-Lamberte, E. 1983. Indicators of upland poverty: A macro-View. Paper presented at the First National Conference on Research in the Uplands, held at the Sulo Hotel, Quezon City, Philippines.

Estioko-Griffin, A. and P. Griffin. 1981. The beginnings of cultivation among Agta hunters-gatherers in Northeast Luzon. In: Adaptive Strategies and Change in Philippine Swidden-based Societies edited by H. Olofson, Forest Research Institute.

Floro, M.S. 1980. Economic effects of resettlement on the organization of production, consumption, and distribution in Barrio Villarica, Pantabangan, Nueva Ecija. Paper presented at the Seminar-Workshop on the Socioeconomic and Institutional Aspects of Upland Development, sponsored by the Program for Environmental Science and Management, U.P. at Los Baños, Laguna, August 4-6, 1980.

Food and Agriculture Organization of the United Nations (UN-FAO). 1979. Economic analysis of forestry projects. FAO Forestry Paper No. 17. Rome .FAO.

Hayami, Y. 1978. Anatomy of a Peasant Economy . A Rice Village in the Philippines. International Rice Research Institute, Los Baños, Laguna.

Hufschmidt, M.M., et al. 1981. Benefit-Cost Analysis of Natural Systems and Environmental Quality Aspects of Development. A Guide to Applications in Developing Countries. Honolulu . East-West Environment and Policy Institute.

Hyman, E.L. 1981. Tree farming from the viewpoint of the smallholder. An ex-post evaluation of the PICOP project. Paper presented at the National Conference on the Conservation of Natural Resources, sponsored by the Natural Resources Management Center and the National Environmental Protection Center at the Philippine Plaza, Manila, December 9-12, 1981.

Kummer, D.M. 1984. Social forestry in the Philippines . A macro view. Policy Paper No. 11. Forestry Development Center. College, Laguna.

Lee, L. and W. Stewart. 1983. Landownership and the adoption of minimum tillage. AJAE pp. 256-264.

Lee, L. 1980. The impact of landownership factors on soil erosion. AJAE pp. 1070-76.

Lopez-Gonzaga, V. 1983. Peasants in the Hills. A Study of the Dynamics of Social Change Among the Buhid Swidden Cultivators in the Philippines. Quezon City . University of the Philippines Press.

Lynch, O., Jr. 1983. Withered roots and landgrabbers. A survey of research on upland tenure and displacement. Paper presented at the

First National Conference on Research in the Uplands held at the Sulo Hotel, Quezon City, Philippines, April 11-13, 1983.

Mindajao, N.M. 1978. Smallholder forestry and rural development . A case study of the PICOP project in Bislig, Surigao del Sur, Philippines, Unpublished Ph.D. Thesis, University of Minnesota.

National Irrigation Administration (NIA) and Engineering Consultants, Inc. (ECI). 1979. Pantabangan Watershed Management Report. Draft Feasibility Report, Erosion Control Study, Vol. I-IV.

National Irrigation Administration (NIA) and Mandala Agricultural Development Corporation (MADECOR). 1979. Pantabangan and Magat Watershed Management and Erosion Control Projects, Feasibility Report, 1979.

Nelson, G.C.N. 1984. The impact of government policies on forest resource utilization. Working Paper 84-04. Philippine Institute for Development Studies.

Nguu, N. and E. Corpuz. Resources, production activities and financial status of a Kaingin Farm. Annual Report Upland Hydroecology Program, UPLB.

Ono, M. 1982. A preliminary study of income-in-kind underreporting in the quarterly integrated survey of household (ISH). Journal of Philippine Development 16: IX, 1 & 2, 80-98.

Paris, T.B., Jr. 1987. Systems analysis and simulation of upland rice-based cropping systems in the Philippines. Unpublished Ph.D. Dissertation, Michigan State University.

Revilla, A.V., Jr. and M.C. Gregorio. 1983. The COSTEF system: Cost-effective reforestation agroforestation strategy. Policy Paper 9. Forestry Development Center. UPLB.

Revilla, A.V., Jr. 1984. The need for a forest renewal trust fund in the Philippines. Policy Paper 15. Forestry Development Center. UPLB.

Samonte, V. 1980. Social profile of upland farmers in barrio Villarica, Pantabangan, Nueva Ecija. UPLB Upland Hydroecology Program.

Segura-de los Angeles, M. 1983. Economic and social impact analysis of agro-forestry development projects in Villarica, Diadi, and Norzagaray. An Economic and Social Impact Analysis/Women in Development Project. Micro Component (ESIA/WID) Report, Working Paper 83-08. Philippine Institute for Development Studies.

1984. Economic and social impact analysis of an upland development project in Nueva Ecija, Philippines. Paper presented at the EWC/FAO Workshop on the Socio-Economic Aspects of Social Forestry in the Asia-Pacific Region, Bangkok, Thailand, September 18-22, 1984.

Seitz, W. and E. Swanson. 1980. Economics of social conservation from the farmer's perspective. AJAE 1084-1088.

Sumitro, A. 1984a. Review of tropical forest administration and revenue system: Their role in resource management, growth, and equity. Bureau of Planning, Department of Agriculture. IDRC/USAID Indonesia.

1984b. Social impact of commercial timber extraction upon local communities in Indonesia's tropical rain forests. Paper, FAO/EWC Workshop on the Socioeconomic Aspects of Community/Social Forestry in the Asia-Pacific Region. Bangkok, Thailand. September 18-22, 1984.

Tapawan, Z.G. 1981. Economics of farming systems in the upland areas of Hamtic, Antique, Philippines. Unpublished M.S. Thesis, University of the Philippines at Los Banos.

Taylor, R.C. and K.K. Frohberg. 1977. The welfare effects of erosion controls, banning pesticides, and limiting fertilizer application in the corn belt. American Journal of Agricultural Economics 13-26.

Upland Hydroecology Program (UHP). 1976. Annual Reports, UPLB-DNR-NIA-FORD Foundation.

1977. Annual Report. UPLB-DNR-NIA-Ford Foundation.

1979. Annual Report. UPLB-DNR-NIA-Ford Foundation.

1978. Policy Report. UPLB-DNR-NIA-Ford Foundation.

Wade, J.C. and E. O'Heady. 1979. Controlling non-point sediment sources of cropland management: A national economic assessment. American Journal of Agricultural Economics. 13-36.

Warner, K. 1981. Swidden strategies for stability in a fluctuating environment: The Tagbanwa of Palawan. In: Adoptive Strategies and Change in Philippine Swidden-based Societies. H. Olofson, ed. Forest Research Institute.

Weiss, C. 1972. Evaluative research. Methods of Assessing Program Effectiveness. Prentice-Hall.

7

FINDINGS FROM EIGHT CASE STUDIES OF SOCIAL FORESTRY PROJECTS IN THE PHILIPPINES

Filomeno V. Aguilar, Jr.

Philippine forestry has traditionally placed more importance on trees than people. When man has been included in the forester's framework, he has usually been portrayed as the enemy of the forest. Compounding this bias is the forester's traditional training which is oriented towards commercial timber extraction. Subsistence agroforestry in upland communities has not been a part of such training. As such, forestry has basically served the interests of extractive industries, while, concomitantly, seeking the eradication of upland communities through the government's classification of forest occupants as squatters.

Recent developments, however, have shown indications of a new orientation within the field of forestry. This orientation, called "social forestry", stems from the realization that punitive measures cannot effectively remove the forest dweller. Instead, the forest resident must be treated differently to transform him from an agent of forest destruction to a partner in forest development and conservation. This "new" forestry maintains that people are an integral part of the upland ecosystem and that a symbiotic, mutually sustainable relationship can exist between forest dweller and environment. The goal of social forestry is thus socioeconomic development of upland communities whose basic needs can be met through judicious, community-based use of forest resources.

BACKGROUND OF THE CASE STUDIES

This developing trend in forestry is seen in the Upland Development Program of the Bureau of Forest Development (BFD), an agency of the Philippines' Ministry of Natural Resources. Assisted by the Ford Foundation, the BFD is undertaking research, training and experimentation to build its expertise and capacities for effective participatory approaches to upland development. In an initial phase, the program called for studies of eight pilot project sites throughout the country. Four studies were carried out by the Institute of Philippine Culture (IPC) of the Ateneo de Manila University and another four by the Integrated Research Center (IRC) of De La Salle University. Representatives of the two research organizations also became part of the BFD Working Group, a loosely constituted body of administrators, foresters and academicians created in October 1980.

Selection of Projects for Study

Upland projects undertaken by the BFD, other government agencies and private organizations were chosen for documentation and analysis. Selection was based on: 1) the projects' having a reputation for some participatory or social component, specifically projects which called for the target beneficiaries to be organized into local associations, 2) the projects' reputation as "successful", 3) adequate representation of different categories of project proponents, and 4) the inclusion of at least one project involving a cultural minority. These considerations led to the selection of the following projects.

1. The Kalahan Educational Foundation (KEF) is a private organization in Nueva Vizcaya. The KEF was the first of only two groups contracted by the government to administer a lease agreement on behalf of a tribal minority, the Ikalahans, with financial assistance from various local and international organizations.

2. The Communal Tree Farm (CTF) in the Narra Settlement at San Pedro, Laguna, was a BFD-sponsored project ranked as one of the 14 best CTF sites in 1979.

3. The Rinconada-Buhi Agroforestation/Watershed Development Project was initiated by the municipal government of Buhi and was implemented as an inter-agency project under the supervision of a regional development body, the Bicol River Basin Development Program Office (BRBDPO). Funding and consultancy support came from the United States Agency for International Development (USAID).

4. The Forest Protection Incentive Plan (FPIP) was implemented mainly in the highland Tinggian communities of Abra province under the initial sponsorship of a private sector logging and pulp-making company, the Cellophil Resources Corporation (CRC), and its sister company, the Cellulose Processing Corporation (CPC).

MAP 6. FOUR SOCIAL FORESTRY PROJECTS

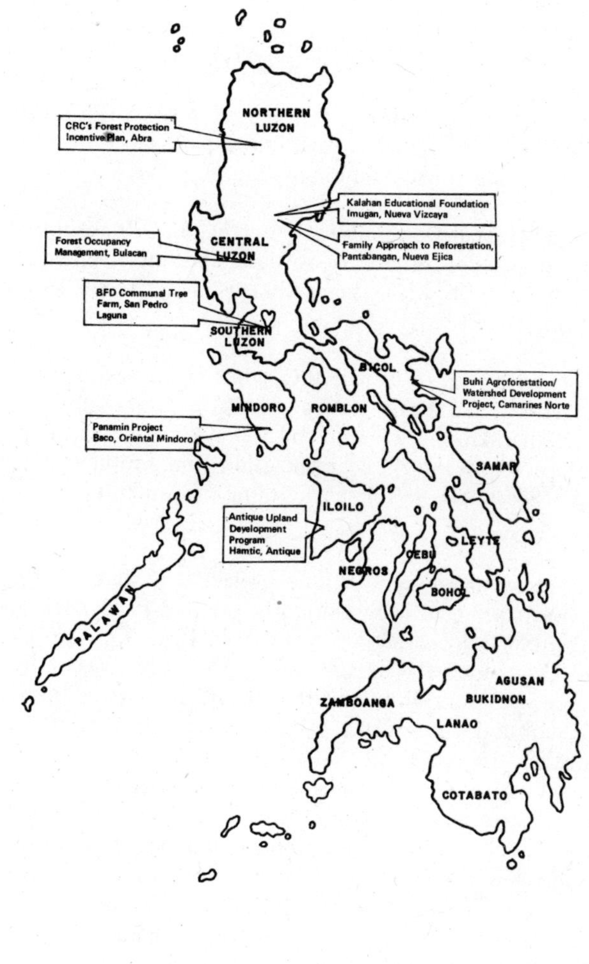

5. The Forest Occupancy Management (FOM) project in Kabayunan, Doña Remedios Trinidad, Bulacan, was one of two pilot sites initiated by the BFD immediately following the issuance of Presidential Decree 705, the Revised Forestry Code of the Philippines, which called for "managed forest occupancy."

6. The Family Approach to Reforestation (FAR) project in Pantabangan, Nueva Ecija, was the second such project of the BFD patterned after the Burmese *taungya* scheme. This project has been turned over to the National Irrigation Administration, which holds jurisdiction over the Pantabangan Dam.

7. The Presidential Assistant for National Minorities (PANAMIN) project in the Lantuyan Settlement was with a community of Alangan Mangyans in Baco, Oriental Mindoro.

8. The Antique Upland Development Program (AUDP) was initiated by the provincial government of Antique and implemented as an inter-agency project with financial assistance from the Ford Foundation.

The first four were studied by the IPC, while the other four by the IRC. The studies covered three projects undertaken initially by the BFD, one by another national government agency, two by local governments, one by a non-governmental organization, and another by a logging company. (See table 1 for background information on these projects.)

Except for the Mindoro and Antique projects, all were located in Luzon island. Most projects were started in the mid-1970s. The areas covered by the projects are variable, ranging from 45 ha from the CTF in Laguna, 1,117 ha for the FOM in Bulacan and to nearly 100,000 ha for a logging company in Abra. In three sites, the project participants were predominantly cultural minorities (Ikalahan, Tinggian, Alangan Mangyan); and another site, participants represented a cultural minority (Dumagat) mixed with a heterogeneous population of lowland migrants. Two projects were located in resettlement areas, one for squatters from the Metro-Manila area, and the other for families displaced by a hydroelectric dam project. Two other projects were located in upland areas with relatively long-established lowland populations (Buhi and Antique). The projects represented a variety of socio-cultural and management conditions although the representation was not exhaustive.

Coverage of the Case Studies

The case studies focus on three major areas of investigation. First, the socioeconomic and cultural conditions of the community and of

Table 1. Selected background information on sample social forestry projects.

Name of Social Forestry Project	Location	Date Established	Area (ha.)	Hshlds. and Sample	Ethnic Composition
IPC Study Sites					
A Kalahan Educational Foundation Agroforestry Dev.	Sta. Fe, Nueva Vizcaya	5/74	14,730	368 (n=38)	Ikalahan
B Communal Tree (CTF)	San Pedro, Laguna	5/79	45	172 (n=40)	Mixed
C Rinconada-Buhi Agroforestation/ Watershed Dev.	Buhi, Camarines Sur	3/79	90.5	134 (n=40)	Buhinon
D. Cellophil Resources Corp/Cellulose Proc. Corp.-F.P.I.P	Abra Province	10/77	99,990	n/d (n=0)	Tinggian
IRC Study Sites					
E. Forest Occupancy Management (FOM) Pilot Project	Dona Remedios Trinidad, Bulacan	April	1,117	253 (n=20)	lowland migrants, Dumagat
F. Family Approach to Reforestation (FAR) Project	Pantabangan, Nueva Ecija	3/76	185 360 102 282 1,139	30 70 40 100 199* (n=36)	lowland settlers displaced by dam
G. PANAMIN's Lantuyan Settlement	Baco, Oriental Mindoro	7/81	402	46(?) (n=31)	Alangan Mangyan
H Antique Upland Dev. Proj.	Hamtic, Antique	76	2,500	698 (n=62)	Antique

Some families participated in more than one phase. Source: Aguilar, 1982 Bernales & Dela Vega, 1982.

the participating households are described. Where data are available, the site's biophysical conditions are described. The circumstances surrounding the formulation, introduction, and establishment of the project are traced. Household and community data and background events provide the context in which project structure, strategies, management, and impact are assessed. The extent of popular participation in the decision-making, implementation, and sharing of benefits is analyzed, and the relationship between participation and project results is explored. The case studies were presented as reports to the BFD Working Group. The IPC report is contained in one volume (Aguilar, 1982), while the IRC prepared a separate report for each case (Bernales and De la Vega, 1982). This chapter presents a summary of the case study findings.

Data Gathering Procedures

Different methods of data collection were employed to meet the objectives of the case studies. Community data, household information, and beneficiaries' opinions concerning the project were obtained through a sample survey of project participants. A questionnaire was administered to respondents, although the instruments of the two research institutions differed. Each project, because of its own peculiarities, required a slightly different interview schedule.

Unstructured interviews were conducted with a variety of key informants (including village leaders and elders), project staff officials and employees, project participants, barangay and municipal officials and BFD field personnel. These interviews helped in reconstructing events leading to the project and the problems encountered during the initial stages of its implementation. The informal interviews also yielded information about different perceptions of the project, which could not be easily articulated in a structured interview setting. Unstructured interviews also revealed information which would not normally appear in official reports.

Field observation provided insights into project dynamics and the target community. Research entailed about a month-and-a-half in the field (except for the Abra case study where direct field observation and the sample survey could not be undertaken due to "peace-and-order" problems). A supplementary study of existing literature, documents, project reports, and minutes of meetings was conducted. Using these various approaches, data-gathering for the case studies was undertaken mainly in 1981.

The Sample Respondents, Access to Land, and Socioeconomic Conditions

The sample survey was to cover about 10% of the participating households in each project site. For the IPC, this gave a sample size of 38 respondents for the KEF study. For the other project sites, however, the sampling rule was changed since the small number of participants resulted in very small samples. Thus, for the later case studies, a standard sample size of 40 was used, representing 23% and 30% respectively of the CTF and Buhi project participants. Sample sizes in the IRC's studies varied from 20 to 62 respondents. In two cases, the sample represented less than 10%, and in the other two, more than 10% of the participating households. Respondents were selected randomly from a list of project participants and in direct proportion to the number of participants from each *barangay* (village). Inaccessible villages were not included in the sampling frame owing to time constraints. Villages where the number of participants was too small for a sample to be drawn were also excluded.

Male respondents predominated in the KEF and Buhi sites while females constituted a sizeable minority in the CTF site. On the average respondents were about 40 to 42 years old. Only about one-third of the KEF respondents completed an elementary education. Among Buhi respondents, the number rose to 47.5%, while in the CTF site as much as 85% of the sample had completed at least elementary school. Although some respondents had little formal schooling, most were literate and could speak two or three languages.

The mean number of owned or claimed parcels per household numbered from about 1.5 to 2, with the aggregate land area varying from 456 square meters in San Pedro to 2.7 ha in Buhi to 6.9 ha among the Ikalahans. Although areas with a gradient of 18% and more are not alienable and disposable according to current Philippine law, private titles exist, with the incidence ranging from 82% in Buhi to 36% in the Ikalahan region to 5% in San Pedro. Land tenure was varied and included traditionally respected use-rights among the Ikalahans and share tenancy among the Buhi residents. Access to land in San Pedro was mainly through the CTF project. Everyone, however, had access to land for which, despite legal impediments, almost all wanted private title. Indeed, most upland residents were convinced they had rights to the land they till.

Annual household cash income was lowest in Buhi (P2,089), a figure that doubles for the Ikalahans (P4,218). The highest income was registered among San Pedro households (P10,669), most of whose members are industrial and city wage workers who commute to the Manila Metropolitan Area. The Gini coefficient of income inequality is about 0.4 for the Ikalahan and San Pedro respondents, but worsens to 0.6 in Buhi, indicating more pronounced disparities where average incomes are lower. On the other hand, the Gini coefficient of land inequality is about 0.4 which approximates the degree of income inequality among Ikalahan and San Pedro residents, and points to a land access situation better than the income disparity characteristic of the Buhi sample.

Upland communities are not homogeneous within and between sites. Wage employment accounts for a major segment of upland households' income, affording better income opportunities to residents of settlements accessible to industrial and urbanized zones. Most communities, however, depend on agriculture, especially swidden cultivation, to meet subsistence requirements. In some areas, agriculture adds to household income through the marketing of surplus crops, livestock and handicrafts. Cash incomes generated are utilized for the acquisition of commodities from the lowlands. Table 2 gives information on respondents, land access, and income for 3 sites.

Table 2. Selected background information on case study respondents, their access to land, income data, for three social forestry project sites (Aguilar 1982)

	Respondents		
	Ikalahan (N = 38)	San Pedro (N = 40)	Buhi (N = 40)
Personal data			
Number of respondents	38	40	40
Percent male	92	55	83
Percent female	8	45	17
Mean age (years)	40	42	42.5
Percent at least elementary school graduate	33	85	47.5
Land access			
Mean number of owner/ claimed parcels per household	2.0	1.4	1.5
Gini coefficient of land inequality	0.41	0.44	0.42
Mean area (ha) owned/ claimed per household	6.9	0.046	2.7
Mean area (ha) cultivated land/household	1.4	0.043	2.1
Percent of owned/claimed parcels w/ private title	36	5	82
Respondent's access to parcels			
Percent use-rights	47	5	—
Percent government program	—	86	—
Percent inheritance	18	—	12.5
Percent purchase	14	2	18
Percent tenancy	—	5	50.0
Percent borrow/others	21	2	19.4
Household income			
Mean annual household cash income (1980 P)	4,218	10,669	2,089

Percent from:

Crop production	7.8	1.6	17.0
Livestock production	19.6	7.9	14.6
Handicraft production	19.0	–	0.9
Sale of forest products	0.5	NIL	1.1
Wage employment	41.5	70.8	58.0
Self-employment	12.1	14.7	6.8
External assistance	–	5.0	1.7
Gini coef: cash income ineq.	0.40	0.39	0.63

BASIC FEATURES OF SOCIAL FORESTRY PROJECTS

Impetus for Project Development

The projects were conceptualized and introduced to different communities in response to outside agency perceptions of problems and in response to problems perceived by the residents. The proponents (agencies) were usually concerned with environmental issues. For instance, about two-thirds of the 10,290 ha in the Lake Buhi watershed is denuded. While the lake now provides irrigation water for 3,620 ha of riceland and is capable of irrigating an additional 10,027 ha, its capacity is imperilled by soil erosion and landslides. Soil losses are estimated at 1.4 tons/hectare/year. If such losses continue, the lake is expected to be silted up within 30 to 50 years. In 1978, Buhi's town council recognized the environmental issue in terms of lives, property, and livestock lost as a result of landslides since the 1950s. The municipal government called for the protection and reforestation of Lake Buhi's watershed and petitioned the national government for assistance. The call was heeded by the Bicol River Basin Development Program office (BRBDPO) and its funding agency, the United States Agency for International Development (USAID). An inter-agency committee was created at the regional level, and the pilot agro-forestry project for the Buhi watershed was formulated.

The reforestation project in Pantabangan was established for similar concerns. The Pantabangan Hydro-Electric Dam, completed in 1974, is capable of generating 100 megawatts of electric power and of irrigating 80,000 ha. The capacity and lifespan of the reservoir are dependent on the 115,819 ha watershed, of which only about 24% is forested. Consequently, two years after the dam's construction, the local BFD district office initiated activities for the watershed's protection and rehabilitation. Similarly, in Dona Remedios Trinidad, Bulacan, the pilot project was set up to halt the degradation of the Angat Watershed Reservation. The area covers 55,440 ha that act as buffer to the Angat Hydro-Electric Project, composed of the Ipo and Bustos Dams. These latter two dams supply part of the electricity and water needs of Metro Manila and surrounding areas. The impetus for project formulation in the other sites was also environmental: frequent flashfloods and landslides in Antique, denudation in a district of San Pedro, Laguna, and wildfires destructive of Abra's primary forest cover.

Only in the Lantuyan Settlement in Mindoro and the Kalahan Educational Foundation in Nueva Vizcaya were the overriding concerns other than environmental. The project in Mindoro was intended to provide assistance to cultural minorities with a goal of "integrating" them into the mainstream of Philippine society. The agro-forestry project in Nueva Vizcaya was the outcome of a seven-year struggle for land security. The conflict started in 1968 when a group of government officials registered 198 ha of the Ikalahan reserve as private property. As a result, 41 families faced possible ejection. After initial judicial attempts failed, a volunteer lawyer filed a civil case on behalf of the affected minorities. Despite political pressures, the case was pursued for about three years, and in August 1972, the titles were nullified. However, apprehension over possible landgrabbing caused some Ikalahan leaders to seek a more secure land tenure. Thus, another court case was lodged to invoke a 1964 amendment to the Public Land Law which granted preferential land rights to tribal occupants. The court's requirement of a sketch map to cover 4,000 ha of individual claims delayed the case for about two years.

In the meantime, sympathetic forestry officials suggested the establishment of a civil reservation, but a local pastor, having witnessed what had happened in other reservations, expressed the concern that the Ikalahans would lose their dignity and autonomy in the process. Even private titling of land was reconsidered since private titles could simply turn out to be an indirect way for the Ikalahans to lose their lands. At about the same time, an effort was started to establish a high school, culminating in the formation of the Kalahan Educational Foundation. The UNESCO office in the Philippines provided aid and, in an interagency meeting it organized, the land problem was highlighted. Present at that meeting was the Director of the BFD who suggested leasing the ancestral lands to the minorities. Thus, the 25-year Memorandum of Agreement (renewable for 25 years) was conceived. Signed on May 13, 1974, the memorandum provided for the gratuitous lease of 14,730 ha of public land to the KEF. In accordance with the agreement, three years later the KEF submitted an agro-forestry development plan to the BFD.

Only in the last case was the land issue responsible for the formulation of a socially oriented forestry troject. Most projects derived impetus from environmental concerns which, in many pilot areas, are intertwined with the aims of servicing lowland agriculture and securing investments in infrastructures and in a logging venture, in one case. Except for the KEF, the initiative in practically all projects was taken by outsiders. Residents in the project areas were aware of this, although they may have differed in their perceptions of who the actual initiators were. Such outside initiative has invariably led to a definition of community needs that differs from the people's own felt needs. In the formalization of project objectives, however, there were attempts to balance environmental and social goals. While projects may be concerned with reforestation, arresting the degradation of critical watersheds, or developing a balanced ecosystem, the strategies to attain these were also designed to improve living conditions, provide livelihood, augment income and create self-reliant and self-sufficient upland communities.

Project Strategies for Upland Development

Each social forestry project studied had its own set of strategies for upland development. In the Ikalahan case, the lease agreement mandates the KEF to convert the area into a viable, multi-purpose forest zone. An agroforestry development plan was devised, stressing fire control, watershed preservation, and reforestation. The KEF initially constructed firelines along its boundaries, but fireline construction proved to be expensive and ineffective in preventing fires from entering the leased area. Consequently, the KEF revised its policy in favor of greenbreaks (fire-fighting teams of locally-hired forest guards) and firelines in fire-prone areas. Ten-meter greenbreaks of *manguey*, giant *ipil-ipil*, and *falcatta* will take a few more years to develop. Swidden cultivation was prohibited in two forested watershed areas of 703 ha. The KEF also conducted reforestation activities and erosion control measures, such as contour belting and riprapping in denuded watersheds, steep areas subject to sheet erosion and the sides of major gullies. In many instances, students of the Kalahan Academy were paid for reforestation work to help meet tuition and other educational expenses. In addition to these projects, local leaders constituting the KEF's Board of Trustees evolved 17 agroforestry regulations for residents within the leased area. These regulations limited access to the area's resources to bona fide residents, specified restrictions on swidden site selection and cultivation, and provided guidelines for resource use.

New areas could not be opened for cultivation without the Board's approval; preferred areas were those previously cultivated and over which residents have acquired communally respected use rights. Whenever a newly cleared area was burned, a 10-meter fireline had to be constructed around the intended swidden to reduce the risk of wildfires. The elders revived an old practice of *gaik* or fireline construction, which many residents had abandoned in retaliation against government personnel who in the past harrassed them for being squatters. The KEF also encouraged the indigenous technology of *gen-gen* construction, which consists of uprooted camote leaves and stalks buried in piles following the contour of a field. Finally, the KEF took steps to rationalize land claims within the leased area. Due to past attempts seeking the release and private titling of land, residents have claimed large tracts of land. The claims are now being redrawn to conform to the KEF's 15-hectare limit or replaced by parcels in other areas to protect critical watersheds. In lieu of titles, "Land Claim Registration" papers are being issued. These not only allow residents use of the land, but also bind them to the Foundation's agroforestry regulations.

The BFD's FOM project in Bulacan had a similar but less broad strategy for upland agriculture. Labelled as "management-in-place" (in contrast to resettlement), the project sought to regulate swiddening by ensuring that only early occupants (established through a census) utilize the area's resources, and by designating parcels for permanent cultivation. Parcels were identified from previously occupied areas, each not to exceed seven hectares. The land was made available through a two-year Forest Occupancy Permit issued by the BFD. The permit stipulated conditions with which the permit holder must comply, including non-expansion of the cultivated area, planting of agricultural and forest crops, and fire protection. Agroforestry and bench terracing were pursued as soil conservation measures.

Land development through bench terracing was more of a focal point in the Buhi pilot project. This project also involved orchard and firewood (giant ipil-ipil) lot development. The Buhi project gave cash incentives on a staggered basis to some cooperators for undertaking these activities. This arrangement was presented as a subsidy scheme, wherein 50% of the labor cost for land development activities is covered by the project. The amount, however, was limited to a maximum of P455 at 0.75 ha per farmer-cooperator, ideally distributed as follows: P300 for 0.15 ha of bench terracing, P100 for 0.50 ha of fruit trees, and P55 for 0.10 ha of ipil-ipil. The incentive was released in four payments, with corresponding output expectations. Farmer cooperators underwent a farmers' training seminar as a prerequisite to project participation. The week-long seminar was designed to "educate" the residents about upland development, arouse the residents' interest in the project, and to equip them with skills deemed necessary for project implementation. No eligibility criteria were followed during the first seminar, but problems necessitated the specification of criteria for subsequent seminars. These criteria were permanent residence in the project site, landholding within the area under whatever tenure, and willingness to complete the seminar. Farm visits were made by the project staff following the seminar to determine activities best suited for each plot. This was followed by the negotiation and signing of the contract between landowner, tenant and the director.

The Antique pilot project, although lacking the mandatory training seminar, paralleled Buhi in emphasis on soil conservation (bench terracing, strip cropping, contour farming) and reforestation through orchard (mango) and woodlot (giant ipil-ipil) development. Coffee production was promoted. The Antique goal was integrated home lot-farm lot-pasture lot-wood lot development for household economic self-sufficiency. A two-hectare model was designed to accommodate four complementary productive areas: 0.8 ha of wood lot, 0.5 ha of pasture lot, 0.5 ha of farm lot, and 0.2 ha of home lot. A livestock breeding farm was established to upgrade native cattle and goats. The project's nursery propagated high-yielding rice varieties, fruit, and commercial forest trees. Like the KEF, the Antique project relied on students for some of its reforestation work. Unlike the KEF, however, tree planting was enforced as a graduation requirement for students.

The other projects lacked intensive land development activities such as in Buhi and Antique. The Lantuyan Settlement project thrust was the intercropping of giant ipil-ipil, coconut, coffee and other tree species in the upland farms of the Mangyan settlers. While payment in kind was given for tree planting, the project had no clear strategy for resource management. Another project where land rehabilitation was not emphasized is the BFD's Communal Tree Farm (CTF) in San Pedro, Laguna. Although "communal", the CTF allocated plots of a minimum of two hectares to individuals. The land allotments in the San Pedro site measured only about 200 square meters. Land was to be made available to persons meeting three conditions: annual income not exceeding P6,000, availability to work on the project at least three days per week and preferably an occupant of the area. Twenty-five-year CTF certificates were awarded to qualified individuals after a Memorandum of Agreement was executed between the BFD and the municipal government acting on behalf of the participants. The land could then be planted with short, medium, and long-term tree species. Giant ipil-ipil, however, was the usual crop grown in San Pedro. The inter-planting of food crops was allowed, although should not exceed 20% of the area.

A different approach was observed in the BFD's Family Approach to Reforestation (FAR) project in Pantabangan. Instead of hiring laborers, as is done elsewhere in the reforestation of the Pantabangan watershed, the FAR project contracted a family for a range of tree plantation activities. Land parcels ranging from three hectares to five hectares were allocated per family, and both parties signed a contract written in Tagalog which stipulated the responsibilities of each party. One requirement for participants was training in nursery and plantation management. Unlike other projects, however, the land was to be turned over to the BFD for full government supervision when the contract expires after two to three years. To compensate participation in the project, cash payments were to be made in six to ten installments, depending on the contract's duration and only after each contracted activity was accomplished. Payment to participants did not exceed P0.70 and P0.90 per surviving seedling grown within a two-year and three-year period, respectively. The contract also required a survival rate of 80% before the turn-over period; otherwise, the last payment was withheld. Mainly giant ipil-ipil and gmelina were planted. Later, cashew and fruit seedlings were inter-planted, and teak seedlings were introduced. Fireline construction was a major activity during one of the projects.

Eligibility criteria for FAR participants were followed in a later project stage. Criteria included bona fide residence, physical fitness and availability for reforestation work, four unmarried dependents, income of not more than P2,000 per year, and possession of tools for project work.

The Forest Protection Incentive Plan of the CRC was unlike the other projects in many respects and was designed to stem the antagonism generated by the entry of a timber company into Abra's highlands. The Tinggian communities expressed several concerns: 1) as a result of the company's operations, they would lose their communal forests and pasture lands, 2) the water system feeding the ricefields carved out of the mountainside would dry up, 3) the river providing their protein source would be polluted. The Tinggians demanded their constitutionally guaranteed first priority rights to the lands, waters, and other natural resources of the area. When the demands failed, many resorted to protest burnings of forests. To control the fires, the company decided to solicit the assistance of some local residents. The Tinggians were organized into Forest Protection Units (FPUs), each with a membership of at least 10 households living within a contiguous area of from 2,000 ha to 5,000 ha. The FPUs were made responsible for any fire within their area and were tasked to contain fires, apprehend any violators, and submit findings of investigations to the company. A P200-monthly incentive pay was awarded for the work, capped by a P1,000 yearly bonus. The incentive and bonus payments, however, were subject to an adjustment factor based on the extent of damaged area.

Common Elements in Project Strategies

Despite apparent differences, the projects have a number of things in common. An incentive scheme is a feature of many projects, although the incentives take different forms: monetary rewards, access to land or free seedlings and other planting materials. Direct monetary payments are offered by the Abra, Pantabangan, and Buhi projects. Monetary gains accrue more indirectly to participants in the CTF (through the sale of tree plantation products) and Lantuyan Settlement (through payments in kind) projects. Land for cultivation is a powerful incentive for participants in the Ikalahan and CTF projects. Free seedlings also serve as an incentive in six of the eight projects studied. In combination with such incentives, projects employ negative reinforcements such as downward adjustments in cash rewards depending on performance (Abra, Pantabangan, and Buhi projects), cancellation of permits (the FOM project) and the fines for violations of agro-forestry regulations (Ikalahan). In five of the eight projects, the implementation of resource management strategies is made binding through a written agreement executed by the project participants and the implementing agency.

Another point of comparison across projects is the mobilization of beneficiaries through community forestry associations. The Ikalahans had the KEF and its 10-man Board of Trustees elected from among village elders for fixed 5-year terms. The KEF also had a Board Secretary and Treasurer, two positions of which may be occupied by non-Trustees. In the Buhi project, there were two uplanders' associations organized separately by the village at the end of the farmers' training seminars. Each of the Buhi associations had an elected president, vice-president, secretary, treasurer, and has five planning committee representatives. Linking the cooperators and project management staff, the farmers' associations were designed to help upland farmers make use of their potentials, appreciate their duties and responsibilities, and foster a more effective participation in the management of their upland resources. The view that participation enhances project implementation was shared by the Antique project where there are now ten Upland Farmers' Associations organized into a Federation and nine Young Farmers' Assocations. Participants in the Pantabangan Family Approach to Reforestation project have their own association, as do the CTF beneficiaries. Some 32 Forest Protection Units have been organized in the CRC project. In the Lantuyan Settlement, an unsuccessful attempt was made to establish a cooperative.

Another organizational complement common among projects is a project staff, which is charged with surpervisory and management functions delegated by the proponent agency. The CTF Development Staff had a national coordinating body and a unit paralleling each level within the BFD's organizational hierarchy. The Bulacan FOM project was not only linked to a major division within the BFD, but was also attached to a regional entity, the Upper Angat River Basin Multiple Use Management District. For projects initiated by local government, the project staff was a special creation e.g., Buhi's and Antique's project management staffs. The private KEF had a number of project staffs which oversaw its different projects, including agroforestry. The CRC, on the the other hand, reorganized its pool of forest guards into a Forest Conservation Coordinators' group.

A number of projects have changed agency affiliation. The BFD's Pantabangan FAR project is now under the National Irrigation Administration (NIA). The Buhi pilot project has been turned over to the BFD. And the CRC is now under the government's National Development Corporation (NDC). In Buhi, at least, the change has disrupted project operations and has caused serious delays in fund releases.

Members of the project staff with technical training were usually recruited from outside the project locality. Only in the KEF, where all members of its staff were locally hired, was there a conscious effort to develop the community's manpower resources. The Buhi Upland Extension Leaders were drawn from among project participants. Other project staff came from the town of Buhi, and the project head was appointed by the implementing agency. The project staff of the different projects undertook various support activities, including nursery development, construction of demonstration farms, and extension of technical assistance.

To summarize, upland development project strategies involve the introduction of labor-intensive outside-packaged technologies and only rarely feature the promotion of indigenous practices and techniques. But considerable variations exist from one project to another. While one project regulates and modifies the manner of swidden cultivation, another may stress reforestation. While one project may have fire control as its sole objective, another may be concerned mainly with soil-conservation and rehabilitation. Some projects attempt a multi-pronged strategy which may include livestock production and the delivery of social services (e.g., education, health). In general, the strategy adopted is seen as capable of meeting both environmental and social goals.

Regulations concerning resource utilization are often legally formulated and agreed to in writing. Violation of such contracts results in penalties and other punitive measures. Positive reinforcements are also employed, and these include land access, monetary rewards and free planting materials. A project staff oversees each project and undertakes support activities, such as nursery development and the provision of technical assistance as well as technology diffusion through farmers' training seminars, demonstration farms and upland extension services. In order to effect popular participation in project implementation, the beneficiaries are usually organized into associations. Enlistment in the project, however, is frequently limited by eligibility criteria (such as bonafide residence, a specified annual income and willingness to work with the project), but these criteria are either waived or evolved in the course of project implementation.

COMMUNITY RESPONSES TO SOCIAL FORESTRY PROJECTS

Enlistment in the Project

The Pantabangan FAR case sheds light on the process of introducing a project to an upland community, initial resistance by the people, and eventual project acceptance. The project staff first called a meeting of upland residents to explain the project and to gather support. Only 10 out of the expected 100 family heads attended. These 10 tried to convince other members of the community, resulting in 30 household heads participating in the first project phase. The reluctance to join the project stemmed from the frustration and mistrust acquired from past experiences with government agencies. The people displaced by the dam project had been promised new housing and farm lots, but the houses were haphazardly built, and the farms were located some 13 kilometers away from the resettlement area, causing many to abandon rice farming. Others in the community became interested, however, when the first 30 families began to receive their incentive pay after the first year. In a second project phase, 70 families enlisted. The third phase, however, witnessed a reversal of the trend, and participation dropped to only 40 families. This decrease was attributed to, among other factors, ill-timed reforestation activities which coincided with the more lucrative option of onion-growing. This led to the scheduling of the fourth phase at the end of the onion season. Consequently, the project succeeded in mobilizing 100 families.

In Buhi, the assistance of the barangay chairman was sought to facilitate negotiation for a nursery site, to explain the project to local residents, and, when the first farmers' training seminar was ready, to encourage residents to participate. The barangay head responded, compelling 50 members of his barangay to attend. Not surprisingly, the sample survey reveals that 18% said that project enlistment was compulsory. Because of material incentives, residents of the adjoining barangay became eager to join. This group then constituted a majority of the 51 participants of the second training seminar.

The case studies illustrate that the incentive schemes were generally effective in eliciting participation from upland communities. In the San Pedro CTF site, 50% of the respondents said they got involved because of their desire to gain access to cultivable land; some 23% thought the project would augment income; 23% wanted to enjoy whatever benefits the project could offer to improve their lives. Participants in Buhi held expectations of finding employment as casual laborers (29%), receiving free seedlings (29%), deriving material assistance (19%) and gaining access to the subsidy scheme (13%).

Evidently, concern for the ecosystem is not a strong motivation for project enlistment, even among those who have undergone training seminars. Basic subsistence, cash incomes, remunerative employment, land security and better sharing arrangements with landlords emerge as more pressing needs. In joining projects, participants inevitably expect immediate and tangible rewards. Since the environmental benefits from labor-intensive land development activities can be enjoyed only in the longer term, these fail to draw the interest of the participants. Incentive schemes, on the other hand, become major points of attraction because they have the potential to address the community's felt needs more directly.

An unfortunate consequence of the incentive system is that the project can become the focus of competition for scarce resources, especially if cash rewards are involved. When people's expectations are not met, the project itself is placed in jeopardy. To illustrate, the often temporary and sporadic employment opportunities offered by the Buhi project raised hopes among many participants. Resentment developed when the project staff's recruitment process was perceived as unfair. While some participants tried to win the staff's favor, others distanced themselves from the project. In either case, project credibility suffered, and people started to view the project as politically motivated and tied to the electoral ambitions of town officials. Delays in fund releases and, thus, in subsidy payments, compounded the problem and discouraged participants from completing the contracted land development activities.

After the San Pedro CTF project was well underway, the project staff realized that the site, a resettlement area, was under the jurisdiction of another government agency. Negotiations with the latter were started, but with no conclusion in sight, CTF certificates could not be issued. The continuing uncertainty over land tenure wore out the interest of many participants. Fortunately, tree crops had already reached maturity and needed little care. Meanwhile, other participants severed their ties with the project due to another land issue. Project implementation had proceeded without full clarification of land claims, a fact hidden by the project staff in initial reports. Land disputes, including ones generated by the project, remained unresolved. These disputes resulted in blocked entry to some parcels, threats to life, poaching, and even intentional burnings. Many enlisted residents became disheartened.

While project incentives can draw people's interests, the same source of attraction can become sources of conflict due to poor planning and management. Project involvement becomes dispirited, eventually leading to cessation of activities in the worst possible case.

The Association of Project Beneficiaries

Awareness of and membership in community forestry associations in the KEF and Buhi projects were nearly universal. In contrast, as much as 87.5% of the San Pedro CTF respondents were not aware of the existence of their organization, and membership was limited to only 10%. At all sites, rates of participation in the associations were rather low. Only 47% of the respondents in the KEF and only one respondent in San Pedro participated in the election of officers. In Buhi, 85% of the respondents participated at least once. However, this was because elections were held before the end of the mandatory farmers' training seminar. In one Buhi association, only 32% of the respondents participated in all three of its elections. In the other association where two elections were held, 43% participated in both.

These low rates of participation may be traced to the manner in which such associations were organized. Since its inception in 1979, the San Pedro CTF was coordinated by the heads of the barangays at the project locality. Although appointed as village representatives, they had neither definite positions nor clear-cut functions in the project, other than to serve as liaison between the proponent agency and the project participants. This loose structure persisted until a handful of participants gathered to elect the chairman and vice-chairman of the CTF organization. Established because the CTF plan called for such an entity, the San Pedro organization of beneficiaries was little more than a paper organization.

In Buhi, each association was formed at the end of a farmers' training seminar. At that time, the participants were expected to have realized the importance of banding together. Informants revealed, however, that the associations were formed out of a sense of obligation, while others stated they were simply cautious not to earn the staff's displeasure. Moreover, after the associations were formed and the officers elected, objectives and by-laws of the associations were presented by the (project) staff and adopted by the farmers after some discussion. Apparently, much was accomplished in too short a time. In evaluating the effectiveness of association officials, 40% of the Buhi sample said they were effective; 34% claimed they were ineffective; 26% said they were neither. The unstructured interviews revealed similar opinions about the associations. The complaints were centered not so much on the officers themselves, but on what the participants perceived as the project staff's attitudes. On a number of occasions, the associations submitted suggestions concerning project personnel and other matters. These were turned down by the staff, causing the impression that associations were futile and that they were established merely as a show for government agencies and foreign visitors.

Although the associations were conceptualized by proponents as vehicles for popular participation, the data indicate that they are largely ineffectual and, in some places, were organizations in name only: Participants were disturbed by their groups' lack of efficacy, resulting in dwindling and unenthusiastic support for such organizations. In other project sites, as in the KEF, the village organization could be a truly functional one. However, the circumstances surrounding the introduction of the project to the community (the signing of the lease agreement in the KEF's case) precluded lengthy discussions with all concerned, leaving the upland residents little opportunity to opt individually for the project. This situation has spawned its own problems, particularly in the crucial area of decision-making.

Participation in Decision-making

In San Pedro, the CTF project staff was identified by 52.5% of the respondents as the usual convener of meetings, while barangay officials were mentioned by 27.5%. Only 45% attended more than half of the meetings called during the six months preceding the interview. Association officials in Buhi were cited as responsible for calling meetings (68%), while project staff members were mentioned by only 26.5%. During the six months prior to the interview, however, only 20% attended more than half of the meetings called. The data signify low rates of attendance in meetings, especially in a site where formal organizations were set up early on in project implementation.

Moreover, major decisions concerning the project were made exclusively by the project staff according to 72.5% of the respondents. Most participants (60%) did not perceive themselves as having any share in decision-making. Despite the somewhat higher rates of attendance in meetings in San Pedro, 85% of the sample saw the project staff as the prime decision-maker, and 55% denied any share in the process. However, 85% of the San Pedro and 90% of the Buhi respondents asserted the importance of having opportunities for making suggestions and voicing opinions about the project.

The Board of Trustees was identified by 87% as the main decision-maker in the KEF case, but resentment exists towards some of the Board's decisions, particularly regulations concerning obtaining permits from the agroforestry staff and the imposition of fines for violators. (Ikalahans, however, show no opposition to the requisite 10-meter fireline around swiddens; they also expressed support for the KEF's policy of encouraging gen-gen construction). Pursuant to the lease agreement, the Board banned the sale of certain forest products, such as orchids and lumber. This action was referred to by a number of residents as unfairly restrictive. Rumors thus spread that some residents wanted to secede from the reserved area. These problems may be attributed to the limited consultation between staff and participants at the project's inception and to the residents' distance from the local center of power. Board meetings were usually held in the central village of

Imugan, and residents of the outlying areas had little opportunity to attend. Exacerbating this was the failure of many trustees to communicate the issues discussed and the decisions made during the meetings to their respective constituents. Respondents felt, however, that consultation with the people prior to decision-making was of utmost importance.

Evidently, upland communities value participation in decision-making, since this gives them a sense of being part of the project. However, in the study sites, direct and effective channels for such participation have been highly wanting. Existing structures, such as uplanders' associations, were shown to be ineffective. In the minds of many, the project staff controls decision-making and is loath to share it. The physical inaccessibility of decision centers in some sites further alienates project participants, and resentments build up. Although the situation seldom leads to direct confrontation, many problems could be obviated if more decision-making opportunities were shared with participants.

The only area in which participants enjoy some leeway is in choice of crops to plant. Most respondents in San Pedro said that crop selection was made by both respondent and project staff. Most respondents in Buhi said they made the choice. Participants were usually constrained, however, by the project package itself. Only a quarter of the respondents in San Pedro would have chosen to plant the giant ipil-ipil the project nursery propagates. The majority would have planted fruit trees instead, or vegetables, root and cereal crops. The Buhi participants had similar preferences, and they succeeded in altering the contents of the project nursery. Initially, the project staff stocked various seedlings deemed environmentally suitable and acceptable to the participants. When the available seedlings were not utilized, the staff asked the cooperators for their preferences and changed the nursery's collection.

Participation in Project Benefits

Reports of receipt of benefits from projects were uneven. About 79% of the Ikalahan sample stated they obtained some benefits from the lease agreement, the percentage rising to 89.5% for benefits from the Foundation itself. Receipt of benefits in San Pedro was claimed by about 67%. In Buhi, only 46% of the respondents claimed to have been benefited. Those who had gained from the project mentioned benefits varying from site to site. Due to the lease agreement, land for cultivation was made available to the Ikalahans, as 73% testified. According to 78%, residents benefited from the medical care program from the KEF. Firewood supply was mentioned as a benefit by 63% in the CTF project; increased income was cited by 48%. Additional income was named by 91% in Buhi, and free seedlings by 73%. Despite these benefits, only a handful (around 15% to 16%) reported their expectations were fully or mostly met. On the contrary, as much as 50% to 55% reported that very little or none of their expectations were met.

Land Area and Security

The CTF was designed to allocate a minimum of two hectares per participant. In San Pedro, however, the actual mean area allocated measured only 280 square meters, decidely inadequate according to 69% of the sample. The decreased allocation was due to waiving of the qualification criteria for beneficiaries. Thinking that all 225 families that showed interest in the project were poor, the project staff decided not to exclude anyone. In accommodating all families, however, the land allotments were radically reduced. This need not have been the case since the survey data indicated that only about 36% of the respondents would have been eligible using the project's annual income criterion of P6,000. (It is not clear whether the income criterion refers to the participant's or his household's annual income). The acceptance of all who showed interest and the consequent reduction in plot size was made more problematic by the site's poor soil, which prevented the cultivation of .crops other than ipil-ipil. As a result, the projct staff decided to trade off greater numbers for smaller, uneconomical lot sizes. In the long term, the decision proved to be counterproductive as it led to loss of interest given the minimal potential benefits from the limited farm size.

Moreover, because CTF certificates could not be granted, the number of years that the land would be available was unclear, engendering insecurity of tenure. The majority (74%) felt that the land could easily be taken away from them, especially by the government. In Buhi, despite the different land access situation, at least one-third of the respondents felt the same way as the San Pedro respondents. The lease agreement in the KEF case had given a measure of land security to a substantial number of Ikalahans, with at least 20% saying that the acquisition of private title was no longer important. Nonetheless, unstructured interviews revealed a continuing sense of insecurity among some who questioned why there were restrictions on the use of the land if they really owned it. Many expressed the fear that success in reforesting the area would only invite the government to take over the land and eject them.

Income

The mean income derived by Buhi participants from the project was P881.71. In Pantabangan, the mean amount was P562.50. In contrast to these two projects which offered direct monetary rewards, participants in the San Pedro CTF reported a mean income from the project of only P26.86. The respondents' assessment of the projects' income effect was gauged by their perception of the proportionate share of income thus earned to total household income. Income from FAR participation in Pantabangan accounted for only 10.5% of average annual income. In both the Buhi and San Pedro sites, only 14% of the respondents considered income provided by the project as constituting more than half of total household income. Sixty-four percent in San Pedro and 42% in Buhi said that the additional income constituted only a very small proportion of total annual household income. A further consideration was equity. The participant's income in Buhi was thought to be one criterion for choosing subsidy recipients. The case study, however, revealed that the median annual household cash income of primary cooperators amounted to P1,024.50. This was significantly higher than the non-subsidized cooperators' median income of P284.00. Since the detailed income data showed that the difference could not be accounted for by the subsidy money, it may be argued that the already relatively well-off residents were those selected, albeit unintentionally, to benefit from the subsidy. Even if that were not so, the data indicated that the project heightened or perhaps created socioeconomic disparities at the village level. Thus, on adequacy considerations, the income contribution of projects was perceived to be minimal. On equity grounds, the projects may have had negative ramifications for upland communities in terms of inequitable distribution of benefits.

However, this is not to deny that projects can improve the income of upland residents. For instance, the provision of land for cultivation and other agricultural purposes in the Ikalahan area did improve living standards. The annual household cash income of the average Ikalahan respondent was P4,217.55, more than twice that of the average Buhi respondent. Of the Ikalahan mean income, 39% was derived from livestock and handicraft production. These activities were possible because of the availability of grazing grounds within the leased area and the cultivation of tiger grass in swidden patches for the village broom-making industry. Despite unequal exchange with the lowlands, the commercial linkages offered opportunities for augmenting household cash incomes.

Product Marketing

The minimal income contribution in some pilot areas was due to the inability to sell products from the project's farm lots. In San Pedro, 30% of the respondents were able to sell once and 2.5% more than once, but 67.5% were not able to sell even once. The situation in Buhi was worse. There, only about 5% were able to sell once or twice; 95%

did not have a chance at all. Part of the problem in San Pedro was the poaching and burning of tree crops. In Buhi, no tree crop was harvested due to limited wood lot development. In the CTF project, the staff did provide marketing assistance by identifying outlets and providing instructions on harvesting techniques.

Overall Assessment

The available data indicate that few respondents saw positive returns from their involvement in the projects. Only 17.5% in San Pedro and 19% in Buhi thought that project benefits outweighed the necessary effort. The rest said that the benefits either merely equaled or might not have even compensated for the required effort. Based on this intuitive analysis, few involved themselves actively in the projects. Further, few respondents said that the assistance provided was adequate (25% and 21% in San Pedro and Buhi, respectively). According to the majority of Buhi and San Pedro respondents, project benefits were not equitable.

PHYSICAL ACCOMPLISHMENTS OF THE
SOCIAL FORESTRY PROJECTS

Soil Rehabilitation

The mean reported swidden cycles were relatively short: 4 years of cultivation and 5.8 years of fallow for a cycle of about 10 years among Ikalahan respondents; 2.2 years of cultivation and 2.2 years of fallow, for an even shorter cycle of 4.4 years among Buhi respondents. (See table 3 for information on swidden cycle, crops grown, perceptions of obligations and privileges, and participation indicators.) Swidden site selection for Ikalahans was based on soil conditions and slope. Areas where the soil is black owing to humus were preferred, while steep mountainsides where the soil rolls off at the touch of a dibble stick were avoided. The Ikalahan swidden land was allowed to fallow when yams became smaller and as watery as turnips and when cogon grass began to grow. For Buhi upland agriculturists, the fallow stage set in when yields declined and the soil showed signs of deterioration, becoming rocky, dry, or sticky. The abandoned field was cultivated again when green grass, small trees, and other vegetation appeared. Apparently, conditions were better on Ikalahan farms, where the swidden cycle allowed for better soil revitalization. However, it is difficult to cite, the extent to which the projects influenced these swiddening practices. For the Ikalahan, the relatively free access to a sufficiently large forest area, coupled with sound traditional practices and local regulations, produced better results.

Table 3. Swidden cycle, crops grown, perceptions of project obligations/prohibitions/privileges and selected indicators of participation in project implementation in 3 social forestry project sites.

	Respondents		
	Ikalahan (N = 38)	San Pedro (N = 40)	Buhi (N = 40)
Mean years swidden in cultivation	4.0	No fallow periods	2.2
Mean years of fallow period	5.8		2.2
Crops planted/grown by respondent-households:			
Rice	28	—	—
Root crops/root crop and abaca	11	—	21
Root crops, fruit trees and others	17	5	—
Fruit trees only/fruit trees and tiger grass or rice	42	—	23
Fruit trees and forest trees	2	12	23
Fruit trees, forest trees & others	—	53[a]	20[a]
Forest trees, rootcrops & others	—	10	12
Forest trees only	—	19	—
Perception of project obligations			
Percent saying tree-planting	80	45	80
Percent saying forest fire prevention	63	—	—
Perception of project prohibitions			
Percent saying tree cutting	54	82.5	72.5
Percent saying improper burning	77	17.5	—

	Respondents		
	Ikalahan (N = 38)	San Pedro (N = 40)	Buhi (N = 40)
Perception of allowable things under the project			
Percent saying cultivation of food crops	97	10	—
Percent saying tree planting on land allotment	—	57.5	25
Percent saying harvest trees at maturity	—	25	—
Selected upland farming practices — Percent reporting:			
Fire control: fireline construction	95	15	57
Reforestation: tree planting at least before fallow stage	97	100	62
Soil conservation:			
composting	49	50	15
gen-gen	43	—	—
contour belting	32	—	—
bench terracing	—	—	59[b]
Assessment of vegetative cover on a 6-point scale-Mean scores:			
Past state	No data	1.4	2.8/3.4/3.0[c]
Present state	No date	5.2	2.5/3.0/3.0
Percent reduction in forest fires	80-90	No data	No data

[a]Others, in San Pedro, refer to rootcrops/vegetables, while in Buhi, to abaca.

[b]Based on agency reports; refers to all project participants as of 1980.

[c]The series to mean vegetative score for land parcels in flat rolling, and steep slopes, respectively.

Source: Aguilar, 1982.

For the Ikalahan swiddeners, *gen-gen* is an indigenous technology that was encouraged by the KEF. Usually constructed by women, the *gen-gen* maintains soil fertility by preventing the erosion of the top soil, maintains moisture within and between piles of burried stalks and leaves, and enriches the soil since it is basically a compost pile. Some 43% of the Ikalahan sample reported the practice in addition to 49% reporting regular composting. Moreover, 32% reported contour belting, utilizing tiger grass or fruit trees. Known as *balkah* ("belt"), the practice stabilizes slopes and controls erosion. Fifty percent of the San Pedro and 15% of the Buhi respondents reported composting to maintain soil fertility.

One of the Buhi project's major activities was bench terracing to control erosion. From the first group of participants, 20 farmers were expected to construct bench terraces over 30,000 square meters. However, only 14 of the farmers participated, and only 10 constructed the terraces over the assigned area. In all, the area terraced amounted to only 16,800 square meters, about 44% short of the target. For the succeeding group, 49 farmers were expected to do bench-terracing to cover 60,000 square meters. However, 55% participated and only 39% accomplished what was expected. Only 39,545 square meters were terraced, 34% short of the goal. For the succeeding groups, the shortfall in the target area for terracing was relatively smaller. But farmer participation declined to only 55%. In all, 59% of the target farmers constructed bench terraces over 63% of the expected area.

Reforestation and Vegetative Cover.

Polycultural production systems were preferred and practiced in most project sites. Respondent-households adopted cropping mixes of root corps, fruit trees, forest trees, and commercial cash crops. Such agroforestry practices provided for household subsistence and other needs and served as conservation measures well-suited to the uplands. The growing of tree crops may be credited to the social forestry projects in the different field sites. Tree planting was widely perceived as the residents' obligation to the projects, while the cutting of especially young trees was seen as a prohibition. Alternatively, in the San Pedro CTF, tree planting was seen as a privilege. Tree planting, at least before the fallow, was a near-universal practice of Ikalahan and San Pedro respondents; it was also reported by a majority from Buhi.

But reforestation targets in some sites were not met. Although records did not show the number expected to participate in the initial orchard development scheme in Buhi, they indicated that only one cooperator participated. Forty-nine farmers were expected to plant fruit trees in succeeding groups, but only 65% participated. Thus, of a targeted 450,00 square meters, only 112,972 square meters were planted, yielding a deficit of about 75%. As for Buhi's firewood lot development, none of the expected first group of farmers participated. Only 29% of the later groups planted ipil-ipil, resulting in a short fall of as much as 82% in the expected area. Overall, there was no improvement in the site's vegetation. On the contrary and according to respondents, a further reduction in tree volume has taken place, with even more parcels having very few or no trees at all.

Better results were observed at the San Pedro CTF site. There was practically no vegetative cover on the respondents' parcels before the project was introduced. At present, however, the project site teems with giant ipil-ipil. The area indeed has been transformed. Nonetheless, only 26 of the planned 45 ha were actually revegetated. In Pantabangan, a cumulative area of 1,139 ha were reportedly reforested, but its significance could not be assessed in the absence of a project target.

Control of Forest Fires

One of the most notable accomplishments of the KEF has been the reduction in forest fires. Although there are no adequate data on the hectarage burned from year to year, rough estimates indicate the magnitude of change. The local district forester approximated an 80% reduction in brush fires from the level prevalent before the lease agreement was instituted. An Ikalahan councilman and the resident pastor both cited a 90% reduction. Indiscriminate burning has been reduced, if not eliminated. The sample respondents cited fire control as an obligation of utmost importance, while fireline construction was considered vital and is nearly universally practiced. With wildfires kept to a minimum, natural revegetation has proceeded relatively unhampered.

The number of forest fires within CRC's concession area has declined since the start of the company's fire protection scheme in October 1977. In the 1976-77 fire season, 148 fires were recorded. Fires declined to 81 in 1978, went up to 86 in 1979, and was down to 34 in 1980. Nonetheless, a different trend was observed in hectarage burned. The total damaged area at the end of the 1977 fire season was 3,482 ha. In succeeding years, the recorded burnt hectarage was 2,173 in 1978, 10,126 in 1979, and 750 in 1980. The dramatic rise in 1979 coincided with the rise in reported cases of arson, which in earlier years had not figured prominently. Arson was the primary cause of fires during the project's implementation period. This phenomenon can be related to hostility towards the company. The fires declined in 1980 when the company's forest activities abated due to financial difficulties. Consequently, no causal relationship between the incentive plan and fire occurence could easily be established. In fact, the company increased its forest guards from 23 in 1981 to 27 in 1982. Thirty-four fire fighters and 46 protection laborers were also dispatched to strategic points within the concession area. CRC's plan to replace forest guards with the incentive scheme was apparently not workable.

To summarize, different levels of ecological success were attained by the projects studied. Fire control, soil rehabilitation, and improved upland agricultural practices were outstanding Ikalahan accomplishments, while giant ipil-ipil trees mushroomed in San Pedro. Large shortfalls were registered in Buhi's land development targets. A lower incidence of forest fires was recorded in Abra's highlands. Accounting for these are various intervening variables that are project and site-specific: 1) the greater measure of land security and availability and the stiff penalties for infractions of the locally-devised regulations in the Ikalahan area, 2) the settlers' desire to gain access to agricultural land in San Pedro, 3) the reduction in company activities which engendered less resistance in Abra, 4) the minimal substantive participation and the residents' subsistence level of living in Buhi.

SOME LESSONS FROM EXPERIENCE

The Use of Punitive Devices and the Role of Incentives

The case studies bring to the fore issues which need to be resolved in pursuing genuinely community-based strategies for upland development. One basic issue is the social forestry model (implicitly) adopted by project proponents. This new thrust is said to be a different approach to upland residents whose behavior has been viewed as intractable despite punitive legislation. Thus, the social forestry approach allows upland cultivators to remain within forest zones and utilize the available resources. As the case studies show, however, this accommodationist stance has restrictions, the government's policy being "managed occupancy." Most projects, particularly those of the BFD, stipulate regulations on resource access and use in a manner reminiscent of the former policy of exclusion. It appears that the old forestry code's penalties of incarceration and ejection have been merely translated into individual contracts that bind the upland resident to the implementing agency. The violation of the new rules merits cancellation of the permit and possibly ejection from the site. Thus, the projects' transformative process is one in which the "enemy" of the forest is forcibly converted, through legal means, to a "partner in forest conservation." From all indications, government imposition of such punitive devices usually does not generate the desired results.

This is not to say that forestry regulations with concomittant penalties are unnecassary to effect improved resource utilization and management practices in upland communities. Such regulations may be indispensable although not enough evidence is available to support this strategy. But if such regulations are to be adopted and enforced, they might best be evolved by the community residents themselves. Despite limitations, the KEF illustrates the possibilities inherent in a community-based set of norms and prohibitions, the enforcement of which is more seriously, and at the same time more sympathetically, pursued. The Abra Tinggians' town councils formulated forestry rules to preserve their forests and protect their watersheds. These local ordinances, which the Tinggians cited in initially rejecting CRC's forest protection scheme, may be credited with the preservation of as much as 46% of the concession area as virgin forest before commencement of the company's operations. Indeed, forestry regulations stemming from local village initiatives tend to produce the targeted results.

That upland residents nonetheless participate in projects which impose sanctions is attributable to the innovative use of material incentives and other rewards for project involvement. These reinforcement schemes, however, must be properly planned and monitored; otherwise, as the case studies show, the project itself is jeopardized when administration is faulty and people's expectations are not met. Indeed, the reinforcement schemes generally have not been optimally used by the projects. This is probably because the staff or the bureaucracy to which they belong did not comprehend, or could not respond to, the crucial role of incentives.

Land Security and Land Adequacy

Another primary issue in the uplands relates to land. The case studies reveal that upland cultivators believe that they are the rightful owners of the land they till, causing them to seek formalization of their rights through private titles. Tribal communities in particular resent being called "squatters" and would go to the extent of indiscriminate use of resources and protest burnings of forest to vent their resentment, especially in the face of displacement threats. In such situations, land insecurity can be counter-productive not only in terms of inflicting environmental damages, but in the uplander's reluctance to plant perennial crops for fear that once the area is reforested, there would be even stronger grounds for the government to reclaim the land as forests. The need for land drives some lowland migrants to risk whatever government action may later confront them. Many thus proceed to plant tree crops despite uncertain tenure. Nonetheless, the general trend is for upland cultivators to want to have a sure footing on the land. If unresolved, uncertain tenure leads to dampened interests in forestry activities.

Albeit indirectly, most projects deal with the land issue. The Ikalahan case illustrates that granting land security, especially by means other than private title, requires intensive consultation with the affected residents. The KEF case also presents the issue of whether individual rather than communal tenure may not be the most suitable form of land tenure for cultural minorities. Indeed, the lease agreement exhibits closer affinity to tribal concepts of communal ownership of property which allow for individual use-rights backed by cultural norms. Individual tenure may, however, be more appropriate among lowland migrants. Community initiative and consultation remain imperative, particularly in settling problems such as boundary disputes.

The other projects touch on the land issue through direct allocation of parcels to project participants. The process normally involved stipulation of land area, ranging from a minimum of 2 ha in the CTF, to 3 ha to 5 ha in Pantabangan, 7 ha in the FOM, and 15 ha in the KEF. The actual land allotment in San Pedro turned out to be an inadequate 200 square meters. While conditions differ from one project site to another, there is a need to establish a standard for the minimum economical size of an upland farm lot. The standard itself may have to be adjusted depending on the project site's aggregate area, biophysical conditions, the needs of the participating households, and the project's specific thrust or resource management strategy. Land distribution need not be guided by an apparently arbitrarily set minimum or maximum, but by the concern that the land can yield adequate economic returns for the workers. There also must be sufficient guarantees of security and the enjoyment of the fruits of the land.

Basic Needs and Target Participants

Although the impetus for project development has often been environmental, the projects studied commonly enumerate goals that balance ecological and social concerns. Some projects adopted strategies entailing highly labor-intensive land-development activities, which were viewed as eventually benefitting participants. In a few years, forest products could be gathered and sold, and restored soil fertility would raise yields and overall productivity. As the case studies show, however, some upland areas are more economically depressed than are others, with daily subsistence being a constant struggle for many. In such communities, residents are not inclined to participate in labor-intensive schemes characterized by long gestation periods. The more immediate concern is to satisfy the household's need for food. While a demonstration in technology was being conducted in one project site, a farmer remarked that one day's work making a contour ditch was one day lost in feeding his family. Certainly, poverty and malnutrition do not provide upland cultivators with enough energy to undertake intensive forestry activities and to still look after the daily needs of their families. It must be recognized, then, that in practice, social and environmental goals are not always compatible. Faced with the social condition in the uplands, a few projects have wisely diversified to include delivery of social services (e.g., health and education) and the promotion of income-generating activities such as livestock production.

Since upland conditions are variable, there is need to match a community's particular socioeconomic status and the specific resource strategy espoused by a particular project so that the most appropriate means can be employed for the desired ends. Some projects attempted to arrive at a such matching by setting eligibility criteria, in effect choosing participants that would fit a pre-determined resource management strategy. There are a number of reservations concerning this approach. For one, the income criteria of P2,000 per year used by the FAR project and the CTF's P6,000 appeared to have been arbitrarily established, there being no prior data used for planning. Such criteria are often vague and difficult to apply, resulting in frequent waivers of the stipulated qualification requirements. A more suitable approach would involve withholding the imposition of a predetermined resource strategy and the attendant eligibility criteria while evolving strategies with the upland community. The community response would be based on the residents' own priorities and felt needs. The residents could even set their own eligibility criteria, if necessary, and monitor their own compliance. Appropriate support services (such as seedlings and markets) could be extended as the project progressed.

Technology Transfer and Indigenous Technology

The core of most projects is the diffusion of new technologies to the uplands. This is undertaken through training seminars, demonstration farms, legal contracts, and extension services. Transferred are technological packages which, from the program planners' view, contain suitable answers to environmental problems. The usual implicit assumption is that the upland cultivator does not possess the technical know-how to exploit the resource system without harming it. On the contrary, the case studies show, that upland communites, especially cultural minorities, have practices geared towards safeguarding the land from overcropping, ensuring fire protection, and hastening soil rehabilitation and natural revegetation among others. Even among non-traditional hillside farmers with no agricultural practices developed over centuries, there is on-going self-generated experimentation on techniques to ensure their survival and make their efforts sustainable and profitable. In the process, they have arrived at their own strategies for controlling erosion and the like.

Projects that stem from community initiative, as the case studies indicate, are better able to tap this pool of local knowledge. Other projects, however, promoted technologies that are inappropriate to the needs of the "target" beneficiaries, who then view these as requiring much effort for little gain. People may not be receptive to the new technology, which is readily attributed to indolence or lack of education. Proponent agencies need to realize that the upland cultivator follows a rational, intuitive framework for judging the extent of project involvement. The participation of upland communities in projecting the different aspects of development is indispensible for uncovering existing practices that can be adopted or improved and for arriving at more suitable technological mixes for implementation.

Popular Participation and Community Forestry Associations

Discussion has highlighted the need for greater popular participation in upland development projects if these are to address the community's pressing needs while being better managed and implemented. Upland residents value participation, especially in the decisions that affect their lives and well being. The case studies reveal that decision-making is usually dominated by project personnel. However, this may not be deliberate on the part of the staff. In fact, most proponent agencies recognize the value of popular participation, as indicated in project plans that call for the organization of beneficiaries into community forestry associations. The reality, however, is that such organizations are generally ineffective and often do not function as intended.

Despite awareness that project success is contingent upon the active role of the intended beneficiaries, most proponents face the problem of how to actually mobilize the community to meaningful levels of participation. Most projects have groped for the means by which participatory approaches can be implemented. The possible answer lies in the experience of private voluntary organizations whose proven community organizing techniques in urban and rural lowland areas can be pilot tested in upland communities.

The Project Staff and the Forestry Bureaucracy

Because of the difficulties associated with the participatory approach, there is an urgent need for committed and trained field staff. As the case studies show, personnel are usually recruited from outside the project locality. As outsiders, they consequently need to establish credibility with the project beneficiaries. Most projects, especially government, fail in this regard. Periodic meetings with the participants are erratic and do not allow for a meaningful exchange between people and staff. Some projects attempt to overcome this problem by hiring residents in the project or nearby areas, a strategy the case studies show as having potential for making the project more acceptable to the community and for developing local skills for project management.

However, the quality of the field staff is often a reflection of the larger bureaucracy to which they belong. On the whole, the actions of the government's forestry bureaucracy reveal an indecisiveness concerning social forestry. While it is willing to extend all possible support services for certain projects, less resolve is shown for others. The Bureau's accomplishments in one place are negated by contradictory actions in others, which thus do not augur well for an agency whose credibility is already wanting. What is essential, therefore, is for the entire forestry bureaucracy to muster the policial will necessary to pursue a social forestry program beyond pure rhetoric.

REFERENCES CITED

Aguilar, Filomeno V. Jr. 1982. Social Forestry for Upland Development: Lessons from Four Case Studies. Quezon City: Institute of Philippine Culture, Ateneo de Manila University.

Bernales, Benjamin C. and Angelito P. De la Vega, 1982. Case Studies of Social Forestry Projects (4 Volumes). Manila: Integrated Research Center, De La Salle University.

8
UPLAND DEVELOPMENT IN CALMINOE: THE ROLES OF RESOURCE USE, SOCIAL SYSTEM, AND NATIONAL POLICY

S. Fujisaka and A. Doris Capistrano

Calminoe is an upland swidden (*kaingin*) agricultural community located in an area previously forested, then logged, and now settled by migrants from all over the Philippines. Interdisciplinary field research was conducted in order to understand local social, economic and ecosystem dynamics and then to design a low-cost, locally appropriate, and participatory development project. Analysis of the local situation, however necessary, turned out to be insufficient. The major impediments to local upland development included the ambiguity of national policies on upland resources, confusion of national institutional responsibilities and of land jurisdictions, and the current national economic situation.

Ideally, development projects in upland *kainginero* communities should progress from community appraisal, to production and environment problem assessment, to design of technical strategies for improving incomes and protecting the environment, and, finally, to implementation, with the people's participation, of those input strategies. The situation in Calminoe was not, however, ideal. An interesting combination of local factors — land, wood, tomatoes, charcoal, factions, and community conflict — combined with the aforementioned national factors have complicated local upland development. We believe that this situation is far from unique and should, therefore, have useful implications for other development projects. National policy confusion alone is certainly a relevant problem in all upland development or "social forestry" projects in the Philippines.

We discuss the local ecosystem and resources, resource utilization and management, social organization, and policy and institutional confusion concerning these resources. We examine the interactions of ecosystem, social system and confused policy. Because our aim was to improve local resource management, the paper concludes with suggestions for this improvement and for avoiding possible stumbling blocks.

THE RESEARCH CONTEXT

Our initial goals were to study local patterns of migration and settlement, examine local social organization and analyze extant systems of cooperation and conflict, examine strategies of resource management and analyze the economics of resource utilization, and study the local ecosystem itself and the effects on it of human activities associated with it. We then tried to develop limited cost, locally appropriate, and participatory "social forestry" or upland development strategies. We have gone a step further in providing some cautionary and hopefully instructive warnings — based on our findings — for the continued implementation of such projects in the Philippines.

"Social" or "community" forestry has been described as "forestry for community development" (Alvarez, 1982). It has developed in India and Nepal somewhat differently than in Thailand, Indonesia, and the Philippines (Fujisaka, 1984). Community forestry in Nepal and India generally involves long-settled, relatively well organized communities (e.g., traditional *panchayats*) which depend on their forests primarily for fuel and fodder (Manandhar, 1982). In the Philippines, social forestry projects involve swidden agricultural migrants to the upland forest areas. Governments in both regions have begun to recognize the problems of deforestation and watershed destruction and to grant responsibility for the management of local forest resources to the users in order to minimize the undesirable effects of common ownership. One result of this has been the development of social forestry programs (FAO, 1977a, b; Allen and Barnes, 1982; Brokenshaw, 1984). Programs in Nepal and India contributed to the early popularity of the approach (e.g., HMG/UNDP/FAO, 1982). Over the last several years the Bureau of Forest Development and other government and non-government agencies have experimented with "social forestry" in the Philippines (Magno, 1981; Aguilar, 1982; Bernales and De La Vega, 1982a, b, c, d, and e). Encouraged by the apparent successes of initial efforts, the BFD combined its approaches into the Integrated Social Forestry Program (Cortes, 1982. BFD, 1983).

Specific aspects and problems encountered in the implementation of social forestry projects led to (or were supported by) further research. Interdisciplinary preliminary field research has been found essential to appropriate project design (Harwood, 1979; CIMMYT Economics Program, 1980; Carruthers and Chambers, 1981; Collinson, 1981; Hildebrand, 1981; Working Group in Agroforestry, 1982a; Fujisaka and Capistrano, 1984a, b). The tropical forest ecosystem itself has posed many problems not encountered in temperate climates, and continued research on ecosystem dynamics has been found essential to improving management strategies (Myers, 1980; Gwyer, 1979; Sajise, 1981, 1982). Land tenure for forest lands has proven complicated, involving indigenous or native users in some cases, migrants or "squatters" in others, and general government control over "public" lands (Adeyoju, 1976; Capistrano and Fujisaka, 1984; Cernea, 1981; Lynch, 1983). The lack of knowledge about sociocultural factors (rather than sociocultural "peculiarities" or traditional "resistance to change") impedes upland development efforts (Conklin, 1967; Schmidt, 1973; Campbell, Shrestha, and Stone, 1979; Schlegel, 1979; Noronha, 1980; Olofson, ed. 1981; Raintree, 1982; Olofson, 1983). Similarly, understanding local social organization has been found to facilitate project participation and technology adoption (Grandstaff, 1980; Gow and Morss, 1981; Raintree, 1983). The upland development technologies themselves need further development and continuous assessment for local appropriateness and economic viability (Laarman, Virtanen, and Jurvelius, 1981; Jacalne, 1982; Capistrano and Fujisaka, 1984).

After the initial enthusiasm for the "new" idea — manifested in projects giving some land tenure security to community residents in return for their participation in tree farming, non-shifting agroforestry, or reforestation — came evaluation and constructive skepticism (Alvares, 1982; Fontanilla, 1982; Ganapin, 1982; Fujisaka, 1983; Mahati Team, 1983). Several authors (De Los Angeles, 1982; Hyman, 1983; and Nelson, 1983, 1984) have recently examined the broader, non-local aspects of forest resource utilization, and have suggested that social forestry projects might result in further migration to the uplands, and thus be self-defeating. Our case study has confirmed this possibility.

ECOSYSTEM AND RESOURCES

Calminoe is located between the lower slopes of Mt. Banahao and the southwestern flank of the Sierra Madre mountains, near the Laguna-Quezon provincial border. The area immediately surrounding the community was, until the early 1940s, primary tropical dipterocarp forest, but was logged continuously from the 1950s into the 70s. With the clearing of most of the forest lands, they became available for cultivation. Today, the area includes human settlement, small and isolated residual timber stands, secondary forest regrowth, *kaingin* plots at various stages, ranging from the newly opened and productive to the depleted and fallowed, and more degraded cogon *(imperata)* and brush lands. The rolling to steep lands (elevation about 400 m) are part of the Caliraya Watershed, which currently provides water for

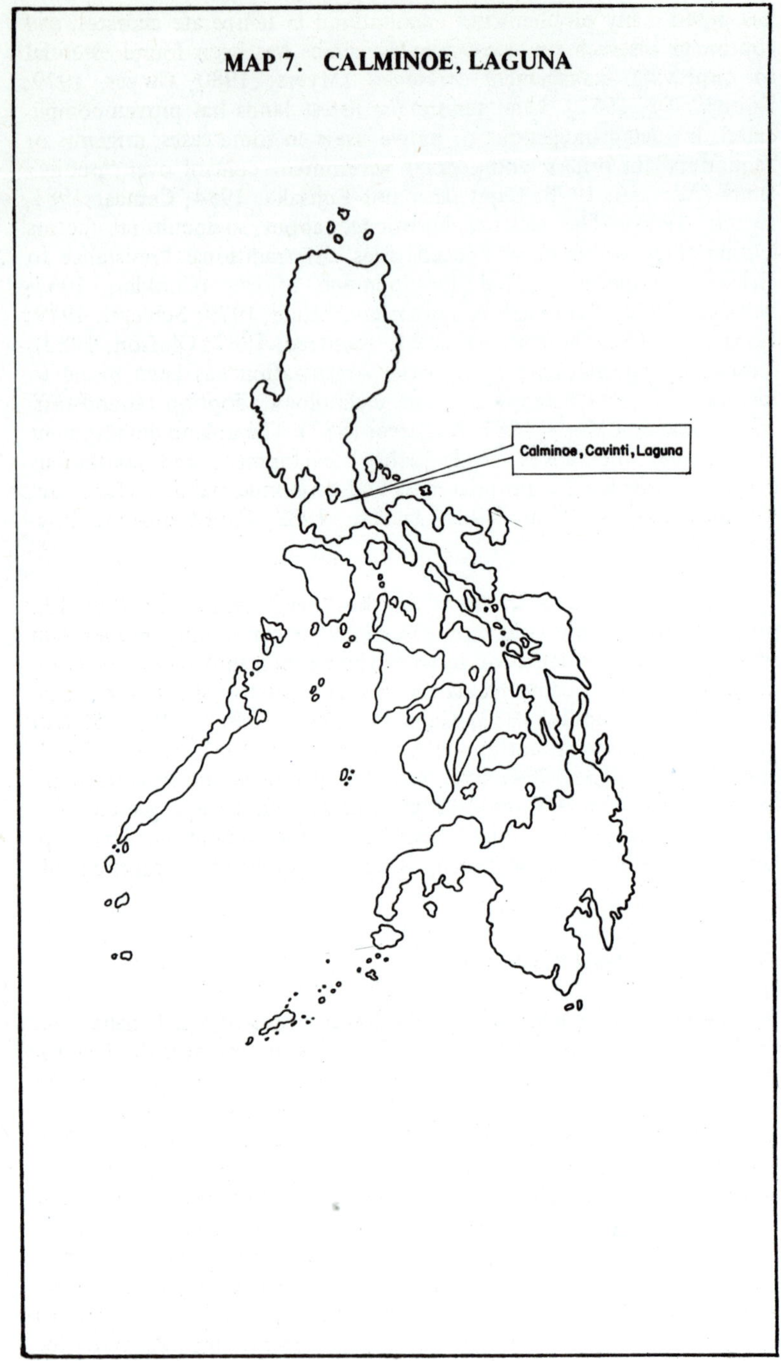

MAP 7.   CALMINOE, LAGUNA

hydro electric power, irrigation, Laguna de Bay, and for such rivers as the Pagsanjan (featured in the movie "Apocalypse Now"). In the last four to five years and since the termination of legal logging, migrants from all over the Philippines have settled in the area. These settlers farm and use the remaining forest products for building materials, fuel, and for sources of cash income. They are steadily increasing in number and, of course, are utilizing more and more resources.

The area's climate lacks distinct rainy and dry seasons (Type IV climate according to Corona's classification), and has a total annual rainfall of from 2500 mm to 3600 mm. The relatively high rainfall is due to moisture from the Pacific ocean rising upslope and condensing and falling with increasing altitude and decreasing temperature. Area soils are of volcanic origin; they have weathered rapidly, with latisol and ultisol formation due to the combination of high temperatures and heavy rainfall. Such soils are characterized by low silica and base content (because of heavy rainfall), by high iron and aluminum oxide contents, and by 1:1 lattice clays. They are red or pinkish in color and are low in organic content because of nutrient leaching from heavy rainfall and high temperatures. The heavy and well distributed rainfall, however, permits the growth of abundant vegetation. Imperata grasslands are eventually taken over by softwood tree species of the fig family (Moraceae). Currently, a few small patches of such young secondary forest growth have spread out from the wetter gully areas into the surrounding grasslands.

The tropical evergreen rainforest was commercially logged until 1979. The land was left bare except in scattered patches where trees were either too young, damaged, or of unwanted varieties. Also left scattered about were large quantities of waste slash and the stumps of the cut trees. Important by-products of logging were a road leading to the area and the timber workers who saw possibilities in the land for settlement. Some of these workers brought their families, mostly from nearby Quezon and Laguna, and settled on lands that had been a part of the (Interwood) concession. At about the same time, the national government established a resettlement community, Caldong, not far from Calminoe for *kaingin* farmers from Mt. Makiling. The access road was somewhat improved, and publicity increased. Other settlers, especially Seventh Day Adventists, followed not only from Laguna and Quezon, but from Bicol, the Visayas, and Mindanao. The settlers built their homes with wood from the forest, planted home gardens — often of considerable diversity, farmed the surrounding hilly lands — and have been exploiting the remaining forest resources.

Land Use: Home Gardens

Unsurprisingly, most of the settlers come primarily for land — for home and home-garden lots, and for farm plots. Home gardens are cultivated next to the house on the same piece of land. Such plots range from approximately 15 x 20 m root crop gardens up to half-hectare, very diversified, multi-storied, and intercropped plantations. Most families initially interplant the subsistence staples of cassava, camote, and gabi, which they harvest as needed. Many of the residents have diversified and intensified their home garden cropping over time. Indeed, a few of the gardens now resemble a sort of agronomic botanic garden, with (depending upon time of year) interplanted root crops (cassava, camote, gabi, ube), vegetables (bush beans, pole beans, soy beans, mung beans, squash, eggplant, garlic, ginger, tomato, peppers of various types, onions, others), tree crops (coffee, cacao, papaya, citrus of various types, lanzones, jackfruit, mango, avocado, rambutan, and others), coconut, sugarcane, rice, corn, and the current cash crops of tomato, pineapple, and banana. Local farmers experiment not only with species of cultigens, but also with different varieties, fertilizer applications and schedules, and various cultivation techniques. Many of the local farmers are quite innovative in their farming practices (and are certainly not "tradition bound" in this respect). Some of them rely on their farm lots for both subsistence crops and possibly cash incomes, while others with smaller, cultivated farm plots rely more on home gardens for subsistence and limited cash incomes. The home gardens are permanently cropped, rainfed, intensively cultivated, and, in their current range of forms, appear well adapted to local conditions. The home gardens are very important to future development in that they represent the type of farming possible throughout the area after the surrounding lands are depleted for *kaingin* tomato production.

Land Use: Kaingin Farming and Commercial Tomato Production

Almost all of the settlers have (or are in the process of obtaining) *kaingin* plots located in the wide area surrounding the community. These plots, of which a farmer family may have up to four, range from about a half hectare to three hectares (although a farm lot is locally defined as three hectares). Plots are either cut from the secondary or residual patches of forest (better soils but more labor) or reopened on previously cropped and/or fallowed areas (poorer lands, less labor).

The *kaingineros* generally clear and prepare land in January and February. For new plots, underbrush and smaller trees are cut, and many of the medium and larger trees (if there are any) are felled. Almost all work is done with bolo and ax. Slash is left to dry, and wood of suitable size and quality cut into beams and removed. The slash is burned at the end of the dry season, mostly during April. Because the dry season is short, burning is usually slow and incomplete with smoky, smouldering fires, rather than fast, robust blazes. Fire breaks are not needed and are not constructed. After a first burning, the wood is gathered and re-fired. Larger logs not used for timber, are left in the fields to rot, although more recently farmers are converting more of the suitable wood slash, both from *kaingin* plots and from logging, into charcoal. Fallowed or previously cropped plots are "undergrassed", that is, cleared of brush and grass and burned prior to planting.

Farmers plant from March through May, using zero to little tillage with digging or dibble sticks, bolos, and hoes. Most of the newly opened plots are planted for commercial tomato production. Tomatoes can usually be successfully planted on a given plot for two years. Depending on condition, older farm plots are either fallowed or planted to subsistence root crops combined with more permanent combinations, such as pineapple and banana (both local cash crops), coconut, coffee, and fruit trees. All of these require fewer inputs and less care than tomatoes. The older farm plots resemble the home gardens in cultigen types and diversity.

Transplanting, practiced by some of the farmers, of tomato seedlings starts towards the end of June. Chicken manure, 40-0-0 (urea), and 14-14-14 (complete) and pesticides (Decis, Thiodan, Folidol) are applied to the plantations according to individual farmer schedules and their respective abilities to obtain inputs they feel are needed. Weeding takes place as necessary, and larger growers might hire local labor. Tomato harvest takes place between July and August. The farmers sell the produce either to consumers in nearby towns or to local or outside traders.

Pineapple is the other major local cash crop. Once established, pineapple requires periodic weeding and little else although it brings less income than tomato. The fruit is harvested, and suitable tops planted in June and July. Bananas are harvested in July and August. Root crops require little labor and are harvested continually as needed. Vegetables, usually grown on the home garden plots, are also cultivated throughout the year on farm plots. Several pole bean farm plots were observed for the first time in 1984. In May and June, 1984, several farmers planted rice for the first time on rainfed plots: yields were low, and pests and weeds were problems. Harvest times of other local crops vary, and most of the local coffee and fruit trees are not yet bearing fruit. None of the coconut trees are of bearing age.

All agriculture is rainfed, but a few very small *palayans* (rice paddys) have been constructed along existing stream beds. The settlers have little livestock, perhaps a few chickens, fewer pigs, and some dogs (occasionally eaten). A local river and the nearby small lakes supply a few fish.

Because of hilly terrain, irregular shape, stumps, the ever-present logging and *kaingin* slash and lack of traction animals, farm plots are not plowed. Contour planting or other forms of erosion control is essentially nonexistent. Little composing or use of local organic materials for soil enrichment is done. If they can afford it, most settlers use commercial fertilizers.

Carabao Logging

Small patches of fairly mature secondary growth and residuals remained scattered throughout the area after legal logging terminated in 1979. Many of the Calminoe settlers and some outsiders have been exploiting (commercially and privately) the area's remaining wood resources, an activity directly related to (increased) cutting of new *kaingin* plots for tomato production.

In carabao logging, large trees of commercial value are cut either during the clearing of new *kaingin* plots or specifically as timber. Employees of outside buyers, local middlemen, or occasionally the local farmers first select and fell the tree, and then cut the main trunk into 100 board feet beams (14" x 12" x 15'). Most of these beams are then dragged over mud grooves (paths) by carabao to a local assembly point where they are cut into (2" x 12" x 15') boards by crews of chainsaw operators. The boards are loaded into trucks and hauled either to buyers or directly to builders in Quezon, Laguna, and Metro Manila. Some of the beams are used in local building; these are sawn into planks with two-person manual cross-cut saws. Most of carabao logging's benefits accrue to the better capitalized outsiders. The locals have received much less through wage labor — for felling, beaming, beam hauling with carabao, "milling" with chainsaw, and loading of lumber into trucks, and as suppliers for the much smaller local wood demand.

Carabao loggers working for outsiders often select and extract trees without the permission of local (non-legal, de facto) authorities or landowners. Local loggers may take logs from unclaimed areas from their own lands, or from — usually with permission — the lands of other residents. Some of the carabao logging is strictly local: workers are hired by either the wood contractors or middlemen, and some of the middlemen are local residents.

Evolving Resource Use: Charcoal Production

There has been a recent and widespread move in the community to produce charcoal from the remaining wood slash left over from *kaingin* plot cutting and from logging. Production has followed demand. Demand for charcoal greatly increased because of recent, steep fuel price increases and the decreasing availability of commercial timber. The rapidly increasing charcoal price and stable lumber prices, the simple technology required to produce charcoal, and the benefits that accrue relatively equally to all participants have helped push the move. Charcoal is being produced in small- to medium-sized pits lined with grass and covered with clay. In the last two or three months, as many as up to 1000 sacks per week have been shipped out of the area.

A few other forest resources, including rattan and pandan, are used by the local residents. Rattan is gathered, stripped, cut into thin strips, bundled, and hauled to outside buyers. Pandan is similarly processed, but is locally woven into mats, baskets, and hats for local use or for sale nearby.

THE ECOSYSTEM: DEPLETION

The resource utilization patterns described above are probably not sustainable. Tomato production has become widespread in the surrounding area within the last few years and is rapidly increasing. Outside buyers are coming to Calminoe and nearby communities, and local middlemen are extending credit for production inputs. New *kaingins* are being cut at an accelerated rate. A few non-settler outsiders have opened new tomato plots for use for only a season or two. These farmers clear, plant, and harvest, and leave the area with their profits. Tomato farming succeeds economically in the short run, but it requires high natural soil fertility, which can only be supplied by newly cleared lands. New lands are disappearing, and, unfortunately, practices (on the farm plots) of the *kaingineros* have not been soil-maintaining. Most of the tomato farmers rely on purchased inputs — chicken manure, fertilizers, and chemical sprays. They do not control erosion nor use green manures or composts. Farm plot soil nutrient losses have been significant. Tomatoes, profitable for the short term, are hastening local resource depletion.

Nutrient losses from cropping were documented for the area. Soils from a newly cleared (from secondary forest) plot and from a plot that had been cropped for four years were sampled and qualitatively tested. Samples were taken from upper and lower slopes of each field, from below the surface (0 cm to 2 cm depth), and from the subsurface (10 cm to 12 cm depth). The cropped soils are more acidic than the "new soils, and are lower in N and P. Since the new soils show medium levels of N, nitrogen leaching is assumed for the cropped areas. Phosphorus, on the other hand, is easily fixed chemically into the soil particles. Phosphorus levels declined with soil depth on the upslope and increased with soil depth downslope in the newly opened area. The opposite was observed for the four-year cropped area. This might be explained by leaching, which concentrated phosphorus on the subsurface upslope, and by soil erosion, which deposited phosphorus on the surface downslope. Crop extraction might account for the decline in phosphorus level in the subsurface soil downslope (Samson, 1984).

Carabao logging is also depleting resources and degrading the environment. Outside capital using local labor means much more wood is extracted than could be by the settlers alone. Most is cut for and financed by outsiders, and most of the profits leave the area. While such logging is less destructive than mechanized logging, it too scars and erodes the landscape and lessens the effectiveness of the watershed.

Charcoal making, as currently practiced, results in little ecosystem damage. It requires wood waste, labor, and little more than a shovel and bags for hauling it. However, the slash and wood waste that seemed abundant in 1983 has already diminished visibly. If the settlers start to cut trees for charcoal production, very serious denudation, erosion, and watershed destruction could result.

SOCIAL ORGANIZATION AND RESOURCE USE

The local social system consists of competing factions in conflict over resources. Because of diminishing availability of resources and the diversity of the community, mistrust, poor communication, and conflict between groups characterize the local social system. As of September 1984, the conflicts extended to the national level where institutional and policy confusion contributed to the problem.

Since settlement by the first timber worker families in the late 1970s, in-migration rates have steadily increased. The early settlers were mostly independent Catholics, many of whom remain nonaligned with local organizations or factions. Seventh Day Adventists began to settle in the community in 1980 and now represent about half of the population. The settlement of Adventists was accelerated by their national communications network, which advertised to Adventists that Calminoe was ideal for them. The Adventists are so well organized. they can assist some of their members in pioneering a new community.

Competition among interest groups developed in the community as more settlers arrived. The competing (often overlapping) groups reflect splits between Adventists and Catholics, between Bicolano and Visayan-Mindanao Adventists, between the main Adventist group and a smaller Adventist group now allied with some of the local Catholics, between different language speakers, between different political groups, and, recently, between local settler-residents and a group planning to move into the area. The major factions are: 1) early Catholic settlers, 2) Adventists, mostly from Bicol, 3) Adventists, mostly from Visayas and Mindanao, 4) Adventists and Catholics combined, and 5) outsiders planning to settle in the area and a few local residents. Political and language affiliations further divide these groups. Each of the major factions has created its own leadership and sought outside recognition. For example, the larger Adventist group has sought the community's recognition as a *barangay* from the Sampaloc, Quezon mayor's office and has tried to discredit the Catholic-Adventist alliance. That alliance, meanwhile, sought recognition as *barangay* officials from the Cavinti, Laguna, mayor's office. Each group has sought a legal grant to their organization for the community's sorrounding lands and applied to different national institutions. In the meantime, each group "grants" land to its respective newcomer settlers.

Among the issues contested are: 1) the jurisdiction over the land — Bureau of Forest Development (BFD) or Bureau of Lands (BL), 2) the location of the community — whether in Quezon or Laguna, 3) the corresponding municipal and provincial jurisdictions involved, 4) the legitimacy and land granting rights of each of the local power structures, 5) both farm and home lot boundary disputes, 6) and rights over timber. That is, local group leaders reflect national confusion in having to guess which national agency has jurisdiction over the area. Local leaders are forced to debate issues that can only be resolved at the national level. Factions are a symptom of increasing competition for resources, and conflicts are rapidly worsening with attempts by faction leaders to enlist local residents and to ally factions with nearby municipal political machines. Leaders try to bring legal action against rivals by pleading their cases with different municipal, regional, and national authorities. For example, the leader of the main Adventist group prevailed upon the mayor of Sampaloc to briefly incarcerate — for impersonating an official — the leader recognized by the mayor of Cavinti. To increase their power, major functions are recruiting loyalist settlers for in-migration.

So far, violence has been avoided, and most conflicts are being carried out through a war of words, gossip, claims, and counterclaims. Although some "strong-arm" tactics and threats have been reported, the factional leaders seem to emphasize recruiting locals and petitioning of outside authorities. The local leader typically promises his followers that he will be able to procure their exclusive rights over the land, and he collects funds for doing so ("following up papers at Malacanang"). He takes advantage of a real or supposed information advantage over the local farmers. Amidst such official confusion, even supposed knowledge is sought and paid for.

NATIONAL POLICY, THE INTEGRATED SOCIAL FORESTRY PROGRAM, AND UPLAND DEVELOPMENT

National policy ambiguities concerning public lands and resources are a problem not only for Calminoe but for other upland communities. as well. Local uplanders ask:

1. Who has legal and administrative jurisdiction over the land, the Bureau of Forest Development or the Bureau of Lands? Recently, the leader of the community's major Adventist organization was reportedly granted 1000 ha — i.e., the whole area currently used by the settlers — by the BL. The Regional BFD office said it is fairly certain that the land falls under BFD jurisdiction, and they have seen no evidence supporting the Adventist's claim. The regional BFD office did not have an up to date map showing jurisdictional boundaries and did not know the BL's official position on these lands. In effect, no one knows who has legal jurisdiction over the Calminoe area, and the unfortunate local settlers have become polarized according to their bets on the eventual outcome. If the land is granted by the BL, it would be privately titled and could be legally bought and sold. If it is under the BFD's jurisdiction, the land could be leased to settlers for use in a social forestry project.

2. Is Calminoe in the province of Quezon or Laguna and which municipal authority has jurisdiction over the community — Sampaloc, Quezon, or Cavinti, Laguna? The Adventist leader reported to have the BL grant of 1000 ha is also claiming that the area is a part of the province of Quezon and the municipality of Sampaloc. If under BFD jurisdiction, the area would probably be administered through a BFD social forestry project, and the provincial (and municipal) jurisdiction would be largely irrelevant. But if granted by the BL, the land could then have Alienable and Disposable status, and the provincial jurisdiction would be important because of local taxes and regulations, and because the community would become incorporated in a nearby politico-administrative unit. The provincial boundary was established in the 1920s and changed in the 1970s, but up-to-date boundary maps are not available.

3. Who, if anyone, has rights over the remaining timber? The loggers tell the locals that they (the loggers) have timber rights and that the settlers are illegal squatters. If the Calminoe area is under BFD jurisdiction, then both local logging and settlement are illegal. But the BFD is powerless in the face of logging's potential profits. In Calminoe, the individual farming family generally does not know whether or not the loggers have a legitimate claim to local timber. This ignorance has worked to the farmers' disadvantage because outside loggers have exploited it. They have taken logs and, in the process, damaged *kaingin* plots and the watershed itself. The logging continues with both local and outside participation. Timber is treated as an open access resource, and as long as most residents do not know who is acting legally or illegally, they have fewer incentives for improved resource management.

4. What is the Integrated Social Forestry Program of the BFD and what is its intent? The BFD has recently recognized the futility of trying to evict upland settlers and is facing the management problems of **de facto** common resource ownership. Hence, the Integrated Social Forestry Program (ISF). Individuals within project communities or entire communities are granted renewable 25-year stewardship contracts (similar to a lease) to maintain the land through agroforestry and *non-shifting* cultivation (or face loss of the land lease). The ISF seeks sustainable resource use through limiting *kaingin* shifting and further in-migration, in exchange for reasonable land tenure security. Project participants are required to have settled in the uplands prior to December 1981. The program is not supposed to engaged further migration to the uplands.

However, the leaders of a relatively new group in the area have recently claimed BFD sanction for a new IFS project covering 300 ha of relatively *unsettled* lands slightly overlapping and adjacent to the lands farmed by the Calminoe settlers. These leaders are certain that they hold a legal grant. Therefore, they are surveying the land, recruiting members mostly from outside the area for settlement in the project, and collecting fees for both surveying and for "following up papers at Malacanang". The BFD says that the group received sanction only for feasibility investigation from Minister Pena's office (Ministry of Natural Resources, the parent ministry of the BFD). It seems fully possible that this project will be carried out and that the ISF program will, by lack of action or protest, have *encouraged* upland settlement. In fact, the group's president has stated, "We should take advantage of the ISF program since it is a program for the benefit of all Filipinos, especially those who would like to come to the uplands to make their home."

CONCLUSIONS

Any government upland development or social forestry project introduced in Calminoe will face many obstacles. Current resource utilization combined with population increases have already damaged the ecosystem. Ambiguities in national jurisdiction over the area lands and confusion over the policies such as those of the BFD's Integrated Social Forestry Program now threaten to render resource management efforts futile.

The very resources important to maintaining the ecosystem are being used and not renewed. The forest cover is being expoited for land and timber, and, thereby, the Caliraya watershed damaged. The farm plot lands, including some that are relatively steep, are being cultivated without protection against soil erosion. Already portions of the area consist of unproductive cogon (imperata) grasslands.

Incomes supporting in-migrating settlers until now may dwindle with the resources as the ecosystem changes. Some initial outside support to the more organized Adventist settlers, returns from tomato production on previously uncultivated lands, incomes from carabao logging, and, now, returns from charcoal-making possibly allowed settlement of more families than can be supported in the future after such resources are depleted. Many of the migrants were informed by earlier settlers that the area afforded ample income, which may have been correct as long as such resources were available.

Effects of local over-population have been worsened by the community factionalization. The existence of factions has increased rates of resource depletion and ecosystem degradation because settlers do not attempt to improve local resource management. On the contrary, they compete for as much as possible as fast as possible. And faction leaders have brought in more settlers to strengthen their respective groups through greater numbers.

Problems of lack of enforcement and ambiguity of national policies on resource use must be resolved to curtail local ecosystemic degradation. Extensive land clearing for tomato cropping and illegal carabao logging have continued, in part due to the settlers' lack of rights over such local resources and in part due to ignorance regarding who does have those rights. Confusion over the jurisdiction over the land (both in terms of national agencies and provincial boundaries) must be eliminated; it fuels speculation and factional conflicts, further polarizes the factions, and sustains a situation rendering future intra-community cooperation more difficult. The intent of upland development programs must be clarified, that is, they are not intended to encourage further in-migration. Problems would be aggravated if they inadvertently encourage further in-migration. The relatively new group planning to move to the Calminoe area under the aegis of the Integrated Social Forestry Program of the BFD is a discouraging example.

On the Positive Side

Despite the many obstacles to be overcome, there is significant potential for healthy upland development. The people, far from being tradition bound and tied to inappropriate technologies, are agriculturally innovative. While the current situation seems to dictate the relatively uncontrolled exploitation of the commons, the local evolution of the home gardens illustrates that local residents may be developing technologies for future use in a poorer environment. Home gardens provide basic subsistence (root) crops, are highly diverse in cultigens and now include more and more permanent crops (many of which would be encouraged by an agroforestry project), host farmer experiments and trials, and are essentially permanently cropped even though they provide much less income than does tomato production of fresh lands. Future researchers and development workers in Calminoe may witness a race between ecosystem destruction and expansion of the home garden agricultural practices. The nature of the population that could be supported under such a system is not known, and out-migration from the area may take place before a new dynamic equilibrium is reached. And, of course, if the residents attempted such a short-term total exploitation as to make charcoal out of all the remaining vegetative cover, the ecosystem would be destroyed.

The factions themselves provide another possible source of hope. While inter-group cooperation is essentially non-existent, each group is becoming more organized. Organizational skills may become an asset for dealing with outsiders. In this, the perhaps 25% of the local residents who are not aligned may be at a distinct disadvantage.

Development Implications

The implications for introduced upland development in Calminoe and similar communities are straightforward. (By "introduced" we refer to the type of project that might be implemented if the area turns out to be under BFD jurisdiction. If the land turns out to be under the BL and available for conversion to alienable and disposable status, the natural evolutionary development described above would be hoped for, and could be aided by outside agencies.) First, as the ISF program recognizes in its granting of 25 years "stewardship contracts", local settlers need to be responsible for the local resources. Whether the contracts are awarded individually or communally should be based in

Calminoe on the rights of individual non-aligned families or of families aligned with the weaker local organizations. These might be better protected through individual leases. If an ISF project were implemented, the allocation of lands through either arrangement would be problematical. A fair distribution system should respect current ownership claims, but counter those that are blatantly speculative or power-seeking.

Upland development in Calminoe can be facilitated by some of what the settlers are doing, for example, developing their own "agroforestry" technologies (new annuals and perennials and combinations of cash and subsistence crops). But we were discouraged to find a strong orientation toward commercial inputs (especially for tomato production) and a lack of soil erosion control measures. Easy-to-learn-and-implement erosion control measures should be introduced — especially contour ditching and planting, as should composting and mulching. Participatory development would necessarily follow the current socio-cultural "contours". Impartial work with the existing groups and with the non-aligned individual families would be called for.

Policy Implications

Local development can be obstructed by non-local factors, which can worsen local problems. The major non-local problems include: 1) On the institutional level: agency and politico-administrative jurisdiction over the land was not known at the time of writing (now settled as BFD lands, provincial boundary still unsettled). 2) On the policy level: the Integrated Social Forestry Program of the BFD needs to define its intentions and establish practical guidelines for implementation of its projects. 3) On the national economy level: the current worsening of economic conditions in the Philippines has exacerbated local problems by, for example, pushing migration to the uplands and raising the demand for charcoal.

The specific implications are easier to name than to accomplish. The BL and the BFD must meet to decide who has jurisdiction over the land. If the land is BL land, an equitable system must be created for land titling and distribution, while ensuring protection of critical watershed areas. If the land is under the BFD, an ISF project could be implemented. Representatives of the provinces of Quezon and Laguna must meet to decide the provincial boundary. If the land is under BL jurisdiction, this will determine the area's political-administrative status. If BFD land, the decision would influence eventual provincial cooperation with any BFD efforts in the area. Obviously, the establishment of national jurisdiction would go far in decreasing current conflicts and polarization in the community and might slow down the exploitation of remaining resources. Some locals may lose land and/or resources with the decisions, but for the community and the nation in the long term, these decisions are essential.

Social forestry policy questions must be resolved and clarified to the public. Eligibility for ISF project participation, resources to be used, and conditions, responsibilities, obligations, or limitations for participants need to be ascertained for specific, real-life situations. We believe that under no circumstances should such a program directly or indirectly encourage further migration to the uplands. Our current experience with bureau employees and officials indicates that they share this viewpoint, but policy ambiguity and, perhaps, some "weak spots" in the process by which such projects are eventually proposed, approved, and implemented may be responsible for current migration into uninhabited upland areas.

We propose local responsibility and management, not only of the land, but of the remaining timber resources. (This is not allowed under current BFD policies.) The trees are now treated as common property, and the benefits from timber extraction are mostly leaving the community. If logging must continue, more benefits should accrue to the community. A well managed ISF program, or conversion of lands to A and D status combined with equitable distribution, might be considered as future possibilities for Calminoe.

National economic conditions are, of course, largely beyond the scope of this discussion, but Calminoe is not atypical. Demands on local resources and resulting ecosystem damage have increased with worsening economic conditions in the lowlands. Thus, any plan for the uplands must consider the larger economic context.

Calminoe's future may include further in-migration, clearing, cropping, logging, and charcoal making, even denudation and eventual depopulation. Or the settlers may modify their current resource management strategies and successfully adapt to changing conditions. They may reach a new equilibrium, one that lacks the quick returns now provided by uncleared land and tomatoes, timber, and charcoal but has stability that may be reached with the assistance of outside agencies.

REFERENCES CITED

Adeyoju, S.K. 1976. Land tenure problems and tropical forestry development. FAO: FP: FDT/76/5(b).

Aguilar, F.V. 1982. Social Forestry for Upland Development: Lessons from Four Case Studies. Quezon City, Institute of Philippine Culture.

Allen, J.C. and D. Barnes. 1982. Social forestry in developing nations. RFF Paper D-73. Washington: Resources For the Future.

Alvares, C. 1982. A new mystification? Some social forestry is old practice rhetorically dressed up. Development Forum (UNDP) 10:1:3-4.

Alvarez, Jr., Jesus. 1982. Social forestry in the Philippines: prospect and restrospect. Participatory Approaches to Upland Development Seminar Series II. Manila: Integrated Research Center, DLSU.

Bernales, B.C. and A. De La Vega. 1982a. Case study of the Antique upland development program. Manila: DLSU-IRC.

1982b. Case study of the family approach to reforestation program in Pantabangan, Nueva Ecija. Manila: DLSU-IRC.

1982c. Case study of the forest occupancy management program in Dona Remedios, Trinidad, Bulacan. Manila: DLSU-IRC.

1982d. Case study of the Peace Corps Volunteers' Pondasyon ng Bagong Buhay ng Gubatnon Mangyan, Inc. project in Occidental Mindoro. Manila: DLSU-IRC.

1982e. Case study of PANAMIN's lantuyan settlement in Baco, Oriental Mindoro. Manila: DLSU-IRC.

Brokenshaw, D.W. 1984. Social and community forestry. Development Anthropology Network: Bulletin for the Institute of Development Anthropology 2:1, 2: 6-11.

Bureau of Forest Development, Philippines. 1983. Strategies for implementation: Integrated social forestry program. BFD-ISF Paper.

Campbell, G., R. Shrestha, and L. Stone. 1979. The Use and Misuse of Social Science Research in Nepal. Kathmandu: Center for Nepal and Asian Studies.

Capistrano, A.D.N. and J.S. Fujisaka. 1984. Tenure, Technology, and productivity of agroforestry schemes. Paper for Philippine Institute of Development Studies Seminar, Economic Policies for Forest Resources Management.

Carruthers, I. and R. Chambers. 1981. Rapid appraisal for rural development. Agricultural Administration 8:407-422.

Cernea, M. 1981. Land tenure systems and social implications of forestry development programs. World Bank Staff Working Paper 452.

CIMMYT Economics Program. 1980. Planning Technologies Appropriate to Farmers: Concepts and Procedures. El Batan, Mexico: CIMMYT.

Collinson, M. 1981. A low cost approach to understanding small farmers. Agricultural Administration 8:433-450.

Conklin, H. 1967. Some aspects of ethnographic research in Ifugao. Transactions of the New York Academy of Sciences 2:30-99-121.

Cortes, E.V. 1982. Interim guidelines in the establishment and implementation of social forestry projects. Ministry of Natural Resources, Philippines. BFD. Letter.

De Los Angeles, M. 1982. Forest policies for development planning. In: PIDS Surveys of Philippine Developmental Research II.

FAO. 1977a. Forestry for local community development in Asia and the Far East. Asia-Pacific Forestry Commission. Tenth Session.

1977b. Report on the FAO/SIDA expert consultation on forestry for community development. TF/INT 271 (SWE).

Fontanilla, Conrad. 1982. Social forestry is in now. Canopy 8:3:13.

Fujisaka, J.S. 1983. Philippine social forestry: The current participatory approach model. Manila: Integrated Research Center: DLSU.

1984. Issues in forest resource management policies. Report on ADC-JCIE Seminar. Bangkok: ADC.

Fujisaka, J.S. and A.D.N. Capistrano. 1984a. Calminoe: An upland kaingin community case study with implications for development. Unpublished paper.

1984b. Community appraisal for social forestry: Lessons from Calminoe. To be published as PESAM Research Paper.

Ganapin, Jr. D. 1982. Social forestry: The Forester's point of view. Paper. Seventh Social Forestry Forum of the BFD-UDWG.

Gow, D. and E. Morss. 1981. Local organization, participation, and rural development: Results from a seven-country study. Rural Development Participation Review 11:2:12-17.

Grandstaff, T. 1980. Shifting cultivation in northern Thailand: Possibilities for development. Resource Systems Theory and Methodology Series 3. United Nations University.

Gwyer, G. 1979. Developing Hillside Farming Systems for the Humid Tropics: The Case of the Philippines.

Harwood, Richard R. 1979. Small Farm Development: Understanding and Improving Farming Systems in the Humid Tropics. Boulder, Colorado: Westview Press.

Hildebrand, P. 1981. Combining disciplines in rapid appraisal: The Sondeo approach. Agricultural Administration 8:423-432.

HMG/UNDP/FAO. 1982. Community forestry development project progress up to mid 1982. Kathmandu: FAO.

Hyman, E.L. 1983. Forestry administration and policies in the Philippines. Environmental Management 7:6.

International Council for Research on Agroforestry. 1983. Guidelines for Agroforestry Diagnosis and Design. Nairobi: Working Paper 6.

Jacalne, D.V. 1982. Possible technologies for upland development. Paper for workshop, Perspectives in Social Forestry, UPLB-CF and DLSU-IRC.

Kundstadter, P. 1978. Implications for socioeconomic, demographic, and cultural changes for regional development in Northern Thailand. In: Proceedings of Workshop on Agroforestry and Highland-Lowland Interactive Systems. Chiang Mai.

Laarman, J., K. Virtanen, and M. Jurvelius. 1981. Choice of Technology in Forestry: A Philippine Case Study. Quezon City: New Day.

Lynch, O. 1983. A survey of research on upland tenure and displacement. Paper, National Conference on Research in the Uplands. BFD.

Magno, V. 1981. The social forestry program of the BFD. Paper, 71st Anniversary of the UPLB-CF.

Mahiti Team. 1983. Why is social forestry not 'social'?. Paper, Ford Foundation (India) Workshop on Social Forestry and Voluntary Agencies. April.

Manandhar, P.K. 1982. Introduction to policy, legislation, and programmes of community forestry development in Nepal, Kathmandu: HMG/UNDP/FAO.

Meyers, Norman. 1980. Conversion of Moist Tropical Forests. Washington: National Academy of Sciences.

Nelson, G.C. 1983. Approaches to reducing renewable resource depletion and environmental degradation: Social forestry or macroeconomic policies. Paper, Agricultural Economics Society of Southeast Asia Conference, Bangkok.

_______ 1984. The impact of government policies on forest resource utilization. Paper, PIDS Seminar-Workshop, "Economic Policies for Forest Resource Management".

Noronha, R. 1980. Sociological aspects of forestry project design. Paper AGR-World bank.

Olofson, H., ed. 1981. Adaptive Strategies and Change in Philippine Swidden Based Societies. Los Baños, Philippines: Forest Research Institute.

Olofson, H. 1983. Indigenous agroforestry systems. Philippine Quarterly of Culture and Society 11:149-174.

Raintree, J.B. 1982. Readings for a Socially Relevant Agroforestry. Nairobi: ICRAF.

_______ 1983. Strategies for enhancing the adaptability of agroforestry innovations. Agroforestry Systems Journal 1:2.

Sajise, P. 1981. Some facets of upland development in the Philippines. Paper, Participatory Approaches to Development Series II. DLSU-IRC.

_______ 1982. Ecological approaches to managing degraded uplands in the Philippines. Paper, Joint Chinese Environmental Protection Office-E-W EPI Workshop. Kunming, PRC.

Schlegel, S. 1979. Tiruray Subsistence. Quezon City. Ateneo de Manila Press.

Schmidt, D. 1973. Anthropological and ecological considerations regarding the transition of shifting cultivation in the tropics. African Soils 18:2: 59-68.

Working Group in Agroforestry. 1982a. Economic evaluation of agroforestry projects. In: New Directions in Agroforestry. N. Vergara, ed. East-West EPI.

1982b. Initial tasks in agroforestry projects. In: New Directions.

9
AGROFORESTRY SYSTEMS FOR SMALLHOLDER UPLAND FARMERS IN A LAND REFORM AREA OF THE PHILIPPINES: THE TABANGO CASE STUDY

Filemon Torres and John B. Raintree
with
M.V. Dalmacio and T. Darnhofer

The hazards of upland farming in the Philippines, as elsewhere in Southeast Asia, are notorious. Problems exist for the spectrum ranging from shifting cultivation to permanent upland farming. In areas of high population density where permanent arable cropping has been the predominant land use for generations, continuous cultivation without fallows results in soil mining, accelerated erosion, and diminishing yields. Although the system may eventually stabilize at a very low level, it holds little prospect of being able to meet a future of an increased human population that cannot be realistically absorbed into the urban economy. Low cash incomes of peasant farmers and lack of adequate input supply and marketing infrastructures create a situation offering little hope for the improvement of the rural economy through conventional high-input approaches to agricultural intensification.

UPLAND FARMING AND LAND TENURE

Land tenure issues compound the problem. Unlike among shifting cultivators where the main issue is insecurity of tenure over ancestral lands in the face of restrictive forest laws and increasing pressure from land-grabbing outsiders, the main problem in peasant farming is unequal distribution of land and other productive resources between smallholders, often tenant farmers and large, often absentee, landowners. In some cases the traditional systems of rights and obligations safeguarded a reasonably stable symbiosis. With modernization and the breakdown of these relations, however, landlord-tenant relations have become increasingly exploitative in many areas. In response to mounting political pressures, the government has sought to implement land reform through various programmes. One such programme, "Operation Land Transfer," is being implemented in the Tabango case study area on the island of Leyte.

In Tabango, the relationship between ecological and tenure problems in upland farming system is particularly interesting and challenging for developers of agroforestry technologies. There are basically two main forms of land use in the area: coconut-based perennial tree crop systems and permanent upland cultivation of field crops. As presently practiced, the coconut-based farming system, with or without additional tree crops and livestock grazing in the understorey, appears to be a productive and sustainable land use system; permanent upland cultivation is not. Although the tree crop system is feasible for farmers with upwards of 5 hectares at their disposal, the major part of the coconut system is held by a small number of large landowners.

In areas not designated as "land reform areas," large landowners are expanding their tree crop holdings using the labour of landless farmers to plant the trees in return for the right to cultivate annual crops until the canopy closes and forces the laborers out. Eviction, in this case, is accomplished by the tree rather than the landlord. While they have the option of remaining on the large haciendas as wage labourers, most farmers would prefer to be independent smallholders, or at least permanent share-croppers on open land.

To meet the small farmers' need for land, the government land reform programme, Operation Land Transfer (OLT), first buys lands from the landlord, and then offers the title to the same farmers after a specified pay-back period. That is, the programme offers the tenants their traditionally cultivated lands. Many of the programme beneficiaries face the problem in which the amount of land they can lay claim to is quite small — usually under two hectares and too small — to contemplate a transition to the tree crop systems. This problem is especially true for farmers meeting daily family subsistence needs. They are left, then, with the upland field crop system which, under traditional technology, is only marginally productive and highly subject to severe upland degradation.

In Tabango, we were confronted with the land use paradox of sustainability without equity and equity without sustainability. If land reform is to succeed in Tabango, and in similar areas of the Philippines, technologies must be developed to make smallholder upland farming sustainable. Agroforestry may have a role to play. The specific form which that role could take is examined in this case study chapter.

THE STUDY AREA AND ITS ECOLOGY

The municipality of Tabango is located in the northern part of the island of Leyte, Philippines, about 159 km from Tacloban City, between the municipalities of Villaba and San Isidro. Comprising an area of some 129 square km, the municipality is traversed by unpaved all-weather roads. Coastal areas along the Visayan Sea are accessible by boat. Tabango has a total of 13 local administrative units, or *barangays* Due to the inaccessibility of some of the barangays during heavy rains, the study was confined to five barangays: Poblacion, Catmon, Omaganhan, Tabing and Campokpok.

The topography is mostly hilly with moderate to steep slopes draining to the Visayan sea. The highest peak, Mt. Canturaw, rises to 362 ft. above sea level. Level lowland areas are limited to about 25 percent of the total land area.

The project area is situated in the humid tropics, with an annual average rainfall between 2000 mm and 2500 mm. The water balance is usually positive (rainfall in excess of evaporation) through 9 of 12 months, from May or June to February or March. The growing period is approximately 270 days/yr. Rainfall patterns are strongly influenced by tropical depressions and storms (2-4 per year) such that inter-annual variability is considerable and temporary drought as well as heavy, erosive rains are not uncommon.

The mean temperature at sea level is around 27°C, with little annual variation (2.5°). The absolute maximum and minimum range is between 34° and 22° respectively. The elevation affects the mean temperature with a loss rate of 0.55°C per 100m.

Nebulosity follows the seasonal changes in rainfall distribution and is accordingly high. The average duration of bright sunshine is between 40% and 50% of the astronomical day length. High relative air humidity is also in accordance with the general climate pattern. The prevailing wind directions change from northeast in January to west-south-west in July. The mean windspeed is generally 2.0 m/s.

As of 1980, the total population of Tabango Municipality was 29,356. Males comprise 50.41% and females 49.59% of the population. The annual crude population growth rate is 1.6% and the average population density is 234 square km.

The major municipal products are maize, coconuts, and rice. In terms of employment, 45% are self-employed farmers, 40% are labourers, and 15% are "other employed." In terms of monthly income, 14% make less than 100 pesos, 68% make 100 to 300 pesos, and 18% make more than 300 pesos.

There are two agroecological zones which are determined by 1) level topography (0-5% slope) or Class A cropland which can be cultivated

using ordinary methods of farming and whose erosion hazzard very slight to nil, and 2) sloping cropland (5-605 slope) which should be cultivated using special conservation practices since the soil is subject to moderate erosion hazzards.

THE FARMING SYSTEM OF TABANGO

Farms in the study area can be classified into different farming system types on the basis of land tenure, farm size, and production emphasis. The main classes are as follows:

Local land tenure arrangements include: 1) owner cultivators, 2) beneficiaries of the Operation Land Transfer (OLT) in which, according to the Ministry of Agrarian Reform, about 3000 ha in the area are now under the land reform programme, benefiting some 1,887 tenants, and 3) leaseholders, a category reportedly being phased out by the OLT (the survey team only encountered one farmer in this category). Owner cultivators and OLT beneficiaries constitute about 90% of the farmers. Farm sizes can be classified as less than 2 ha, 2 ha to 5 ha, more than 5 ha, and large hacienda. Most farmers cultivate no more than 5 ha. The majority operate farms of 1.5 ha to 2 ha.

Characteristics

This section summarizes the salient characteristics of a typical farm in the farming system selected for priority emphasis by the diagnostic and design team.

Relevant agroecological characteristics include
— topography, characterized by upland slopes ranging from 10% to 60%, with an average of 30%
— soil, with the following characteristics
    — clay to clay loam, generally deep with a pH of 7.7
    — percent N (total) is 0.06% to 0.14%
    — percent OM is 0.4% to 2.5%
    — P (available) is a trace, particularly in upper slopes, and 22 to 32ppm in the lower slopes.

Sheet and rill erosion are notable on many farms. The farmers in the study area are OLT beneficiaries. Farm sizes are between one and two hectares. The production emphasis of the predominate farming systems in the Tabango study area is indicated in table 1. Table 2 describes the production system.

Table 1. Production emphasis of the predominant farming systems.

	Primary Production Emphasis	Secondary Production Emphasis
Annual Crops	Subsistence staples	supplementary foods & crops
	— rice	— cassava
	— maize	— sweet potato & other root crop
		— bananas
	cash crops	
	— peanuts	
	— mung beans	
Perennial Crops	cash crop	cash crop
	— coconut	— Leucaena leucocephala (ipil-ipil) for leaf meal and fuelwood
Farm Animals	draught power	subsistence consumption
	— carabao (water buffalo)	— chicken
	cash income	
	— pigs	

Table 2. **The role of major crop and livestock components in the selected farming system and their associated cultural practices (where two or more purposes are served, the primary one is underlined).**

Crop/Animal	Purpose	Cultural Practice
Primary		
— maize	subsistence, cash, animal feed	monocrop in rotation
— rice	subsistence, cash, animal feed	monocrop in rotation
— peanut	cash	monocrop in rotation
— mung bean	cash	monocrop in rotation
Secondary		
— sweet potato	subsistence, cash	monocrop in rotation
— cassava	subsistence	monocrop
— banana	subsistence, cash, feed	boundary planting
— ipil-ipil	cash, fuelwood	clusters on farm
Animals		
— carabao	draught animal	grazing on roadside and neighboring coconut plantations
— pigs	cash, social, consumption	scavenging and supplementary feed
— chicken	subsistence	scavenging and supplementary feed

The cropping schedule for the major crops is as follows:
- Maize is planted in April and May, and is harvested in July and August.
- Peanut is planted in September, and harvested in December and January.

Average crop yields are five sacks of maize per ha (250 kg/ha) and 10 sacks of peanut per ha.

DIAGNOSIS OF MAJOR LAND USE PROBLEMS AND POTENTIALS

The results of the diagnostic survey and analyses carried out by the regional D & D team are presented in fig. 1.

Problems in the Household Supply System

Contrary to initial expectations, the diagnostic survey revealed that there are no presently perceived problems in the staple food and fuelwood supply subsystems of households in the target farming system. Domestic needs for fuelwood are reported to be adequately met by the ready supply of coconut fronds in the general area together with farm plantings of *Leucaena Leucocephala*. Although farm-produced staple foods are in most cases insufficient to meet the annual needs of some households, the dominant strategy of the household economy is one of mixed subsistence and cash crop production, supplemented by income from off-farm labour (mainly agricultural labour on other farms in the area). Since cash income is sufficient to meet the needs for supplemental purchases of food, food shortages per se are not felt to be a problem. They are, however, a drain on household cash resources.

Inadequate cash income. Insufficient cash for purchases of other household needs (clothing, educational expenses, and non-food consumer items) and for capitalization of farm improvements emerges, therefore, as the main household supply problem in the survey area. The main sources of cash income are proceeds from the sale of cash crops (mainly peanuts and mung bean), livestock (pigs) and earnings from off-farm employment (table 3).

Table 3. Sources of cash income.

Source of Income	Percentage of Farmers Surveyed (N = 10)
Off farm employment	90
Peanuts	90
Pigs	80
Mung bean	50
Maize	20
Rice	20
Sweet potato	20
Banana	10
Ginger	10
Onion	10

The main drains on household cash resources are food purchases to supplement farm production of staples and hired labour for weeding, land preparation, and planting. Some 50% of the households surveyed own no carabao and must spend an average of 250 pesos per crop (2 crops/year) to hire the services of a ploughman and carabao. This lack of a draught animal is a severe capital constraint which, given low net income, is very difficult to remedy.

Income from sale of cash crops and livestock is generally insufficient to meet the cash needs of the household, and most farmers resort

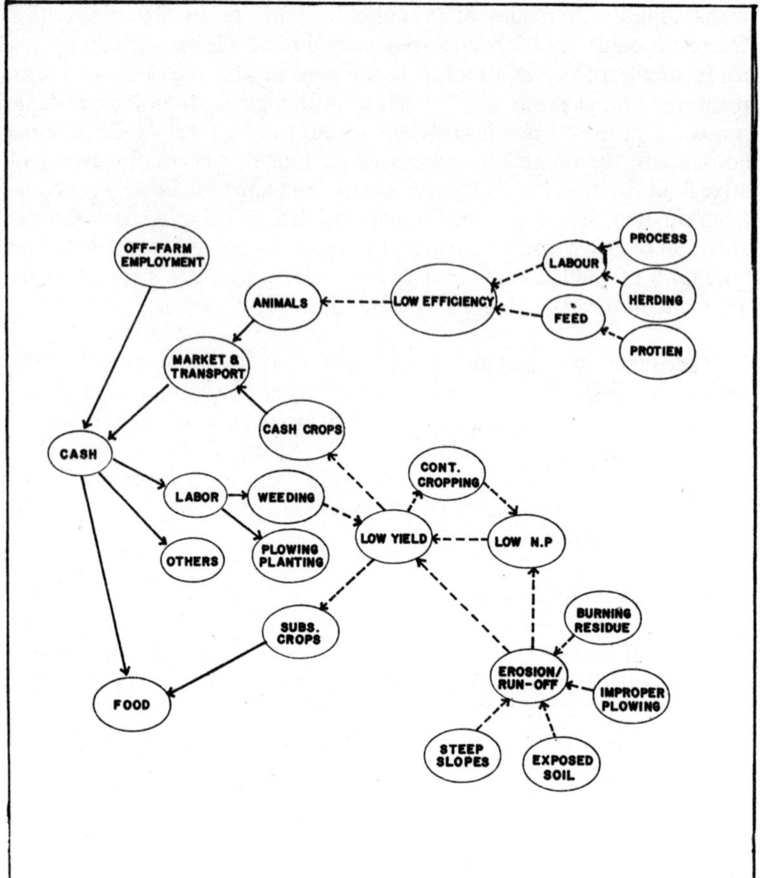

Figure 1. *Combined casual network and flow diagram of the farming system problematique in Tabango, Leyte. Depicted from left to right are the major household supply problems, the flow of related inputs and outputs (unbroken lines), and the casual factors which contribute to the creation of the identified problems (broken lines).*

to off-farm employment to supplement farm earnings. Of the ten farmers surveyed who fall into the category of the target farming system, eight engage in off-farm agricultural labour and one is self-employed as a maker of wooden ploughs. All farmers expressed interest in increased income from farm sources, but the survey team did not detect a notably high level of entrepreneurial aspiration among this group of farmers.

Causal factors responsible for household cash problems. On the income side, the main factors associated with low on-farm cash production are shown in figure 1 as causal complex or "syndromes" responsible for low yield of field crops and low-efficiency of the pig production enterprise.

Low Yield Syndrome

Continuous permanent cropping without mineral nutrient inputs and inadequate soil conservation practices have resulted in low nutrient status of the soils and low and declining yields of both subsistence and cash crops (e.g. 250 kg/ha of maize). The small farm size and low productivity of the land necessitates a permanent cultivation regime with no opportunity to rest portions of the land in order to restore soil fertility, thus perpetuating the low-yield syndrome. The non-use of mineral fertilizers is attributable both to the lack of cash resources and the lack of a local infrastructure for supply of agricultural inputs.

Sheet and rill erosion processes further exacerbate the degradation syndrome on these farms through continued nutrient losses and poor crop stands. Erosion hazards are substantial under the prevailing conditions of steep slopes and cropping practices which leave the soil exposed to heavy rains following field preparation. Improper ploughing practices and the burning of agricultural residues are major contributing factors in soil erosion. The fact that the farming system is still operational at all, even at low levels of yields, is attributable to the depth and only moderate erodibility of the soil. Nevertheless, at an estimated rate of loss of 1 cm of soil per year, this system of poor husbandry cannot be sustained indefinitely.

The prevailing tillage practice is to plough the field in lines perpendicular of the draft animals and not specifically intended to prevent erosion. Although proper contour ploughing would more efficiently satisfy both erosion control and energy conservation purposes, no contour ploughing by smallholders was observed in the entire survey area.

The retention of agricultural residues on the fields would also tend to mitigate both erosion and low-fertility problems, but all farmers surveyed reported a strong inclination to gather up crop residues and burn them. The main reason given for removal of crop residues from the fields was to avoid interference with ploughing operations, since, with the traditional chisel plough used in the area, maize stalks and other residues tend to bunch up in front of the plough, making tillage difficult. The expressed reason for burning the gathered residues was not and conscious desire to release nutrients to the soil, but rather to destroy potential habitats for rodent pests.

All of this is done without any evident consideration for erosion control and maintenance of soil fertility, although both of these factors are acknowledged by farmers as major contributing causes of declining yields. The only notable fertility-conserving measure practiced by the majority of farmers is the rotation of cereal crops with grain legumes.

Low Efficiency of the Livestock Enterprise

The low time-use efficiency of carabaos (water buffalo) as draft animals is a noteworthy latent constraint on agricultural production, but not one which admits of an easily conceivable and feasible solution. These animals, unlike oxen, must be rested and allowed a cooling-off period in water after brief periods of draught labour. Nevertheless, the slow work pace of the carabao seems to accord quite well with the somewhat leisurely attitude to work of the local farmers. There seems to be little scope at present for any attempt to disrupt this congenial symbiosis by substitution of more efficient draught animals.

The main addressable livestock constraint, therefore, is the low efficiency of the pig production enterprise, which is present as a source of cash income and socially expendable wealth (fiesta contributions, etc.) on most farms. Inefficiency arises in connection with the labour requirements of the present system of feeding and the nutritional inefficiency of the present diet. Free-range pig feeding is not possible on most farms because of the danger of damage to one's or the neighbor's crops. Consequently, labour is required to move and tether pigs from time to time. Supplemental feeding of rice bran and shredded banana stalk is commonly practiced, but this does not make up for the lack of high quality protein in the diet. The resulting slow rate of live weight gain yields rather low returns to labour invested in the pig enterprise.

TOWARD AN INTERVENTION POINT

The following list of design specifications may, for convenience, be divided into 1) functional specifications derived from analysis of intervention points in the system suggested by the combined causal network/flow diagram (figure 1) and other design specifications derived from the understanding of resource constraints and farmer strategies gained in the course of the diagnostic survey exercise.

Functional Specifications

In order to reduce the outflow of cash, the need for hired labour can be reduced by reducing the amount of tillage required for field preparation and the level of labour required for weeding. Also, cash expenditures on supplementary food purchases can be reduced by increasing on-farm production of staples, through removal of yield-reducing constraints which can be removed by improving the availability of nutrients (mainly nitrogen, or N, and phosphorus, or P) and by reducing erosion through adoption of proper contour ploughing practices and of a feasible system of organic matter maintenance on the fields.

In order to increase the inflow of cash, new cash crops can be incorporated into the farming system. The efficiency of the main livestock enterprise can be improved by producing a more balanced pig ration from potential farm feed sources and reducing the labour requirements of pig feeding. Labour for feeding can be reduced by introducing improved processing technology and substituting the current tethering practice by an efficient pen-feeding regime.

Additional Design Specifications

To be appropriate and adoptable by farmers in the context of their existing farming system (i.e., without major infrastructural or motivational changes), candidate technologies must: 1) have low capital requirements and must not rely on inputs which are not generally accessible to small farmers in the area, 2) be efficient in the use of available labour resources (i.e., at prevailing levels of effort acceptable to the majority of farmers) and must not exacerbate labour bottlenecks in the farming system, 3) make efficient and intensive use of scarce land resources, 4) not contribute unduly to increased pest problems, 5) be consistent, in the case of new cash crops, with marketing potentials in the area.

ALTERNATIVE TECHNOLOGIES

Cash deficit was identified at the diagnostic stage as the main symptom of the farming systems in the Tabango area, a problem that motivates farmers to look for off-farm income. The causal diagram in fig. 1 indicates that this deficit stems mainly from factors in the operation of cropping and animal production enterprises within the farm. Of the two, a higher priority should be placed on improvement of cropping enterprises, since they constitute the main source of both food (maize) and cash (peanuts).

Low soil fertility and soil erosion were singled out as the main causes of low crop yields, indicating the most promising points for technological interventions in the system. Table 4 summarises some of the alternative practices of an agroforestry and non-agroforestry nature, envisaged by the multidisciplinary team as having the potential to overcome those constraints.

Table 4. Possible non-agroforestry and agroforestry inverventions to solve or mitigate the diagnosed problems.

PROBLEMS	INTERVENTIONS	
CASH OUTFLOW	NON-AGROFORESTRY	AGROFORESTRY
1. Low Yield		
— N.P. deficiency	— Mineral fertilizer	— Legume tree mulching
	— Composting	
	— Green manure	
— Residue burning	— Mould-board ploughing	— Mulching for minimum tillage
	— composting	
	— No tillage	
— Water erosion		
Steep slopes	— Terracing	— Hedges across slopes
Improper ploughing	— Contour ploughing	— Contour hedges
	— No-tillage	
Exposed soils	— Perennial crops	
	— Relay cropping	
	— Mulching	— Mulching
2. Tillage		
— Carabao		
Hired ploughing	— No-tillage	— Mulching for minimum tillage
Working capacity	— Oxen	
	— No-tillage	— Mulching for minimum tillage
— Weeding	— Mulching	— Mulching
CASH INFLOW		
1. Low Yield		
(Peanuts)		
Same as outflow	— Same as outflow	— Same as outflow
2. Swine Production		
— Feed sources	— Balanced ration	— Gabi & leaf meal in hedgerow intercropping
— Feed processing	— Mechanical shredder	
— Management	— Appropriate housing	

3. Cash Sources
— Crops	— Pineapple	— Hedgerow inter-cropping of:
		Pineapple
	— Black pepper	
	— Ginger	Black pepper +
		Ipil[2] Leaf Meal
		Ginger
— Livestock	— Cattle Fattening	

Non-agroforestry Alternatives

Although a thorough discussion of these options vis-a-vis the identified constraints is beyond the scope of this document, it is worth mentioning some of the reasons leading to the selection of an agroforestry alternative, most of them linked to the design specifications discussed above.

In the field of soil nutrient deficiency, the introduction of inorganic fertilizers would most certainly overcome the biophysical constraint. However, such an input would further strain the already scarce cash resources available at the farm level. Even if a credit programme were made available, there is no existing infrastructure for distributing fertilizers. Compost could be an interesting alternative, particularly to avoid residue burning. However, lack of knowledge and labour requirements may prevent its adoption.

Green manuring with appropriate herbaceous species has been shown to improve soil fertility. However, land and labour scarcity and local attitudes against growing a crop solely for the purpose of improving the soil would tend to prevent the acceptance of conventional green manuring practices by farmers.

One of the problems farmers would face in trying to incorporate residues is the type of ploughing equipment in use, which does not have a mouldboard to turn the soil over. Substituting the existing chisel plough by a mouldboard plough would be a relatively costly option, possibly requiring stronger animal traction. It is known that no-tillage methods will improve soil and water conservation as well as reduce traction energy and labour requirements, but the specialized knowledge and equipment required, together with high cost and problems with the supply of herbicide, may again prevent its widespread adoption. There is a need, moreover, to provide for a source of mulch which should not compete unduly with the main crops for light, moisture, nutrients, space, and/or time in the cropping regime.

The investment required for construction of proper bench terraces would seem to rule them out as a locally adoptable alternative. Contour ploughing would certainly be a feasible erosion-mitigating alternative to which farmers ought to be introduced. At the moment there is a kind of "across the slope" ploughing, not precisely as a soil conservation practice but to avoid having the carabao go up and down the slope.

A permanent vegetation cover would certainly decrease the impact of erosive factors. However, establishing perennial crops (like coconuts) would compete with food and cash crops for the use of scarce land, depriving the smallholder family of its livelihood for quite a few years. (It could become a viable alternative, however, for holdings of 5 ha or larger). Relay cropping would also provide a permanent protective cover, but its efficiency may be impaired by the low soil fertility level.

The efficiency of tillage operations could possibly be improved by substituting the carabao (short, effective working time) with oxen. However, this may be a very difficult change to introduce, and it may disrupt a cherished symbiosis between Visayan farmers and carabaos. At any rate, it would appear that in the medium to long-run, land pressure may tend to displace animal traction from smallholdings, particularly if and when free access to grazing under the coconuts of the larger plantation owners is restricted.

Cash income could be increased by substituting alternative cash crops for peanuts, but adoption may be slow under existing marketing conditions. Cattle fattening does not appear as a sustainable alternative for smallholders, since fodder production will compete with food and cash crops for land and labour.

Agroforestry Alternatives

Column 3 in table 4 summarises the interventions of an agroforestry nature which were considered by the team for addressing specific constraints in the diagnosed system. From the discussions at the design stage of the D&D process, there emerged a general design concept for an agroforestry land management system which appears to have the potential to remove or mitigate several of the most limiting constraints. The team decided, therefore, to focus attention on the elaboration of this design concept as a prelude to the planning of a research project to develop and test the proposed agroforestry system. The results of the team's thinking, as developed initially in the rapid appraisal mission and later in the follow-up phase of the D&D process, are presented below.

Design concept: hedgerow intercropping or alley cropping system. The use of mulch from a nitrogen-fixing tree as a source of organic fertilizer for the predominant field crops appeared to the team as a promising technique for restoring and maintaining soil nitrogen at an increased level. The possibility for arranging such trees in hedgerows along contour lines in the field suggested itself as a way of forming vegetative barriers capable of arresting soil erosion which in time would result in the development of natural terraces. The existence of contour hedgerows would also force farmers to plough along contour lines, thus reducing inter-row erosion.

The food and/or crash crops normally grown by farmers could be intercropped between the hedges with the expectation that, if such a system were perfected, the increase in productivity from organic fertilization would more than offset the reduction in cropping area alloted to the main crops. Moreover, if the hedges were composed of leucaena or other fodder-producing trees or shrubs, the tops could be used to produce leaf-meal during those times of the year when the prunings were not being used for mulching. Firewood could also be produced as a by-product of other management practices. It was further envisaged that the space alloted to the hedgerows could be intercropped in a two-storey arrangement with shade-tolerant root crops grown at the side of the hedge, and vine crops (such as black pepper) grown within the hedgerow by planting self-pruning tree species at 10 meter intervals or more in the hedgerow to serve as live supports for the vines (see figure 2 for an artist's conception of the proposed system).

The design concept will be recognized as a variation on the "alley cropping" theme found in the literature (Parera, 1976; Guevarra, 1976 and 1978; Benge, 1979; de la Rosa, 1980; IITA, 1980a and 1980b; Wilson and Kang, 1980; Wilson and Raintree, 1980; Raintree and Turay, 1981; Kang, Wilson, and Spikens 1981; Metzner and Evensen, 1982). As a form of "zonal cropping" (Huxley, 1980), alley cropping is a type of agroforestry system which may allow simplification of management procedures by dealing with each zonally-arranged component largely as a sole cropping enterprise within the cropping system.

Evaluation of different alley cropping systems, however, has shown (Verinumbe, 1981; Balasubramanian, 1983) or predicted Hoekstra, 1983; Torres, in press) variable results in different environments, indicating the need to generalize with caution and distinguish clearly between specific design variations on the general alley cropping theme. The concept of alley cropping or, more generally, hedgerow intercropping implies a "family of technologies" often linked to each other only by the shared attribute of some zonal arrangement of woody and herbaceous components.

The specific hedgerow intercropping system depicted in figure 2 is only a design prototype, intended to convey the general idea of the

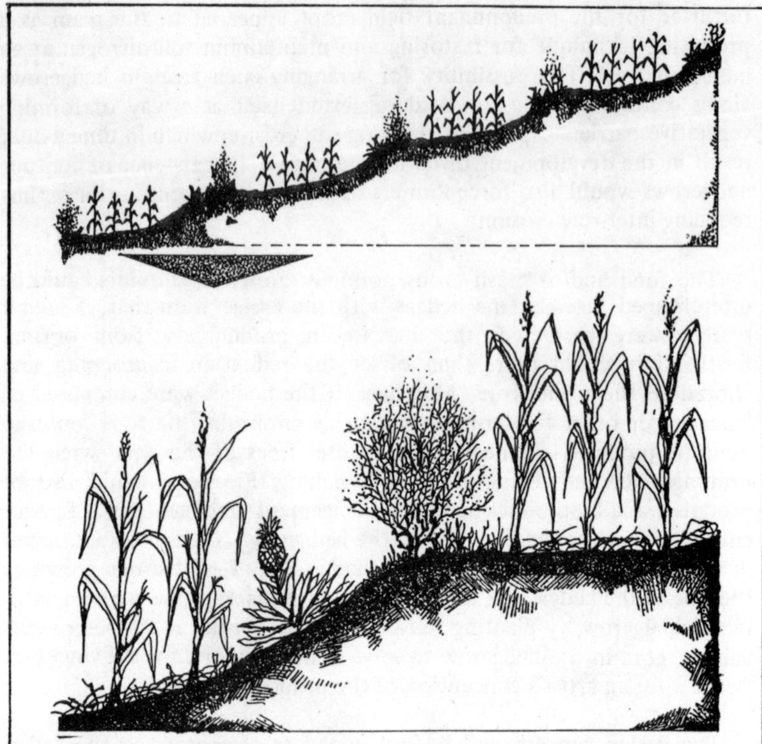

Figure 2. *Artist's conception of the experimental hedgerow intercopping system proposed for research at the Tabango site.*

team's proposal and to serve as a starting point for the research seen as necessary to develop an optimal system for the farming conditions of the Tabango area. The product of such research, assuming the project is successful in developing an optimal site-specific system, would revert itself in the dissemination phase to status of a prototype for adaptive research throughout the potential recommendation domain.

Although various hedgerow intercropping systems of this general type are known to exist in the Philippines and in other forms elsewhere in Southeast Asia, it is doubtful whether any such system has gone through the process of reiterative development and refinement, which would be characteristics of a systematic and cost-effective approach to the development of system-specific agroforestry technology (ICRAF 1983a, 1983b). Such development requires the initial specification of the desirable characteristics (the design specifications), which would make the eventual technological package appropriate and adoptable in the context of the target land use system.

In addition to addressing the functional requirements of soil fertility improvement (mainly increased nitrogen) and erosion control, an appropriate alley cropping system for the target group in Tabango ideally would: 1) be efficient in the intensive use of land, 2) be efficient in the use of labour and not create labour bottlenecks, 3) not increase pest problems beyond an acceptable level, 4) be flexible as regards management for different products on different farms and at different times of the year, 5) allow for the possibility of minimum tillage management in the eventuality of reduced draught power on the smaller farms, and, 6) not involve management complexities which the local farmers, after a sufficient demonstration and trial period, would not be willing to accept.

Plant ideotype for the tree/shrub component of the system. Although maize is virtually a "given" as a component of any widely adoptable alley cropping system developed for the Tabango site, the choice of the most appropriate tree/shrub component for the system is an open question for research. In addition to specifying the desirable characteristics of the proposed system, it is helpful to set forth the desirable characteristics of plant components prior to screening (Raintree, 1981). The following characteristics have been identified as features of the ideotype for the ideal tree/shrub component of hedgerow intercropping systems of the type proposed for Tabango: 1) high N-fixation capacity, 2) fast-growing, 3) able to coppice vigorously and yield required volume of suitable mulch materials under acceptable cutting regime, 4) deep-rooting habit for drought resistance, recycling of subsoil nutrients, and minimal surface root competition with associated arable crops, 5) easy and economic establishment, and 60 high yields of economic by-products (feed and firewood). Vigor and yield requirements will vary with cutting frequency deemed acceptable to the user (which will be affected by labour availability, shading effects, and mulch requirement to achieve desired benefits). Suitability of mulch material will be determined by various characteristics (rapidly decomposing leaves for quick release fertilizer, slowly decomposing leaves for weed and runoff control, desirability of insect repellent properties, nutritional quality and palatability of herbage to be managed for fodder by-product, etc.).

It is difficult to assign priorities to the various characteristics prior to on-farm trials to assess performance in terms of features desired by the end-user. Other more detailed plant physiological characteristics may be identified as research progresses. It may also be necessary to think of satisfying the total package of requirements by a combination of hedgerow species if they are not optimally combined in a single species.

Ex Ante Evaluation of the Proposed System

The basic hypothesis behind the proposal of the experimental hedgerow system is that the increments in crop yield due to organic nitrogen inputs from the nitrogen-fixing woody component would increase the productivity per unit of intercropped land. Available information was analysed to evaluate the potential contribution of leucaena hedgerows and to formulate a more quantitative hypothesis on the impact on grain and fuelwood yields of the intercropping of leucaena hedgerows with maize (Torres, in press).

The use of data on leucaena leucocephala in the following computations is made without prejudice to the eventual selection of other woody components for the hedgerow role. The choice of leucaena as an initial "best bet" component for the prototype system is justified on several grounds: 1) leucaena does possess most of the characteristics of the plant ideotype for the tree/shrub component of the system (although it does not have slowly decomposing leaves, which might be desired if a more durable mulch is required for a minimum tillage variant of the system), 2) the giant "Salvador" type leucaena varieties are a well-established feature of the existing farming system (although not in the arrangement proposed), and 3) sufficient data on the performance of leucaena and associated crops exist in the literature to allow a first order approximation of its performance under the proposed conditions, which is not the case for most of the woody species often mentioned as having potential for agroforestry. The following calculations give an approximate indication of the performance of one variant of the prototype system.

Based on production data from the humid tropics, its was assumed that a linear meter of leucaena closely planted in a hedgerow would produce about 1 kg of "leafy" dry matter per year with a nitrogen content of 3.5% and around 650 g of fuelwood. It was further assumed that an increase of 5.3 kg or 6.3 kg of maize per kg of nitrogen applied as leucaena mulch would be obtained from such a practice. Results of this speculation in terms of kg of maize per ha of inter-cropped land for different baseline maize productions are shown in table 5.

As mentioned above, the existing production level at Tabango appears to be in the order of 250 kg ha^{-1}. Should those relationships be supported by experimental evidence, espacement would still depend on economic considerations, particularly in relation to the labour cost to maize price ratio. The proposed practice would obviously require higher labour inputs than the present system, but at the 150 cm spacing and 1:6.3 ratio, the increase in maize yield would be equivalent to about 60 man days (assuming 1.1 pesos per kg of maize and a salary of 7 pesos day^{-1}) which could be expected to offset the increased labour input.

Table 5. Estimated yields of maize per hectares of intercropped land, as affected by baseline production of maize.

Leucaena inter-row spacing (cm)	Maize baseline production (kg ha^{-1})			
	500	1,000	1,500	2,000
Applying ratios derived from observed yields				
150	2310	2120	2110	2170
300	1405	1560	1805	2085
450	1103	1373	1703	2056
600	953	1280	1653	2043
Applying ratios derived from reduced yields (75% of observed)				
150	1800	1710	1770	1880
300	1150	1355	1635	
450	933	1236	1589	
600	825	1178	1568	

In addition to the potential benefit to be derived from an increase in maize production, a hectare of hedge intercropping at 150 cm spacing would supply enough fuelwood to satisfy annual requirements of 8 persons (table 6). The equivalent value of another 40-man-days of labour could be obtained if one of the hedgerow harvests is sold for leafmeal production (assuming that 50% of the "leafy" portion would be accepted for leafmeal at a price of 500 pesos per ton of leafmeal).

Table 6. Estimated potential fuelwood production per hectare of intercropped land.

Leucaena between row spacing (cm)	Stem Production	
	kg ha-1	m ha-1
150	1610	3.50
300	805	1.75
450	537	1.17
600	403	0.88

The potential of this technology must still be tested under field conditions before being considered for dissemination to farmers. Such testing is part of the research proposed in the next section of this chapter.

A RESEARCH PROPOSAL

Accepting that the quantitative hypothesis formulated in the previous section is based upon some experimental evidence, the designed technology prototype could already be tested on-farm, while more controlled experiments are carried out concurrently to test hypothetical relationships (to allow for a better understanding of fundamental processes) and to try alternative components.

It is, therefore, proposed that research activities be carried out on farmers' fields in a local field station (on land offered by the Tabango Mayor) and at the VISCA campus. In general, information emerging from these sites will reach the development stage (the period of elapsed time before they are ready to be disseminated by the extension system) in the short (2 years), medium (4 years) and long (6 years) term, respectively.

In this context, it is suggested that the following activities be implemented at the different sites.

Intercropping with Leucaena Hedges: Site 1

The site for this research will be the farmer's field. The period will be immediate (short term). Problems to overcome will include low nitrogen (N) and phosphorus (P), soil erosion, and improper ploughing. The general objectives will include utilizing leguminous trees in croppings with contour hedges to reduce water run-off and increasing N and P availability to the interplanted crops. Specific objectives will be. 1) to examine the effect of between hedge distances on the yields of crops established on farms with different baseline productions and slope gradients, 2) to examine the effect to cutting frequencies on the leucaena yield and labour input, and 3) to examine the overall effect of the proposed system on water run-off, erosion, and soil and chemical properties.

The technology components and management practices will be as follows:
- Leucaena cultivars, local giant K8 or K28, will be considered as "best bet."
- Height of cutting will be 15 cm as a best bet and after allowing 6 months for establishment.
- Lemon grass is to be planted on the uphill sites of hedges, and crop cultivars will be local.

The treatment variables will be as follows:
- For the distance between hedges, three inter-row spacings are to be tested on different farms according to existing maize produc-

tion levels. Farms with 250 kg ha^{-1} will have 1, 3, and 5 maize rows between hedgerows. Farms with 500 kg ha$_{-1}$ will have 1, 2 and 3 maize rows between hedgerows.
— Slope gradients will be 20% and 40%.
— The tested frequencies of cutting of hedges will include the farmer's way and six cuttings as shown (first cutting 10 days before planting maize):

May Jun Jul Aug Sept Oct Nov Dec Jan Feb Mar Apr May

   .   .   .   .

  1   2   3   4   5   6   7

There will be three replications.

Criteria for farm selection will include field extending from ridge to lower slope, uniform major slope, and minimum width of 50 m. A total of 12 farms will be required.

Parameters to be monitored will be: 1) crop yield (monitored by the field researchers), 2) leucaena yield (field researchers), 3) labour input (field researchers), 4) farmer's observations (field researchers), 5) rainfall measurement (field researchers), 6) soil chemical and physical properties (VISCA scientists), 7) run-off and erosion control (VISCA scientists).

Intercropping with Leucaena Hedgerows: Site 2

The site in this case will be the field station in Tabango. The time period will be medium term. Problems to overcome will be low N and P, soil erosion, and improper ploughing. General objectives will include the utilization of leguminous trees in croppings with contour hedges to reduce water run-off, and the increase of N and P availability to the interplanted crops. Research will look specifically at the leucaena cultivar, the height of cutting hedges, within-row tree spacing in hedges, and the rate of green manure application in terms of leucaena leaf production, crop yield, labour input, and nutrient (N and P) cycling.

Research design will have the same technology components and management practices as those for the work on the farmers' fields.

Treatment variables will include 1) leucaena cultivars (giant and local), 2) height of hedge cutting (2 treatments), 3) within row spacing of leucaena trees (15 cm, $<$ 15 cm and $>$15 cm), 4) rate of green manure application (three treatments of mulching with a separate leucaena stand to supply needs for additional mulch), 5) replications (2).

Parameters to be monitored will include 1) crop yield (to be monitored by the field researchers), 2) leucaena yield (field researchers), 3) labour input (field researchers), and 4) nutrient (N & P cycling (VISCA scientists).

Hedgerow Intercropping of Cash Crops: Site 3

The third site will be the field station in Tabango. The period will be medium term. The problem to be overcome will be cash shortages. The general objective will be to examine the feasibility of introducing cash crops as components of hedgerows. Specific objectives will be to examine the effect of cash crop inclusion in the hedgerow on labour input on the yields of leucaena agronomic crops and cash corps.

Technology components and management practices will be similar to the previous two research trials. Lemon grass is to be planted on the uphill side of the hedgerows. For pepper, 10m distances will be used between supporting poles. All cash crops will be planted on the downhill side of the hedgerows.

Treatment variables will include types of cash crops (black pepper, (*Jacaranda copaia* and *Gliricidia, Erythrina,* and *Sesbania* as poles at intervals within the hedgerow, ginger, pineapple, and gabi). Parameters to be monitored will include labour input, leucaena yield, field of cash crop and agronomic crop yield.

Mulch Farming with Contour Hedges of Other Species: Site 4

This site will be the research station (VISCA Campus). The period will be long term. Problems to overcome will be low N and P, soil erosion, and labour requirements for cultivation and weeding. The general objectives will be the utilization of trees in farming with contour hedges to reduce water run-off and tillage and weeding while increasing N and P. Specific objectives will be: 1) to examine the possibility of using other N-fixing and coppicing woody perennials as hedge species for providing mulch and nutrients, 2) to determine the most suitable cutting height and within row spacing for sufficient leaf production, and 3) to determine the required quantity of mulch for control of weeds and soil erosion.

The technology components and management practices will entail species selected on the basis of: 1) N-fixing and P-pumping capability,

2) broad and slow decomposing leaves (non-deciduous), 3) deep rooting, and 4) Fast growing and good coppicer (e.g., Erythrina, Gliricidia and Sesbania).

Treatment variables will include height of hedge cutting (2 levels), within row spacing of Erythrina trees (2 spacings), and mulch rate of application (3 rates based on potential production).

Parameters to be monitored will be leaf production, content of N and P in leaves, weed growth, crop yield, and erosion.

REFERENCES CITED

Balasubramanian, V. 1983. Alley cropping: can it be an alternative to chemical fertilizers in Ghanna. Paper presented at the Third National Maize Workshop, 1-3 February 1983, Kwadeso Agricultural College Kumasi, Ghana (mimeo).

Benge, M.D. 1979. Use of leucaena for soil erosion control and fertilization. Agency for International Development. Washington. (mimeo).

dela Rosa, J.M. 1980. A study of the growth and yield of corn intercropped with varying population of giant ipil-ipil (Leucaena leucocephala) on a hillside. Unpublished B.Sc. Thesis. Dept. of Agronomy and Soils, Visayas State College of Agriculture. Leyte, Philippines.

Evensen, C.L.I. 1982. Synopsis of leucaena/maize green leaf manuring study. College of Tropical Agriculture, University of Hawaii. Honolulu (mimeo).

Guevarra, A.B. 1976. Management of leucaena leucocephala (Lam. de Wit) for maximum yield and nitrogen contribution to integrated corn. Unpublished Ph.D. dissertation. College of Tropical Agriculture. University of Hawaii, Honolulu.

Guevarra, A.B., A.S. Whitney, and J.R. Thompson. 1978. Influence of intrarow spacing and cutting regimes on the growth and yield of leucaena. Agronomy Journal 70: 1033-37.

Hoekstra, D.A. 1983. Leucaena leucocophala hedgerows inter-cropped with maize and beans: an ex ante analysis of a candidate agroforestry land use system for the semi-arid areas in Machakos District. Working Paper No. 3. ICRAF. Nairobi.

Huxley, P.A. 1980. Agroforestry research: the value of existing data appraisal. In: T. Chandler and D. Spurgeon (eds). International Cooperation in Agroforestry. ICRAF. Nairobi.

ICRAF, 1983a. Guidelines for agroforestry diagnosis and design. Working Paper no. 7. ICRAF. Nairobi.

ICRAF, 1983b. Resources for agroforestry diagnosis and design. Work-in Paper No. 7. ICRAF. Nairobi.

IITA. 1980 a. Alley cropping: an improved bush fallow system. Research Highlights for 1979. International Institute of Tropical Agriculture. Ibadan.

IITA. 1980b. Tree fallow and in-situ vine support. Research Highlights for 1979. International Institute of Tropical Agriculture. Ibadan.

Kang, B.T., G.F. Wilson, and L. Sipkens. 1981. Alley cropping with maize (Zea mayssL.) and leucaena (Leucaena leucocephala Laree legumes: an underdeveloped branch of tropical agroforestry research. In: L Buck (ed). Proceedings of the Kenya National Seminar on Agroforestry. ICRAF. Nairobi.

Raintree, J.B. and F. Turay. 1980. Linear programming model of an experimental leucaena-rice alley cropping system. IITA Research Briefs 1(4): 5-7. International Institute of Tropical Agriculture. Ibadan.

Torres, F. In press. Potential contribution of leucaena hedgerows intercropped with maize to the production of organic nitrogen and fuelwood in the lowland tropics. Agroforestry Systems, forthcoming.

Verinumbe, I. 1981. Economic evaluation of some zero-tillage systems of land management for small-scale farmers in south-western Nigeria. Unpublished Ph.D. dissertation. Dept. of Forest Resources Management, University of Ibadan. Ibadan.

Wilson, G.F. and B.T. Kang. 1980. Developing stable and productive biological cropping systems for the humid tropics. In: B: Stnehouse (ed). Biological Husbandry: A Scientific Approach to Organic Farming. Butterworth. London.

Wilson, G.F. and J.B. Raintree. 1980. The leucaena-maize-yam alley cropping system. International Institute of Tropical Agriculture. Ibadan. (mimeo).

10
PHILIPPINE LAW AND UPLAND TENURE

Owen J. Lynch, Jr.

The importance of land tenure to upland Filipinos cannot be overemphasized. Land tenure was identified as an overriding concern of upland communities in nine case studies commissioned by the Bureau of Forest Development's Upland Development Working Group (Aquino, 1982). These studies reflect a consensus among many upland citizens that long-term occupancy and management of lands should provide sufficient legal bases for the recognition of tenurial secure rights.

This chapter surveys Philippine law and upland tenure, including the clash between the uplanders' viewpoint and prevailing national laws. Five fundamental issues are examined. First, the colonial foundation of national land laws is analyzed. Second, the allocation of jurisdictional or institutional control among various executive agencies over the "public domain" (which includes most of the uplands) is explored in historical and contemporary contexts. This paper will refer to the "public domain" without agreeing that many of the lands and resources so designated are, in fact, public property. Third, the origins and on-going expansion of the classification power of the executive branch are investigated. The classification power, particularly its certification component, has corresponded to a contraction of judicial power to recognize private property rights within the so-called public domain. Fourth, the private property rights of many upland occupants are discussed. This section includes a brief critique of "the political economy of ignorance" which legitimates an indiscriminate prohibition of shifting cultivation. Fifth, local land laws and customs prevailing in many upland communities are briefly surveyed. Finally, ten specific recommendations concerning law and upland tenure are offered.

THE COLONIAL LEGACY

When Ferdinand Magellan "discovered" the Philippines in 1521 and planted the Spanish flag on Mactan Island, he laid the basis for a legal theory known as the Regalian Doctrine. All land was presumed to belong to the Spanish Crown unless a royal grant described in official documentation recognized a contrary property right. Magellan's act, at least insofar as the colonial powers were concerned, converted the indigenous occupants of the unexplored archipelago into "squatters." In 1985, the Regalian Doctrine remained the theoretical bedrock upon which Philippine national land laws were based. Despite the existence of contrary laws, the Doctrine has rarely been challenged. Essentially, contrary national and indigenous laws have been ignored. Land not covered by official documentation is considered part of the public domain, that is, owned by the Philippine government regardless of how long the land has been continuously occupied and cultivated (Supreme Court Reports Annotated, 1972). Indigenous occupants are presumed to be squatters unless the national government specifically decides otherwise. In Indonesia, by contrast, customary property rights or *adat* are recognized, at least theoretically (Gautama and Harsono, 1972). In practical terms, however, there may be little difference. National Indonesian laws recognizing long-term property rights are often not applied to marginalized rural groups (Soeratman, Singarium-bun, and Guiness, 1977; Fulcher, 1982; Dove, 1983; West Papuan Observer, 1983; Lynch, 1984b; Mayo Vargas, 1984).

The legal framework for the allocation and management of the Philippine public domain, including land, minerals, water, flora and fauna, is patterned after U.S. public land and resource law (Coggins and Wilkerson, 1981). But there is a major difference. The U.S. public domain is largely uninhabited by humans while millions of people in the Philippines live on land classified as "public." The Philippine public domain as of 1984 included almost 62% of the nation's total land area (BFD, 1982 and 1983 annual reports). The national government's perennially incomplete census indicated in 1982 that less than 800,000 *"Kaingineros,* squatters, cultural minorities and other forest occupants" live within this vast area. More credible estimates indicate that more than 12 million "invisible" Filipinos, half of them indigenous, reside on and cultivate a significant portion of the public domain (Lynch, 1984a).

An equally significant and obscured fact concerns the fallacy of the Regalian Doctrine. Historical documentation conclusively proves that the contemporary use of the Doctrine is inappropriate. King Philip II of Spain decreed in 1580 that natives had complete liberty in the disposition of their landholdings. Fourteen years later, he decreed that "grants of land to Spaniards be without injury to the natives and those which have been granted to their loss and injury be returned to their lawful owners" (Laws of the Indies 1580, 1594). Other laws which promoted indigenous property rights, at least in theory, were promulgated by Spanish officials throughout the following three centuries (Barrameda, ed., 1983).

United States colonial administrators ignored most Spanish land laws benefitting the non-elite Filipino majority. Instead, the North Americans appear to have resurrected the Regalian Doctrine. In 1900, the Doctrine provided the legal justification for the new colonial claim to ownership of 27,694,500 ha, or 92.3% of the total Philippine land mass (Philippine Taft Commission, 1900). The land was designated as public domain. The ancestral character of a significant portion was not recognized.

The colonial government was authorized by the U.S. Congress in the Philippine Bill of 1902 to allocate and dispose of tenurial rights to natural resources within the public domain (US Act of Congress, July 1, 1902, section 17). Not surprisingly, most concessions were made to people who were familiar with national laws and had the financial resources for the requisite applications. The only significant limitation on the Governor-General's disposition power was the authority of the judicial branch to recognize the existence of private ownership of lands within the public domain which had been occupied and cultivated for a predetermined period of time. This countervailing power, however, has been eroded in recent years.

By 1984, the public domain had shrunk by one-third. It included about 62% or 18.59 million ha of the Philippine land mass, mostly the interior hilly and mountainous areas of the major islands. As a result, legal and policy issues concerning upland land tenure must be considered within the broader context of "public" land law. At the same time, perhaps 10% of the nation's total upland, or sloping, hectarage is covered by private property rights recognized by all branches of the national government and evidenced by official and proper documentation (Del Castillo, 1983, personal communication).

As of 1985, the U.S. colonial allocation scheme for the Philippine public domain remains essentially intact. The authority to allocate public domain upland resources is exercised by Philippine presidents in a manner reminiscent of the U.S. Governor-Generals. But there are significant differences. As a manifestation of expanding presidential powers, the possibility of upland occupants acquiring recognition of private property rights was foreclosed by provisions in the Revised Forestry Code of 1975 (PD 705, 1975; Lynch, 1982a, and reviewed in the following sections). At the same time, displacement of indigenous and migrant occupants is now widespread and increasing (Barrameda, ed., 1983; Lynch, 1983a). This displacement is a major factor promoting the discontent and injustice which prevails in the upland areas.

JURISDICTIONAL CLAIMS AMONG EXECUTIVE BRANCH AGENCIES

Since the U.S. claimed ownership over a vast area, the colonial government needed to devise a system for management, allocation, and exploitation. This system necessarily included the designation of responsible institutions. Act No. 5 of the U.S. Philippine Commission in 1900 authorized the creation of the Bureaus of Forestry and Mines. The Bureau of Public Lands was established a year later by Act No. 218. These bureaus were controlled by the U.S. Governor-General. The Bureau of Forestry was delegated jurisdiction over public forest land, the Bureau of Lands over public agricultural land, and the Bureau of Mines over designated mineral deposits. Agricultural land fell under the jurisdiction of the Bureau of Lands. Land classified as timber land was under the jurisdiction of the Bureau of Forestry. Jurisdiction was initially concurrent over unclassified public land (over one-fifth of the nation's total land mass remains unclassified as of 1985).

Executive branch agencies and their jurisdictional claims have proliferated since 1902. These claims overlapped by at least 1.27 million ha in 1983 (see chart). The major claimants are the Ministry of Natural Resources' Bureau of Forest Development (15,191,000 ha), the Bureau of Lands (1,964,000 ha), the Bureau of Mines (491,000 ha), the Ministry of Human Settlements (1,501,000 ha), and the Ministry of Agrarian Reform (718,070 ha). In recent years, the publicly owned National Development Corporation (NDC) has acquired extensive claims to public land for capital intensive, agro-development projects. The size of the area under NDC jurisdiction, however, is not publicly known. The National Power Corporation, the National Irrigation Administration, the Office of Cultural Communities, and the Bureau of Fisheries and Aquatic Resources also claim jurisdiction over thousands of hectares. In addition, other governmental agencies and institutions, such as the Ministry of Defense, the Development Bank of the Philippines, and the University of the Philippines have been granted by presidential proclamations and executive orders property rights to public lands.

Most proclamations and executive orders allocating intra-governmental bureaucratic jurisdiction are based on table, as opposed to ground, surveys. As a result, national, provincial, and municipal officials are

LAND RESOURCES of the PHILIPPINES

AN EXECUTIVE BRANCH INTERPRETATION*

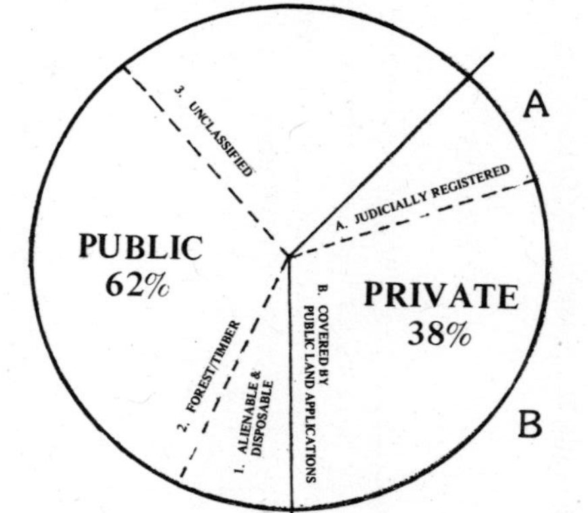

	AREA (in thousand has.)	PERCENTAGE OF TOTAL LAND MASS
I. PRIVATE		
A. Judicially Registered Land	2,624	8.75%
B. Covered by Public Land Applications	8,681	28.94%
SUB-TOTAL	11,305	37.69%
II. PUBLIC		
A. Classification Status		
1. Alienable and Disposable	3,465	11.55%
2. Forest/Timber	9,575	31.91%
3. Unclassified	5,553	18.51%
SUB-TOTAL	18,593	61.97%
TOTAL	29,898	99.66%[1]
B. Jurisdictional Claims[2]		
1. Ministry of Agrarian Reform (Resettlement)	718	2.39%
2. Ministry of Human Settlements	1,501+?	5.00%
3. Ministry of Natural Resources		
a. Bureau of Lands	1,964	6.54%
b. Bureau of Mines	491	1.63%
c. Bureau of Forest Dev.	15,191	50,632%
MNR SUB-TOTAL	17,646	58,80%
TOTAL	19,865	66.19%

[1]Reflects a discrepancy of .34% or approximately 102,000 hectares between Bureau of Lands and Bureau of Forest Dev. statistics if 100% equals 30 million has.

[2]Not all executive agency claims are known or listed.

[3]The jurisdictional claims listed overlap by 1.27 million has. or 4.23%

* Based on statistics presented by the Bureaus of Lands and Mines, Ministry of Natural Resources and the Ministry of Agrarian Reform at the Workshop on Public Land Policy, Development Academy of the Philippines, August 1981. Adjustments were made pursuant to the Bureau of Forest Development 1982 Annual Report and Proclamation No. 2282 of 1983.

often unsure as to the exact location of boundary lines separating the claims of two or more competing agencies. Some lands are claimed by more than one concession. The confusion is exacerbated by inroads made by the Ministry of Human Settlements (MHS) into the Bureau of Forest Development's (BFD) traditional jurisdiction over the public domain.

Proclamation No. 2282 of March, 1984 transferred jurisdiction of more than 1.5 million ha of land from the BFD to the MHS. The proclamation may have effected the largest transfer of public land jurisdiction by the national government since its creation by the Spaniards in the sixteenth century. The proclamation pales next to the 10 to 15 million ha reportedly to be included in the MHS Peoples' Forest Program (according to the Manila newspaper, The Bulletin Today, February 11, 1984).

Despite these inroads, the BFD retains the largest claim to the public domain. Forest reservations, national parks, critical watersheds, and unclassified public land are theoretically under BFD control. The BFD Director estimated in 1982 that approximately two-thirds of the lands under BFD jurisdiction, including most upland areas, were covered by various forest concessions. These concessions include pasture leases, timber licenses, agroforest farms and industrial tree plantations (Cortes, 1982). As of 1982, less than 240,000 ha, or 1.28% of the public domain, were designated for potential distribution of leasehold rights among public land occupants (Bernales, Sagmit, and Bongolos, 1982). Remaining BFD lands are presumably unallocated.

CLASSIFICATION: A HISTORICAL OVERVIEW

The U.S. Congress mandated the colonial government in the 1902 Philippine Bill to classify land according to its agricultural character and productiveness and to regulate the lease, sale, or other disposition of the public lands other than timber or mineral lands. The Philippine Commission accordingly enacted a Public Land Act in 1903 (No. 926) and a Forest Act in 1904 (No. 1148). Public forest and mineral land could not be privately alienated. Public agricultural land was subject to private ownership.

The Public Land Act gave "native settlers" whose occupancy dated back at least six years the right to apply for recognition of ownership. In response, the Chief of the Bureau of Public Lands was empowered to carry out a corresponding investigation as might be deemed necessary. If satisfied as to the application's veracity, the Public Lands Chief was empowered to issue the necessary documentation. But his power was circumscribed by the Director of the Bureau of

Forestry, who was authorized to either certify land as "alienable and disposable" or classify it as "public forest." An interbureaucratic rivalry over classification powers ensued.

To lessen the ambiguities surrounding the classification power, the Philippine Supreme Court ruled in a landmark 1918 decision, *Ramos versus Director of Lands,* that ". . . the presumption should be in lieu of contrary proof that (public) is agricultural in nature" (Philippine Reports, 1918). Besides giving power to the Bureau of Public Lands, the decision bolstered legal rights to private ownership for Filipino citizens who had been occupying the public domain for many years.

A more definitive classification criteria was articulated by the executive branch's chief judicial officer, the Secretary (now Minister) of Justice in 1939. He opined, "In determining whether land is agricultural or not the character of the land is the test. In other words, it is the susceptibility of the land to cultivation for agricultural purposes by ordinary farming methods which determines whether it is agricultural or not" (see also PD 465, 1974). These common sense definitions of agricultural land were increasingly ignored from a legal perspective after 1940. Contrary to the pro-agricultural presumption in *Ramos,* the Republic of the Philippines executive branch began to operate on the premise that public land was to be classified as agricultural only if the Director of the Forestry Bureau did not consider it to be forestal. In other words, the *Ramos* presumption was subtly inverted. Public land was presumed to be forestal unless classified as agricultural. Since vast portions of the nation's total land mass had not yet been (and still are not) classified, the inverted presumption has had a profound impact on the property rights of long-term upland occupants (Lynch, 1982a).

The unheralded policy shift towards the pro-forest presumption provided the executive branch with greater control over public domain resources and shifted the burden of proof to claimants of private lands within the public domain. Illiterate farmers now had to prove that the land they cultivated, and often had inherited, was agricultural in nature. Moreover, these claimants had to do so in an alien legal system. Failure to provide sufficient proof in the manner prescribed meant failure of the application for recognition of ownership.

The Court of Appeals upheld the executive branch's evolving pro-forest presumption in 1966 in *Vicente versus Director of Forestry* (Court of Appeals Reports, 1966). The decision, often cited by executive branch officials, was based on the defunct second "public" land law (Act 2874, 1919) which had been repealed thirty years earlier (Commonwealth Act 141, 1936). The appeals court incorrectly observed that when the *Ramos* presumption was established, a fixed and definite system of classification of public lands did not exist. The courts, therefore, had discretion to decide whether a particular parcel of land was more suited to agriculture or forestry. Apparently, the Appeals Court overlooked the classification scheme in the first Public Land and Forest Acts.

The dissenting opinion in *Vicente* rebuked the majority for refusing to consider properly the evidence submitted regarding suitability of the land for agriculture, as well as large number of fruit trees on the land and agricultural activities dating back at least 30 years. Even more remarkable is that during the same year in which *Vicente* was decided, the Court of Appeals in a related case "reiterated for emphasis" a 1957 decision.

> When the claim of the citizen and the claim of the government as to a particular piece of property collide, if the government desires to demonstrate that the land is in reality a forest, the Director of Forestry should submit to the court convincing proof that the land is not more valuable for agricultural than for forest purposes. . . But a mere formal opposition on the part of. . . the Director of Forestry, unsupported by satisfactory evidence, will not stop the court from giving title to the Claimant (Court of Appeals-General Register, 1957, cited in "Director of Forestry v. Aligno." 1967. Official Gazette 63:34)

Not surprisingly, the Supreme Court never adopted the Court of Appeals' majority rationale in *Vicente*. Besides being poorly grounded in statutory law, *Vicente* represents an abdication by the appeals court of its constitutional obligation to review executive branch decisions which impair private property rights without having first provided due process and just compensation.

Nevertheless, the Supreme Court has appeared reluctant to challenge the executive branch on the classification issue. In a 1963 decision, *Suarez versus Reyes,* the Supreme Court observed that a classification decision by forestry officials has in its favor a "presumption of regularity and observance of due process." The Court did not say, however, that it was without jurisdiction to examine the case (Supreme Court Reports Annotated, 1963). The Supreme Court, held in analogous decisions that findings by executive branch agencies are subject to review if reached arbitrarily, capriciously, or by abuse of discretion (Philippine Reports, 1960). The Supreme Court, however, never considered whether, for example, the 1929 classification of the centuries-old rice terraces in Northern Luzon as "public forest" was arbitrary, capricious, or an abuse of discretion (Proclamation 217, 1929).

Classification: The Certification Component

The Public Land Acts, including the current one enacted in 1936, provide that land classified as agricultural should be certified by the Bureau of Forestry as "alienable and disposable" (A&D). The certification originally did not preclude the courts or the Bureau of Lands from recognizing the private nature of uncertified public land which had been continuously occupied and cultivated for requisite number of years (Borja, 1940).

Today the executive branch insists that A&D certification is a prerequisite to the recognition of private property rights by long-term occupants of public lands. Once land has been certified as A&D, the executive branch admits that rights of private ownership may vest. The Minister of Justice stated in May, 1982, that efforts by the MNR/BFD to lease alienable and disposable land to pre-1945 occupants was "inconsistent with or contrary to the government's intentions" (Opinion No. 67, series 1982). The opinion means the executive branch recognizes that private property rights accrue to long-term occupants of agricultural land within the public domain. But first the land must be certified as alienable and disposable.

It is virtually impossible for most long-term occupants of public lands to have their agricultural land certified as alienable and disposable. The classification of public land as agricultural and its subsequent certification as alienable and disposable is no longer based on local biophysical factors. Instead, by 1984 the classification and certification powers had subtlely merged into a mechanism which strengthens the executive branch's unilateral prerogative to allocate upland tenurial rights to favored individuals or corporations.

The Classification/Certification Muddle

The merger has created a muddle which becomes apparent when the 1974 Ancestral Land Decree and a 1983 Presidential Proclamation are considered. The executive branch certified ancestral land in 27 provinces as A&D in the Ancestral Land Decree (PD 410, Section 8, 1974). The executive action may have been a response to pressure exerted on behalf of ancestral land occupants. From the historical perspective of the executive branch, the A&D certification was a major legal concession on behalf of some indigenous occupants. As noted, the executive branch traditionally concedes that private rights vest in long-term occupants of alienable and disposable public land. The A&D provision in the 1974 Ancestral Land Decree, however, has never been implemented. Qualified citizens on land covered by the decree are still viewed as squatters.

In March 1983, more than 1.5 million ha of public lands thoughout the country were certified by presidential proclamation as alienable and disposable. Jurisdiction over the land was transferred to the Ministry of Human Settlements (Proclamation 2282, 1983). This A&D proclamation also legally strengthened the tenurial rights of thousands of long-term upland occupants within the certified areas. There has been no public acknowledgement of this legal fact, however, by the national government. Indeed, the public rationale for the proclamation did not express a concern for Filipinos already living within the public domain.

The MHS could allocate some of the land to existing upland occupants through its Land Resources Management Program (LRMP), but such an allocation does not appear to be a priority of the program. The purpose of the LRMP is "to transform idle and logged-over areas into land investment trust packages through the financial instrumentation of such areas into equity issues to be known as 'Shares in Kabuhayan' fully guaranteed as to principal and returns by the government." The Shares will provide investors with "undivided proportionate interest" in particular parcels. The most that indigenous and migrant upland families can hope for under the LRMP is to be "conditionally conveyed" a still undefined property right of 5 to 10 ha "upon compliance with" still undefined conditions (Kilusang Kabuhayan at Kaunlaran Land Resources Plan of Operations 1983: 12). The Ministry of Agrarian Reform is also preparing to participate in the LRMP by allocating land rights to qualified lowland citizens.

A related concern is that financial problems may prompt the government to use the LRMP as a means for granting rights over large tracts of land to foreign and domestic corporate investors. Such grants might promote the arbitrary displacement of people living on public lands.

Classification: Assault on the Judicial Prerogative

The expansion of the classification powers of the executive branch is correlated to a contraction of the judiciary's power to recognize long-term private ownership rights within the public domain. The trend accelerated after the declaration of martial law in 1972, and it is best evidenced by an increasing use of the A&D certification power by the executive branch. The Revised Forestry Code (PD 705, 1975), promulgated by martial decree, provides an illustrative example. Section 15 of the Forestry Code declared that "No land of the public domain 18% in slope or over shall be classified (sic) as alienable and disposable." Additional criteria in Section 16 precluded the alienable and disposable certification for areas under 18% slope, but ". . . less than 250 ha and far from or not contiguous with any certified alienable and disposable land." A&D certification was also prohibited for areas previously proclaimed by the President as forest reserves.

The 18% slope rule is predicated on an arbitrary assumption that national environmental stability depends on 40% of the nation's total land area retained as public forest. The 40% figure was reportedly (Cortes, 1982) identified by a forester in 1954. More than 20 years later, the figure became the basis for national policy. The 40% figure would presumably apply on an island to island, as opposed to archipelagic, basis. As of 1981, however, more than 75% of the islands of Bohol, Basilan, Cebu and Negros were identified for classification as agricultural and certification as alienable and disposable. On the other hand, more than half of the total land area in provinces with high concentrations of indigenous occupants was identified for classification as permanent forest (BFD, 1980). These include Agusan del Sur, Ifugao, Mountain, Kalinga-Apayao and Occidental Mindoro.

A reluctance by some judges to invoke the Court of Appeals' 1966 rationale in *Vicente* or to enforce the certification/classification restrictions in the Forestry Code would prompt an executive assault on the judiciary's power to recognize long-term private ownership. A 1977 presidential decree amended the Public Land Act by limiting the power of the courts to confirm private ownership of public lands. This change was accomplished by narrowing the meaning of "agricultural" to include only A&D land within the public domain (PD 1077, Section 4, 1977).

Nevertheless, some judges continued to confirm the private ownership of long-term occupants of agricultural public lands. This judicial independence and other considerations prompted an amendment to the 1973 constitution. Adopted by plebiscite in January 1984, the amendment elevated the certification power of the executive branch to constitutional proportions. The amendment increased the President's power to grant vast tracts of public land and other natural resources. As a result, Section 12, Article XIV of the 1973 Constitution now empowers the executive branch to formulate and implement agrarian reform programs which "may include the grant or disposition of alienable and disposable lands of the public domain to. . . landless citizens." The effect of the amendment on the classification/certification muddle is unclear. Presumably, the amendment is designed to strengthen the executive branch's control over the allocation of public resources. It remains to be seen whether further restrictions of the judiciary's power to confirm private ownership within the public domain will result.

The Amendment also raises several questions, including whether the Constitution now mandates that, prior to judicial confirmation of private ownership, public land must not only be classified as agricultural, but must also be certified as alienable and disposable. Only the executive branch has the power to certify land as alienable and disposable. An affirmative answer would eliminate the judiciary's constitutional power to recognize private ownership within public lands. Another question concerns the definition of landless citizens. Can corporations be considered landless? They are already legally defined as citizens.

RECOGNITION OF TENURE RIGHTS

There is a widespread belief that Filipinos living within the public domain are squatters unless they possess documentary recognition of tenurial claims. The idea was convenient for colonial officials exploiting natural resources, and it continues to benefit some sectors of the Philippine elite. But the idea is incorrect. National land laws recognize four distinct tenurial rights, as well as a variety of correlative rights, which benefit qualified upland occupants. Most of these rights are predicated on occupancy for a specified period of time. The rights overlap in that some citizens within the public domain possess all the rights, others a lesser number. Persons without at least one of these rights could be considered as squatters in that they possess no legal right to the land they presently occupy and cultivate.

Each of the rights emanates from national laws recognized by the Philippine legal community as valid and in force as of May, 1985. Unfortunately, these laws are often not effective. Tenurial rights which upland citizens acquire by virtue of national laws are often not recognized by national and local governments. Furthermore, the processes of obtaining recognition have been increasingly restricted and are prohibitively complex and expensive for the vast majority of upland occupants.

One of the rights benefits only members of "National Cultural Communities," that is, members of ethnic groups which had not been hispanicized by the end of the nineteenth century. The right developed in response to a concern by the now defunct Philippine Congress for the plight of the so-called ethnic minorities. These groups are among the most politically and legally disenfranchised. Most live on lands within the public domain.

Tenure Rights: Native Title

The strongest private property right to public lands is possessed by Filipinos who, along with their predecessors-in-interest, have continuously occupied and utilized an area since "time immemorial". These citizens possess native titles. They are private property owners. This law originated, in a 1909 decision of the U.S. Supreme Court, *Cariño versus Insular Government* (Philippine Reports, 1918). Mateo Cariño was an indigenous uplander from Benguet Province in northern Luzon. He claimed ownership to land customarily inherited for more than three generations. The colonial and U.S. government claimed the land was public and opposed Carino's efforts to obtain recognition of private ownership and acquire a paper title.

In an opinion written by Oliver Wendell Holmes, the Court ruled that ". . . every presumption is and ought to be against the government in a case like the present. . . when as far back as testimony or memory goes, the land has been held under a claim of private ownership it will be presumed. . . never to have been public land." A Philippine Supreme Court decision subsequently held that property owned by native title could be registered under the Torrens system (Philippine Reports, 1923).

The validity of the time immemorial presumption has been reiterated at least six times by the Philippine Supreme Court, most recently in June, 1982 (Supreme Court Reports Annotated, 1982). These cases reinforce the view that at the very least the *Cariño* decision "carved out an exception to the Regalian Doctrine" (Aranal-Sereno and Libarios, 1983: 431). Since ancestral land has never been public, there is no need or legal power to classify ancestral lands as agricultural, alienable and disposable, forest, timber and/or mineral. Except for Carino's claim, however, the executive branch has not only ignored the time immemorial presumption, it has illegally extinguished the private property rights of many.

Section 8 of the 1974 Ancestral Land Decree reveals the executive's familiarity with the *Cariño* precedent. Contrary to the prevailing myth that ancestral land occupants are illegal squatters, the decree gave ethnic minority ancestral land owners in 27 provinces until March 11, 1984 "to perfect their titles. . . otherwise they shall lose their preferential right thereto and the land shall be awarded to other deserving applicants." Other than the threat of Section 8, however, the decree was a sham. It established an overly cumbersome procedure for obtaining a virtually meaningless Land Occupancy Certificate. Not surprisingly, not one ancestral land owner was able to "perfect" his or her title prior to the expiration date.

Section 8 cannot be implemented according to the constitution unless native title holders are provided due process and just compensation. Nevertheless, the decree can now be used to legalize arbitrary displacement of uplanders throughout the Philippines (Lynch, 1983b). A similar threat to holders of private water rights within the public domain is posed by Section 95 of the Water Code of the Philippines. Promulgated by presidential decree in 1976, it gave citizens two years to register claims and provided that "Any claim not registered within said period shall be deemed waived and the use of the water shall thereupon be available for disposition as unappropriated water" (PD 1067, 1976; Lynch, 1984c). This provision is also unconstitutional if implemeted without according due process and just compensation to holders of private rights.

Tenure Rights: Pre-1945 Migrants on "Agricultural" Land

Filipinos who have been in open, continuous, exclusive and notorious possession and occupation of agricultural land of the public domain since June 12, 1945, are "conclusively presumed to have performed all conditions essential to a government grant and shall be entitled to a certificate of title." In 1980, the Philippine Supreme Court declared in *Herico versus Dar* that when these provisions of the Public Land Act ". . . are complied with, the possessor is deemed to have acquired, by operation of law, a right to a grant, a government grant, without the necessity of title being issued. The land, therefore, ceases to be of the public domain." The Court added that lack of application (for documentary recognition of ownership) does not affect the legal sufficiency of the title. Title over the land is "vested" (Supreme Court Reports Annotated, 1980b). The language is unambiguous; the act applies to all citizens who meet the statutory prerequisites, including owners of native title.

Three years before the *Herico* decision, the Public Land Act was amended to limit its application to land certified as alienable and disposable (PD 1077, Section 4, 1977). The land at issue in the Herico decision, however, had already been classified as alienable and disposable. In fact, the Supreme Court has yet to confront the question whether the vested private property rights of pre-1945 occupants of public agricultural land can be divested by a presidential decree which restricts the legal definition of agricultural land. Can a vested right be constitutionally revoked if citizens adversely affected receive no notice, due process, or just compensation (See U.S. Reports, 1950, 1972)?

If standard norms of constitutional jurisprudence are followed, the answer must be an unequivocal "no". Private vested property rights to lands being occupied and utilized are constitutionally protected and cannot legally be taken away without due process and just compensation. These fundamental, participatory principles are enshrined in the 1935 and 1973 Constitutions. Any law adversely affecting these vested private property rights is unconstitutional, unless due process and just compensation are first provided (Philippine Reports, 1928; Supreme Court Reports Annotated, 1980a).

Tenure Rights: Pre-1955 Tribal or Muslim Migrants

Widespread displacement, particularly of tribal and Muslim Filipinos, captured the attention and sympathy of the legislative branch of the national government in 1963. During that year, an amendment to the Public Land Act was passed which provided that an ethnic minority citizen who ". . . continuously occupied and cultivated by himself or through his predecessors-in-interest, a tract or tracts of land, whether disposable or not since July 4, 1955 shall be. . . conclusively presumed to have performed all conditions essential to a government grant and is entitled to a certificate of title" (Republic Act 3872, 1964).

The executive branch looked with disfavor on the amendment. Its public concern focused on the constitutionality of recognizing private ownership to land within the public domain which had not been certified as alienable and disposable. The directors of two executive branch agencies, the Bureau of Land and the Bureau of Forest Development, referred the matter *in consulta* to the Secretary of Justice in 1966. The Secretary did not rule on the matter of constitutionality. Instead, he said that by complying with the amended provisions, qualified occupants would merely enjoy preferential rights to acquire title if the land was classified as alienable and disposable. That is, the executive branch considered the amendment to be legally insignificant.

The Supreme Court never ruled on the legal validity or significance of the 1963 amendment. In September of 1983, however, the Court of Appeals of *Cario Versus Director of Lands* affirmed a lower court decision which recognized private ownership to public agricultural land never certified as alienable and disposable: "Being a native Igorot, appellee is a member of the national cultural minorities. It is, therefore, immaterial whether the land in question is disposable or not. The fact remains that the property has been devoted to agriculture." (Court of Appeals General Register No. 58347-4). The language in *Cario* is refreshingly unambiguous. Unfortunately, it had little impact on realities confronting most upland ethnic minority citizens.

Tenure Rights: The Amnesty of 1975 and Social Forestry

The executive branch's insistence that nearly 62% of the total Philippine land mass is public has not gone unnoticed in many increasingly crowded lowland communities. One result has been increased migration to public lands including upland areas. The migrants' claims often encroach on the private land of indigenous and pre-1945 occupants. Since the land is allegedly public, migrants often believe that they are entitled to claim a portion, regardless of prior rights. Many migrants are emboldened by the government's inability to police such a vast area and prevent illegal occupancy.

The magnitude of migration, the political, legal, and financial implications of prosecuting and relocating thousands of people and the need to relieve population pressures in many lowland areas were reflected in the Revised Forestry Code. Promulgated by presidential decree on May 19, 1975, the code provided that all those who ". . . entered forest lands before the effectivity of this Code without permit or authority, shall not be prosecuted provided they do not increase their clearings" (PD 705, 1975).

The amnesty was hailed as a partnership between public land occupants and national government. Interest in uplanders and in upland ecological problems was also apparent in the Integrated Social Forestry (ISF) Program, launched in 1983. The program provides that all citizens who resided on land under BFD jurisdiction on or before December 31, 1981 are eligible to participate in ISF activities (Letter of Instruction No. 1260, 1982).

The keystone of the ISF Program is the individual leasing of three to seven ha of public forest or unclassified land, or the communal leasing of a still undetermined number of hectares. The possibility for most uplanders of acquiring any type of lease, however, is remote. The BFD Social Forestry Division was allocated less than 32 million pesos (approximately US$1.7 million) in 1984 but received even less due to national economic problems. The amount allocated for social forestry will be smaller in 1985. Most of the money will be spent on salaries for BFD personnel. The number of indigenous and migrant upland occupants who might benefit from an ISF lease in 1985 will be considerably less than the number of poor lowland farmers who will migrate during the same year to public lands.

Correlative Upland Tenurial Rights

Besides the four specific tenurial rights, there is a variety of correlative rights. In many upland communities the payment of real estate taxes on public land provides a certain degree of local land security. The Philippine National Bank, the Development Bank of the Philippines, and some rural banks have been known to use real estate tax receipts as sufficient proof of ownership to justify a loan. The public land occupant's property right is used as collateral (Lynch, 1982). Payment of real estate taxes for public land is prohibited, however, unless the person is a tribal or Muslim Filipino (PD 705, Section 75). This law is of dubious benefit to minority citizens. Two other laws are not so onerous. Section 120 of the Public Land Act states that any conveyance of property rights by an ethnic minority citizen is invalid unless written in a language the citizen understands, or, if illiterate, unless approved by the Office of Cultural Communities (Commonwealth Act 141, section 120, 1936). The Supreme Court said that conveyances not in compliance with this provision are null and void (Philippine Reports, 1942, Supreme Court Reports Annotated, 1964).

Section (g) of Presidential Decree No. 1414 in 1978 prohibited the granting of forest concessions in provinces containing ethnic minorities unless the Office of Muslim Affairs and Cultural Communities (formerly the Presidential Assistant on National Minorities) certified that there were no ethnic minority citizens living within or having a claim to any portion of the area being applied for. This law, however, is rarely applied.

Shifting Cultivation and the "Political Economy of Ignorance"

Karl J. Pelzer observed in a classic study, **Pioneer Settlements in the Asiatic Tropics,** that, "Judgments on the merits of shifting cultivation have been rather contradictory since the practice first came to the attention of scientific observers" (1945: 21). Nine years later, Harold Conklin documented positive aspects of shifting cultivation among the Hanunoo Mangyan of southern Mindoro in his classic, **Hanunoo Agriculture** (1954). The failure of the national government to consider such thought provoking information is disturbing. Efforts to promote government and uplander partnership must recognize that not all shifting cultivation is bad for the tropical ecosystem.

National law still proscribes swidden agriculture. Shifting cultivators (kaingineros) are liable for up to two years imprisonment and a twenty thousand peso (P20,000) fine (PD705, section 69). A less indiscriminate, but still hostile attitude towards shifting cultivation in Indonesia was recently described in an article, "Theories of Swidden Agriculture and the Political Economy of Ignorance." Dove (1983: 90, 95-96) states:

> . . . pioneering studies of swidden agriculture in Southeast Asia long ago demonstrated that long-fallow forest farming is a highly sophisticated productive use of the environment...
>
> The view of forest farming as the destruction of a valuable natural resource, and the view of grasslands as wastelands, have similarly served. . . to justify the extension of external control into the territories of swidden agriculturists, in many cases resulting in the alternative exploitation of forest land for commercial timber extraction. . . The studied ignorance of the role of swidden agriculturalists in smallholder cash crop production has clearly been in the interest of the alternative development of plantations, whether historically or currently, by private or government interests. . .

In all of these cases, it is patently obvious that the prevailing myths have in no way benefited the swidden cultivators themselves. In most cases. . . it can be argued further that the policies supported by the myths also have not benefited the greater society as a whole..

LOCAL LAWS AND CUSTOMS

The foregoing discussion highlights the complex and sometimes muddled state of national laws and policies concerning upland tenure. Equally complex and considerably more varied are local laws and customs concerning upland tenure, particularly those rooted in indigenous tradition and the local ecology. After surveying Philippine literature on property customs among hunter-gatherers, shifting cultivators, Igorot wet-rice cultivators and Islamicized groups, this author concluded that indigenous attitudes toward land in the Philippines tended to be usufructory. Members of a community traditionally possessed the right to use and occupy uncultivated lands. Time and labor spent to develop land or to cultivate crops usually vested rights of varying exclusivity. Boundaries were not drawn on maps but comprised part of a collective folk geography (Lynch, 1983a).

Traditional practices concerning land frequently provided a form of security to all within a local community. Subsistence security in Southeast Asia, however, was not radically egalitarian. Indigenous social relationships usually implied that:

> . . . all are entitled to a **living** out of the resources within the village, and that living is attained often at the cost of loss of status and autonomy. They work, moreover, in large measure through the abrasive force of gossip and envy and the knowledge that the abandoned poor are likely to be a real and present danger to better-off villagers. These modest but critical redistributive measures nonetheless do provide a minimal subsistence insurance for villages (Scott 1976: 5).

Colonization altered indigenous social fabrics and relationships to land. The effects were felt in virtually every Philippine community. But in much of the uplands, the onslaught was less severe, at least until the twentieth century. In Northern Luzon, despite government efforts to introduce a Western system of land ownership to Kalinga, it is only recently that the drive has been successful. Such change came in the wake of the cash economy's increasing penetration into Kalinga society (Aranal-Sereno and Librarios, 1983; Takaki, 1977). Land is increasingly viewed as a commodity to be bought and sold, or otherwise exploited for the profit of individuals having paper title.

RECOMMENDATIONS

Despite some encouraging rhetoric, little political will exists at the national level to address the plight of both indigenous and migrant upland occupants equitably. The meaningful recognition of the uplanders' tenurial claims might hamper large scale and capital intensive exploitation of natural resources within the public domain. Nevertheless, some general recommendations can be made. Most are, of course, contingent on adequate financial and political support.

First, the colonial nature and framework of the national legal system's dominant disposition scheme should be recognized. Allocation schemes that are flexible, participatory, economical, productive, culturally appropriate, environmentally sound, and equitable should be identified and promoted. Efforts need to include small-scale, participatory, and labor-intensive strategies for upland agricultural and forest development.

Second, the precise nature and boundaries of various jurisdictional claims made by executive branch agencies must be determined. Overlapping claims should be eliminated by transferring sole jurisdiction to an organization primarily responsible for, and capable of promoting, sustainable, productive and equitable upland usage. Although in need of reorganization and financial and educational resources, the Ministry of Natural Resources, in particular the Bureaus of Land and Forest Development, are best suited for primary upland responsibility.

Third, arbitrary classification schemes such as the 18% slope rule should be abolished. Classification should be based on biophysical factors prevailing in a given locale. Unclassified land should not be indiscriminately subject to a pro-forest presumption.

Fourth, the BFD's incomplete census of citizens within the public domain should be completed. Meanwhile credible population estimates should be used in upland development planning.

Fifth, the vested, private property rights of qualified upland occupants which emanate from existing national laws should be meaningfully recognized. In other words, all Filipino citizens who have continuously occupied and cultivated land within the public domain since mid-1945 — either by themselves or their predecessors-in-interest — possess constitutionally protected private property rights which should be acknowledged and safeguarded by the national government. For migrant ethnic minority citizens the cutoff date is July 4, 1955. Recognition should be accompanied by an upland zoning law making it illegal to utilize private lands in an ecologically unsound manner. The law could prohibit agricultural practices which demonstrably promote soil erosion, sedimentation, and water yield reduction.

Sixth, upland occupants who settled on public land after 1945/55 but prior to May 19, 1975, should be given a meaningful opportunity to acquire 25-year renewable leases to land they actually occupy and cultivate, unless their presence is harmful to the local environment. Such leases could be processed through the MNR/BFD Integrated Social Forestry Program.

Seventh, areas within the public domain suitable for occupancy and cultivation should be identified and opened up in an orderly fashion to public land occupants who staked claims after May 19, 1975. These areas could also be made available to landless lowlanders willing to migrate and comply with environmental laws.

Eighth, laws prohibiting shifting cultivation should be amended. Indigenous farming systems which are ecologically stable should not be considered illegal. The desirability of long-term, rotational farming on sloping land not susceptible to permanent agriculture should be explored.

Ninth, forest land which is environmentally fragile and genuinely unsuited to human habitation or cultivation should be identified and protected. Any displacement of private property owners must occur only after they have been accorded due process and have been paid just compensation.

Finally, upland communities should be allowed to choose between individual and communal recognition of their private property or leasehold rights. The communal option could save the national government large sums of money by obviating the need for time-consuming and expensive individual ground surveys. Flexible and participatory bureaucratic procedures should be devised to process communal rights and claims. These procedures must allow local, customary variations and ensure specific definition of the community, the communal perimeter, the legal personality of the landholding entity (corporate or documentary), and procedures for resolving conflict between members of a recognized communal entity (see Fernandez, 1980, 1983; and Lynch, 1984d).

CONCLUSION

This chapter highlights the confusion and confrontation concerning upland tenure and the legal frameworks on the national and local levels in which these dynamics occur. It is hoped that the foregoing information and analysis will assist those striving to develop equitable, environmentally sound, and culturally appropriate laws and strategies for allocating and utilizing natural resources in the Philippine uplands. As displacement intensifies and upland resources continue to be rapidly depleted, the challenges, legal and otherwise, grow correspondingly greater and more urgent.

REFERENCES CITED

Aquino, Rosemary M. 1982. Lessons from nine selected case studies. Integrated Research Center. De La Salle University. Manila.

Aranal-Sereno, Ma. Lourdes, and Roan Libarios. 1983. The interface between national land law and Kalinga land law. Philippine Law Journal 58:2:431.

Barrameda, M. Constancy, ed. 1983. Human Rights and Ancestral Land: A Source Book. Quezon City: Ugnayan Pang-Agham Tao.

Bernales, B., Sagmit, and Bongales. 1982. Social Forestry Projects in the Philippines: An Inventory and a Listing of Communal Forests and Pastures. Manila: Integrated Research Center, De La Salle University.

Borja, Alfonso S. 1940. The Law on Public Land Conflicts in the Philippines. Manila: North Star Publishing.

Coggins, George C. and Charles F. Wilkerson. 1981. Federal Public Land and Resource Law. Mineola, New York: Foundation Press.

Conklin, Harold C. 1954. Hanunoo Agriculture: A Report on an Integral System of Shifting Cultivation in the Philippines. Rome: Food and Agriculture Organization.

Cortes, Edmundo V. 1982. Institute on the legal aspects of the management and development of energy and natural resources. Address. University of the Philippines Law Center.

Court of Appeals-General Register. 1957. Ramos v. Director of Lands, et.al. Court of Appeals-General Register 8749-R.

Court of Appeals Reports. 1966. Court of Appeals Reports 10.

Dove, Michael R. 1983. Theories of swidden agriculture and the political economy of ignorance. Agroforestry Systems 1:2.

Fernandez, Perfecto V. 1980. Towards a definition of national policy on the recognition of ethnic law within the Philippine legal order. Philippine Law Journal 55:4.

1983. The legal recognition and protection of interests in ancestral lands of cultural communities in the Philippines. In: Human Rights and Ancestral Land: A Sourcebook. Barrameda, ed.

Fulcher, Mary B. 1982. Dayak and transmigrant communities in East Kalimantan. Borneo Research Bulletin 14:1.

Guatama, Sudargo and Budi Harsono. 1972. Agrarian Law. Bandung: Fakultas Hukum Universitas, Padjadjaran.

Laws of the Indies. 1580. Laws of the Indies. Book 6. Title 1: Law 15.

1594. Laws of the Indies. Book 4. Title 12: Law 9.

Lynch Jr., Owen J. 1984a. The invisible Filipinos: Indigenous and migrant citizens within the 'public domain'. Philippine Law Register 5: 18-22.

1984b. The ancestral land rights of tribal Indonesians: A comparison with the Philippines. Unpublished paper.

1984c. Looking down a murky well: the legal aspects of water rights in the Philippines. Paper, Seminar on A Survey of Water Rights Claims and Disputes in Selected Areas of the Philippines

1984d. Procedural considerations involving the systematic recognition of communal tenure. Paper, Interprofessional Workshop on Legal and Anthropological Interface Issues Concerning National and Indigenous Laws Affecting Cordillera Land Tenure. Baguio City.

1983a. Withered roots and landgrabbers: A survey of research on upland tenure and displacement. Paper, National Conference on Research in the Uplands. Quezon City.

1983b. Freedom from injustice: towards recognition of the human right to ancestral land ownership. Paper, 27th Annual Baguio Religious Acculturation Conference.

1982a. Native title, private right and tribal land law: An introductory survey. Philippine Law Journal 57:2: 268-306.

1982b. Native title: Its potential for social forestry. Likas Yaman 4:1.

Mayo Vargas, Donna. 1984. The interface of customary and national law in East Kalimatan. Ph.D. Thesis (draft). New Haven Yale University.

Pelzer, Karl. 1945. Pioneer Settlement in the Asiatic Tropics. New York: American Geographic Society.

Philippine (Taft) Commission. 1900. Laws and Regulations Relating to Public Lands in the Philippine Islands. Washington, D.C.: Government Printing Office.

_______ 1903. Laws and Regulations Relating to Public Lands in the Philippine Islands. Washington, D.C.: Government Printing Office.

Philippine Reports. 1960. Maribojoc v. Guzman. Philippine Reports 109.

_______ 1942. Porkan v. Navarro. Philippine Reports 73.

_______ 1928. Balbao v. Farrales. Philippines Reports 51.

_______ 1923. Abaog v. Director of Lands. Philippine Reports 45.

_______ 1918. Philippine Reports 39.

_______ 1909. Philippine Reports 41.

Popkin, Samuel L. 1976. The Rational Peasant: the Political Economy of Rural Society in Vietnam. Berkeley: University of California Press.

Scott, James. 1976. The Moral Economy of the Peasant. New Haven. Yale University Press.

Soeratman, Masri Singarimbun and Patrick Guiness. 1977. The Social and Economic Conditions of Transmigrants in South Kalimantan and South Sulawesi. Jogjakarta; Gadjah Mada University.

Supreme Court Reports Annotated. 1982. Manila Electric Co. v. Republic of the Philippines (Castro-Bartolome). Supreme Court Reports Annotated 114.

_______ 1980a. Development Bank of the Philippines v. Court of Appeals. Supreme Court Reports Annotated 96.

_______ 1980b. Supreme Court Reports Annotated 95.

1972. Lee Hong Hok v. David. Supreme Court Reports Annotated 48.

1964. Mangayao v. Lasud, Supreme Court Reports Annotated 11.

1963. Supreme Court Reports Annotated 7.

Takaki, Michiko. 1977. Aspects of Exchange in Kalinga Society. Ph.D. Thesis. New Haven: Yale University.

United States Reports. 1950. Mullane v. Central Hanover Bank and Trust, Co. United States Reports 339: 314.

1972. Grayned v. City of Rockford. United States Reports 408:108.

11
ORGANIZATIONS BEHIND DEVELOPMENT IN THE UPLANDS

Romulo A. del Castillo and Charles P. Castro

The Philippine uplands, a relatively new focus on national development, are now the concern of an increasing number of organizations, both government and non-government. The uplands comprise nearly 60% of the nation's 30 million ha. The Bureau of Forest Development (BFD) was previously the sole steward and administrator of this area, but its task of managing upland development is now taking on new and more complicated dimensions. The traditional view of forestry that resource management is concerned essentially only with biophysical factors is being challenged by advocates of more socially-oriented approaches. The BFD's area of jurisdiction is being entered by numerous agencies and institutions interested in contributing to a more meaningful and holistic national growth. This trend is due to several factors.

First, in recent decades, population has grown against a backdrop of shrinking resources. The problems of unemployment and underemployment, declining wages, lowland congestion, and poverty are pushing people to the uplands. The view that uplands can provide livelihood, habitation, agriculture, and fuel has persisted. Indigenous upland communities are also poor and have increasing populations.

Second, there is now greater interest in people-oriented approaches to development. Planners are recognizing the potentials of the rural poor themselves in aiding in countryside development. According to this view, the uplanders, like the lowlanders, can contribute to nation-building.

Third, cause-oriented organizations are becoming more numerous. Some have noticed the precarious state of the uplands and the plight of the communities there. The uplands have become an area of concern for development planners, resource managers, environmentalists, demographers, human right advocates, social and natural scientists, and missionaries.

Finally, concerned agencies and institutions are realizing the strategic importance of the uplands in national development, especially in light of worsening global food, wood, water, energy, and environment problems. Aside from sheltering an estimated seven-and-a-half million Filipinos, 80% of whom are reportedly absolutely poor, the uplands contain the country's commercial forests, natural parks, wildlife, and mineral resources. They also serve as catchments for major river systems that support vital multipurpose dams, hydroelectric plants, and irrigation works. Properly managed, the uplands can materially and psychologically enrich human life; poorly managed, they could contribute to the disruption of the social, economic, political, and ecological health of the nation.

AGENCIES AND INSTITUTIONS IN UPLAND DEVELOPMENT

Because of the proliferation of agencies and institutions involved in upland development and the speed with which development plans are formulated and implemented, an exhaustive listing is difficult to make. Even a formal survey can miss the informal organizations of tribal and non-tribal communities, those not registered, and those in which the developmental effects on the uplands are real but incidental to the primary purposes of such agencies.

Appendix 1 lists agencies and institutions involved in upland development through research, training, and other activities. A review of publications (see References) together with various other brochures, pamphlets, primers, and periodic reports allowed for the preparation of the matrix of development concerns in appendix 2.

This chapter first considers agencies and institutions by their mandates. Institutional mandates are either the production and management of commodities or the production and delivery of services. Commodities include upland resources such as wood and other forest products, wildlife, water, land, food, energy, and minerals. Services include the promotion of land security, agricultural productivity, cultural or spiritual upliftment, health, education and research, social and political consciousness, livelihood activities, infrastructure improvement, maintenance of peace and order, and integrated area development.

Agencies and institutions are classified into five categories based on organizational mandates and development priorities: a) commodity-oriented organizations, 2) service delivery institutions, and 3) funding institutions.

Commodity-Oriented institutions

Most of the land-based natural resources of the country are found in the uplands. The development and exploitation of such resources are the major concerns of many agencies. This category includes agencies that have threatened the jurisdictional exclusivity of the BFD in the uplands. For example, P.D. 1515 almost transferred exclusive jurisdiction of watershed areas from the BFD to the National Power Corporation, and the Kilusang Sariling Sikap program of the Ministry of Human Settlements nearly removed 1.5 million ha of residual forest areas from the BFD's administrative control.

There are at least eight agencies involved in different resource sectors of the uplands. While interests reflect the agencies' respective functions and objectives (see appendix 2), overlaps and conflicts of resource concerns have occurred.

Wood Resource Agencies. The BFD holds most of the responsibility for forest resources management and conservation, particularly for timber and wood. It controls protection and extraction and, to a large extent, oversees processing and disposal.

However, the recent government thrust in energy production using forest biomass, wood charcoal, and fuelwood has caused other agencies such as the National Electrification Administration (NEA) to use the uplands for the production of wood for dendrothermal plants. The NEA, recently given large upland areas for conversion into dendrothermal plantations, started a project in which uplanders plant ipil-ipil *(leucaena leucocephala)* for fuel and the generation of electricity.

The Ministry of Human Settlements (MHS), the Ministry of Agriculture and Food (MAF), and other community and non-government organizations are interested in the production of wood for fuel and charcoal as part of their agroforestry and upland farming systems programs. The Manila Seedling Bank Foundation (MSBF) and private companies such as the Mabuhay Agroforestry Industrial Tree Farming Program (MAITFP) are planting short rotation trees for wood chips and charcoal.

Large industrial firms such as the Paper Industries Corporation of the Philippines (PICOP) and the Nasipit Lumber Company (NALCO), both members of the Philippine Wood Products Association (PWPA), undertake regular replanting and stand-enrichment planting as required by timber license agreements. These processing firms are establishing industrial forest plantations for sustained production of raw materials. Lately, the government, through the National Development Corporation (NDC), organized the National Industrial Tree Corporation (NITC), which is to obtain low-interest, long-term loans for the establishment of industrial tree plantations in partnership with the private sector.

Water Resource Agencies. The BFD still is responsible for most of the upland water catchment areas that feed streams and irrigation systems. Increasingly though, a provision that all "critical watersheds" should be managed by the National Power Corporation (NPC) is changing the situation, especially with new government multi-purpose dams. The NPC, for example, now controls the watershed adjoining Ambuklao and Binga dams in Benguet. Critical watersheds (according to P.D. 705 and P.D. 151) are those supporting large infrastructures such as dams.

The National Irrigation Administration (NIA) was also given rights over sizeable tracts of hilly land surrounding the Pantabangan Dam in Nueva Ecija and the Magat Dam in Isabela. The Ministry of Public Works and Highways (MPWH) was directed to start construction of mini-dams in rainfed areas to serve as a hedge against dry periods. The Ministry of Agriculture and Food (MAF) is interested in watershed and forest areas because these areas provide agricultural irrigation water.

Land Resource Agencies. The BFD seeks to maintain the uplands on a sustained-yield basis. A lack of lowland areas for agriculture and settlement, however, threatens such efforts. The MAF and MHS are viewing the uplands for agriculture, livestock production, and agro-industries. These agencies have developed programs to manage selected upland areas for agricultural production and for expansion zones for settlement and livelihood activities. The MHS, through its Kilusang Kabuhayan at Kaunlaran (KKK), is currently negotiating for at least 1,000 ha per province of forest zones from the BFD to be converted into Kilusang Sariling Sikap (KSS) production sectors.

The National Council on Integrated Area Development (NACIAD) is exploring the possibilities for development programs in the uplands. It uses an "integrated area development" approach to help make rural areas economically productive, environmentally stable, and socially habitable.

The Ministry of Agrarian Reform (MAR), in an effort to extend the benefits of land reform to forest occupants, may also become involved in projects in the uplands.

Food Resource Agencies. The uplands are viewed by some agencies as potential sources of food for lowland populations. The MAF is intensifying studies on upland farming systems, rainfed and hillside agriculture, and highland cropping systems. Under negotiation is a loan from the

Asian Development Bank (ADB) for the development of the Cordillera highlands for the production of highland fruit trees and other crops.

The International Rice Research Institute (IRRI) has studied rice production and varieties in the uplands. The National Science and Technology Authority (NSTA) has studied wild forest plants as alternative food material. The BFD had a Forest Industries Food Production Program (FIFPP) which involved, among other things, the utilization of upland swamps for production of fish and other water-based food. With an interest in inland fisheries, the Bureau of Fisheries and Aquatic Resources (BFAR) could also work in the uplands.

Mineral Resource Agencies. The uplands harbor much of the country's mineral reserves (gold, copper, nickel, manganese, iron, and cement). The Bureau of Mines and Geo-physical Sciences (BMGS) regulates mining and now could become active in the uplands. The lowland migrants employed by the mining companies have entered the uplands, inviting service institutions.

Wildlife Resource Agencies. Management of wildlife, a concern of the BFD, is limited by a lack of material support. International organizations such as the World Wildlife Fund (WWF) and the International Union for the Conservation of Nature and Natural Resources (IUCN) have started to extend assistance. The WWF through the Films and Research for an Endangered Environment (FREE) and the Haribon Foundation (HF) have come to the rescue of the Philippine Eagle *(Pithecophaga jefferyii)* with a captive-breeding project in Mt. Apo, Davao.

Environmental Agencies. A number of agencies and institutions concerned with environmental conservation and protection have development projects, research, training, and public education and support programs. These include the BFD, the Forest Research Institute (FORI), and the Natural Resources Management Center (NRMC) of the Ministry of Natural Resources (MNR); the National Environmental Protection Council (NEPC); the Forestry Development Center (FDC) and the Institute of Forest Conservation (IFC) of the UPLB College of Forestry; and other non-government organizations such as the Society of Filipino Foresters (SFF) and the Philippine Federation of Environmental Concerns (PFEC).

The uplands host a vast wildlife population, a rich variety of flora, and many indigenous Filipino communities whose colorful cultures may be valuable to the tourism industry. As a result the Ministry of Tourism (MOT) may seek recreation development in the uplands through collaborative efforts with the BFD.

Energy Agencies. In addition to wood, coal and water-generated electrical energy, the uplands are being explored for geothermal power. The NPC, with the collaboration of the Philippine Volcanology Commission, is exploring the harnessing of subterranean heat for conversion into electricity.

Service Delivery Institutions

Common sources of error among development planners included assumptions that the uplands are uninhabited or that the uplanders were inconsequential. Inequities in the distribution of benefits from upland resources resulted. Several private voluntary organizations are changing the situation by addressing human rights, socio-economic development, natural resource conservation, and cultural integration. Unlike the commodity-oriented institutions, this category is dominated by non-governmental organizations (appendix 2).

Private Voluntary Organizations. These are non-stock, non-profit organizations usually founded by social scientists serving the poor. Only a few of these organizations are in the appended directory, but, fortunately, they continue to multiply.

An example of a private voluntary organization involved in upland development is the Philippine Business for Social Progress (PBSP). Founded by Filipino businessmen to acknowledge their social responsibility to country and people, the **PBSP** has contributed to the socio-economic improvement of upland communities and to knowledge about social development through an action research program documenting the changes in given communities. It has complemented the government by operating organizations in the uplands that are concerned with agriculture, education, small economic projects, health, and infrastructure.

The Philippine Association for Intercultural Development (PAFID) helps upland communities to gain access and tenurial security over ancestral lands. It is advocating the Communal Forest Lease for social forestry and is organizing indigenous uplanders to build community capability at their own pace and in their own cultural setting. They believe that a strong community of cultural minorities can better contribute to nation-building.

Religious Missions. A number of small religious groups of varying denominations are working for the education, health, spiritual, socio-cultural, and economic upliftment of upland communities. Among these are the National Council of Churches in the Philippines (NCCP) and the

Episcopal Commission on Tribal Filipinos (ECTF). Both have been militant in seeking justice and better government attention for the minority groups.

The Mindanao Baptist Rural Life Center pioneered an increasingly popular upland agricultural technology and other appropriate livelihood systems for hillyland communities. Some Christian religious orders and the Bahai Faith have set about to improve literacy in upland communities seldom visited by government educators.

International Development Volunteers. The U.S. Peace Corps, the Japanese Volunteers Organization, the Organization of Netherlands Volunteers, the World Neighbors, and other international groups have contributed various upland development efforts and to the openness of many uplanders to change agents. Projects include agroforestry and upland development, cottage industries, health and nutrition, and education.

Government Agencies. The Philippine government has also been trying to alleviate problems in upland communities. The Ministry of Social Services and Development (MSSD) is the lead agency for extending social services to uplanders. The Ministry of Muslim Affairs and Cultural Minorities (which absorbed the PANAMIN) is mandated to integrate underprivileged minority groups into the mainstream of Philippine society. The government goal is to preserve the integrity of the cultures and customs of the minority groups and to guard these groups against exploitation and harassment by other Filipinos.

Armed Services. The negative effects of the activities of this group cannot be ignored, since their presence in the uplands has led to a more intensified national focus on that area. Included are the government military forces, the Philippine Constabulary and the Philippine Army, and their opposition counterparts, the New People's Army. The latter includes armed groups involved in community organization and civic action work. The works of these two opposing groups in the uplands have brought about development which otherwise would have been absent and have sped education about the uplanders' particular situation, their role in Philippine society, and particular strategies or technologies to improve their position.

Community Organizations

"Development" has caught the attention of many uplanders who desire better living conditions. Influenced by external change agents, uplanders have realized the advantages of self-reliance and self-government, have adopted lowland mechanisms for survival, and are trying to harmonize these new mechanisms with their indigenous cultures.

Included are externally-catalyzed groups, tribal groups, non-tribal groups, and commodity-oriented groups.

Externally-Catalyzed Groups. These organizations are exemplified by the Kalahan Educational Foundation (KEF) which, while displaying a strong indigenous orientation, was initiated by an outside agent. The community is now a working model for the Communal Forest Lease System of the Integrated Social Forestry Program of the BFD, has a body of officers to decide issues related to the 14,000-hectare forest lease and has graduated from a bargaining group on forestry and agrarian matters into a self-sustaining land community.

Other examples include the Pondasyon Ng Bagong Buhay Ng Gubatnon Mangyan and the Allah Valley Muslim Association. The former was organized through the assistance of a Peace Corps volunteer and was successful in negotiating with the government for a long-term lease of ancestral lands covering 1,340 ha. The lease had terms and conditions similar to those of the KEF. The latter was formed with the assistance of a former peace corps volunteer who played a key role in generating development funds from three different external sources.

Tribal Groups. This category acknowledges the collective efforts of a number of neighboring tribal communities to resist outside forces. An example of the *pagta* system (also called *bodong* by other tribes) now being used by the native communities of the Central Cordilleras to unite against external threats to their lives and culture. A peace-pact custom that protects individuals from harm by other tribes, the *pagta* has evolved into a mechanism to resist "top-down" government policies affecting uplanders — such as the construction of dams, granting of ancestral land to logging or mining concessions, and militarization.

The Igorot Community Assistance Program (formerly the Igorot Mutual Assistance) was formed through the initiative of an Igorot Episcopal priest, who also played a key role in maintaining the viability of the young organization. The two important programs of ICAP concern land tenure and farmers' training. The former assists members in getting titles to their ancestral lands, while the latter improves the skills of selected farmer leader-demonstrators in various farming technologies, including processing and marketing cooperatives.

Non-Tribal Groups. Similar to the tribal groups, lowlanders who migrate to the uplands may unite in the face of common difficulties. An example is the Ilagas in Central Mindanao who were Visayan migrants to lands formerly held by Manobos and the Muslims.

Upland farmer associations in Villarica and Cuyapo in Nueva Ecija, Buhi-Lalo in Bicol, Hamtic in Antique, Balinsasayao in Negros Oriental,

Calminoe in Laguna, and many other parts of the country are organizations of non-tribal upland farmers organized for mutual benefit. These upland farmer organizations have formed an umbrella organization called Pambansang Samahan Ng Mga Magsasaka Sa Mataas Na Lupa (National Federation of Upland Farmers' Organizations).

Commodity-Oriented Groups. These groups may consist of pure tribals or a combination of tribal and non-tribal peoples. The newly created Cordillera People's Alliance is bound by the desire for security of indigenous upland community land tenure and for government recognition of ancestral lands. This alliance consists of Kalinga, Bontoc, Ibaloi, Kankanaey, and Ifugao victimized or threatened by government projects such as the Marcos Park Complex, the Marcos Highway, and the Chico River Dam, as well as by logging and mining companies.

Other examples include commodity associations such as coffee growers, potato growers, and others.

Research and Educational Institutions

The uplands are still relatively unknown to most development planners and students. The First National Conference on Research in the Uplands (April 11-13, 1983, at Sulo Hotel, Quezon City and participated in by over 150 academicians, researchers, policy makers, and upland project officers) demonstrated that there is still much to learn in terms of ethnography, farming systems, land tenure patterns, poverty, and ecology. Many research groups are working in the uplands, and surveys and studies have proliferated. The institutions in this category are classified as academic institutions, research institutions, and policy study centers.

Academic Institutions. Included are educational institutions offering formal and non-formal courses in forestry, upland agriculture, agroforestry, environmental sciences, applied social and management sciences. Although the primary responsibility of these institutions is teaching, most are also active in research and extension, since these functions are mutually reinforcing.

Institutions include the University of the Philippines at Los Baños, Silliman University, De La Salle University, Ateneo de Manila University, Mountain State Agricultural College, Xavier University, and U.P. College Baguio. Most of these have small operational budgets and are thus dependent on external funding for research and extension programs.

Academic offerings at the masters level include the M.S. in Environmental Studies and in Social Forestry at the University of the Philip-

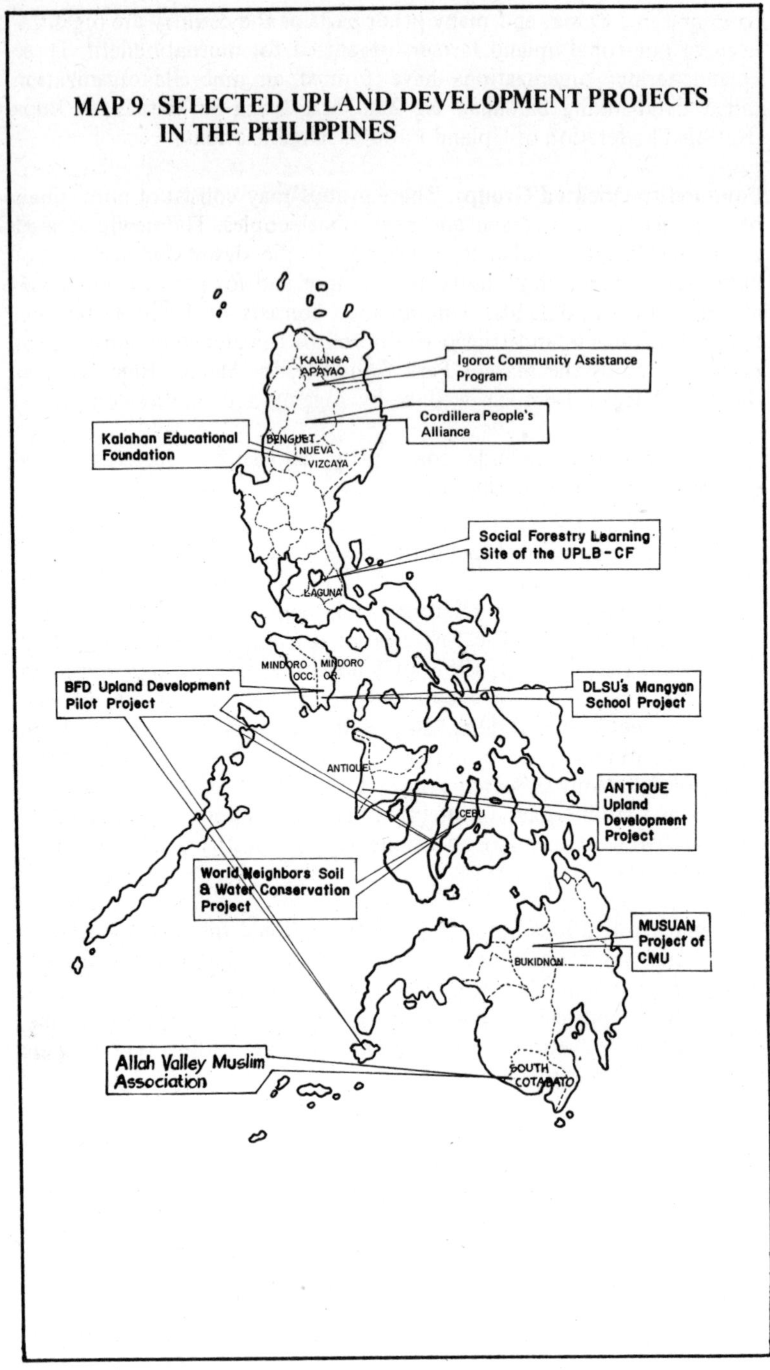

MAP 9. SELECTED UPLAND DEVELOPMENT PROJECTS
IN THE PHILIPPINES
Igorot Community Assistance Program
Cordillera People's Alliance
Kalahan Educational Foundation
KALINGA APAYAO
BENGUET
NUEVA VIZCAYA
Social Forestry Learning Site of the UPLB-CF
LAGUNA
MINDORO OCC.
MINDORO OR.
BFD Upland Development Pilot Project
DLSU's Mangyan School Project
ANTIQUE
CEBU
ANTIQUE Upland Development Project
World Neighbors Soil & Water Conservation Project
MUSUAN Project of CMU
BUKIDNON
Allah Valley Muslim Association
SOUTH COTABATO

pines at Los Baños, in Development Studies at the U.P. College at Baguio, and in Social Development at the Ateneo de Manila University. These programs are now attracting students from the government and the private sectors locally and abroad.

Short training courses in agroforestry, social forestry and related fields are available at some of these institutions, particularly U.P. at Los Baños. The International Institute of Rural Reconstruction (IIRR) is developing rural development workers through various course offerings. IIRR has ventured into action-oriented and participatory research on upland farming systems, the sites of which serve as a field laboratory for training participants.

Research Agencies. Two research institutions are mandated to conduct 'research on the technical aspects of forestry: the Forest Research Institute (FORI) under the Ministry of Natural Resources (MNR) and the Forest Products Research and Development Institute (FPRDI) under the National Science and Technology Authority (NSTA). FORI handles forest production research, which includes biological, physical, socio-economic, and environmental aspects of forestry. FPRDI takes care of forest products utilization research, which includes wood processing, seasoning and preservation, wood products manufacture and finishing, and the various uses of other forest products such as bamboo and rattan. Both institutions are now carrying out research to promote small scale forestry useful for upland community development.

The various colleges of forestry, agriculture, and related fields in applied social sciences are engaged in biotechnical, socio-economic, institutional and management research for upland development. Some institutions are refining field methodologies, developing models, and testing. The Philippine Council for Agriculture and Resources Research and Development (PCARRD) coordinates research and conducts studies of rainfed areas, agroforestry systems, and indigenous forestry technologies.

Policy Study Centers. Through policy analyses, case studies, and other analytical techniques, a policy center should be able to anticipate problems and thoroughly study issues to make possible the formulation and adoption of sound policies. Policy institutions dealing with upland management are the Forestry Development Center (FDC, created by P.D. 1559) and the Institute of Forest Conservation (IFC) at the UPLB College of Forestry, the Center for Policy and Development Studies (CPDS) at the University of the Philippines at Los Banos, and the Philippine Institute of Development Studies (PIDS) of the National Economic and Development Authority (NEDA). The Natural Resources Management Council (NRMC) and the BFD-based informal multi-agency Upland Development Working Group (UDWG) also conduct policy-oriented studies.

Funding Institutions

Any listing of groups working in the uplands will be incomplete without the donor agencies, both local and foreign, who have supported the various programs and projects. Many of the upland development projects could not get started without money to start and to nurture them through various growth stages. This category is divided into private foundations, research funding institutions, bilateral aid agencies, United Nations agencies, and banking institutions.

Private Foundations. A number of private foundations are supporting upland development. Included are organizations that provide technical assistance and/or funding support. Support includes funding for research or developmental activities and, through a line agency, funds that promote multidisciplinary and inter-agency collaboration.

Among the more active of private foundations is the Ford Foundation. One of its current program thrusts in Southeast Asia, particularly the Philippines, is rural poverty and resources, an interest area that includes the uplands and upland development. Present concerns include generating knowledge and expertise in the technological, socio-economic, and institutional aspects of upland development. A recent interest includes the active participation of people in their own development. The following are Foundation-assisted projects in the uplands.

— The BFD Upland Development Program is an agency-based program of research, training and experimentation designed to develop participatory approaches to upland development. It uses a working group approach, combining people of varied disciplines from different institutions to learn and to help the agency deal more effectively with the uplands. Pilot projects are used to test and refine the learnings.

— The Program on Environmental Science and Management (PESAM) is a university-wide program which pioneered an interdisciplinary approach to upland hydroecology and agroecosystems research and training. A newly instituted program leading to a Masters of Science in Environmental Science is a recent PESAM achievement. PESAM maintains a network of upland field projects as a part of its research and training programs, and, through time, has been able to forge strong linkages with line minist ies/ agencies and other policy-making bodies.

— The Highland Research and Development Program (HRDP) is being implemented by the Highland Agricultural Research Center (HARC) of the Mountain State Agricultural College (MSAC) in La Trinidad, Benguet. The program includes participatory research at a Village Research and Demonstration Laboratory (VRDL), and research on farmers' fields. Highland agroforestry technologies are being developed, field tested, packaged and disseminated among Cordillera upland farmers.

— The Cordillera Studies Center (CSC) is a program of the University of the Philippines College at Baguio aimed at developing interdisciplinary research and training. Their work focuses on the lands and peoples of the Cordilleras, the home of one of the largest indigenous Filipino upland populations.

— The Participatory Upland Management Program (PUMP) is building the capability of the De La Salle University Research Center for research and training in participatory approaches to development and the management of upland resources. This program includes two major projects. The Upland Research Center works with institutions in the BFD Upland Development Program and is developing a computerized data bank, clearing house, and a policy study center to aid teachers, students, researchers, policy-makers and others concerned with upland resources. The Mangyan-Appropriate School Project is developing culturally appropriate elementary school education for a group of this tribal people.

— The Social Forestry Program of the UPLB College of Forestry, Department of Social Forestry, is strengthening post-secondary training in social forestry and has a "Learning Site" or field laboratory for its faculty and students.

— Action Research in Upland Agroforestry is a program of the Silliman University Research Center that is developing upland agroforestry models using a multidisciplinary approach and is studying interethnic competition in the use of resources.

— Land Tenure Programs have also received Foundation assistance. The Philippine Association for Intercultural Development (PAFID) assists cultural minority groups seeking secure access to lands handed down by their forefathers and is helping them build organizational capability. The Igorot Community Assistance Program (ICAP) concentrates on a land tenure program in the mountain provinces, particularly Kalinga-Apayao, and provides farmer members security of tenure through the Certificate of Land Title (generally issued on declared alienable and disposable land). ICAP also implements a program of farmers' training with the cooperation of the Mountain State Agricultural College in which selected farmer leaders or demonstrators are trained in upland agriculture, food processing and cooperatives. Another community-based organization, the Allah Valley Muslim Association (ALMUSA), maintains a land tenure program that uses the Association's revolving fund to extend credit to their members and assists them in negotiating and securing (redeeming) their lands from non-Muslim mortgagors.

— The Antique Upland Development Program (AUDP) was initiated by local government. With modest external support, AUDP developed and tested a "Self-Sustaining Small-Time Farmer" model of upland farming, a system said to be environmentally protective. AUDP employed a People's School System in extending the program to province municipalities.

Research Funding Institutions. This group includes the International Development Research Centre (IDRC) and the Agricultural Development Council (ADC) which recently merged into Winrock International. Both provide assistance for research, training, and institutional development, and for the generation of knowledge related to upland development.

Bilateral Aid Agencies. These are agencies of foreign governments which extend funding support to the national government. An example is the United States Agency for International Development (USAID) which has long extended funding to many development projects in the country. A major program currently supported by USAID is the Rainfed Resources Development Program, being implemented mainly by the MAF and the MNR with the support of NEDA, PCARRD and the Office of the Budget Ministry (OBM). USAID also provides support for the Los Banos-based ASEAN-US Watershed Project (AUWP), which covers all the member countries of the Association of Southeast Asian Nations (ASEAN).

United Nations Agencies. Work of the UN in the uplands is carried out by the United Nations Development Programme and the Food and Agricultural Organization. The concerns of these agencies include forestry for rural development. Thus, they are involved in social forestry research and pilot projects, forestry extension, watershed management, timber stand improvement, and other aspects of upland and forestry development. United Nations aid passes through the usual national government channels.

Banking Institutions. Loans to upland farmers contribute to the viability of upland life. The Development Bank of the Philippines supported PICOP's small-holder tree-farming program and, in so doing, encouraged uplanders in Mindanao to go into tree plantations. Other international banking institutions such as the International Bank for Rural Development (IBRD, otherwise known as the World Bank) and the Asian Development Bank (ADB) are showing interest in supporting upland development.

TOWARDS EFFECTIVE INTER-AGENCY COLLABORATION

The list of agencies and institutions is by no means complete. There are surely many more groups and individuals contributing to a national ethic on how upland resources and communities should be dealt with. Most of the agencies and institutions, however, act largely "to each his own." While there may be a convergence of the view that the uplands are critical for national development, overlaps and conflicts can develop in defining jurisdictional boundaries and implementing field activities. Such disharmony results in less work, confusion, and alienation on the part of the affected uplanders. Coordination in the planning and implementation of upland development programs is needed.

Such a proposition could well be wishful thinking given current institutional, political, and technical constraints. Possibly needed is a legislative measure creating a super organization (e.g., a Ministry of Upland Development) to oversee the country's uplands.

Meanwhile, the following are suggested: 1) Speed up research in the uplands. 2) Adopt multidisciplinary, interdisciplinary, and multi-sectoral research approaches. 3) Link research to action. 4) Examine the working group concept for effective interagency collaboration.

The Need for More Research

As pointed out during the First National Conference on Research in the Uplands, the uplands are an open area for research. There is much to be learned about technological, socioeconomic, and institutional aspects of upland management, and individuals, groups, and agencies need to collaborate. Agroforestry technologies, for example, might best be studied if foresters, agriculturists, anthropologists, economists, other professionals, and the affected upland farmers all work together.

A Research Approach

The uplands represent a complex system of social, biophysical, cultural, and political elements. There is corresponding need for upland research to be interdisciplinary and multisectoral. Such research would pave the way for a more holistic understanding. Collective action among participating disciplines and agencies would be possible and conflicting perspectives would decline. Collaborative research would mean better upland development.

Linking Research and Action

Just as research must lead to problem solutions to make it attractive, research results must be useful to make them worthy of the investment. Thus, agency-based research is generally more attractive than those detached from operational realities. Speeding up upland research and making it relevant and useful require that results have immediate and direct application. One way to achieve this is greater collaboration between line agency, researchers, and research institutions.

Such a scheme would assist in achieving relevant, better quality research results, would make results more available and acceptable to the agency, would make possible their immediate use in policy formulation and translation to action, and would decrease long gestation periods in which data eventually become irrelevant and meaningless. Furthermore, bonds between line agency and research institutions would be strengthened, making them more credible to the upland communities, especially if the process involves people's participation.

The Working Group Concept

The difficulties of bringing together professionals and agencies of varying disciplines and interests can be managed through a "working group" approach. The services of a multidisciplinary group from both government and non-government organizations can be harnessed to address specific development problems. For success, it is necessary that members are chosen not only on the basis of professional competence, but also on willingness to experiment and learn together. The working group is better in an informal setting backed by flexible funding and free from rigid and debilitating red tape.

The working group was pioneered in the Philippines by the National Irrigation Administration. It did not function as a permanent or highly formalized body and did not have a fixed membership or staff. To the bureaucrat, such restrictions may not leave enough for an organization, but the working group was effective because it was informal and unstructured.

This mechanism is being tried by the BFD in its Upland Development Program. The BFD Upland Development Working Group is composed of BFD officials involved in the planning and implemetation of the Integrated Social Forestry Program and a few researchers in the bio-physical and social sciences from Ateneo de Manila University's Insti-tute of Philippine Culture, De La Salle University's Research Social Forestry and the Program on Environmental Science and Management, and the Philippine Association for Intercultural Development, with the assistance of the Ford Foundation.

CONCLUDING REMARKS

The efforts of the organizations working in the upland are too many and too varied to make possible a complete listing. The BFD Upland Development Working Group, however, is monitoring such activities for possible lessons that may become the basis for new policies and a more acceptable and relevant national ethic on the use of the country's uplands. The UDWG's goal is to develop effective approaches to parti-cipatory management of Philippine upland areas through research, training, and experimentation. The vehicle for trying out research find-ings is the BFD's Integrated Social Forestry Program, a government program working directly with upland communities on a national scale and on a long-term basis.

APENDIX 1 : DIRECTORY OF AGENCIES AND INSTITUTIONS

A. Government Agencies and Projects

1. Antique Upland Development Program (AUDP)
   Antique Provincial Capitol Building, San Jose, Antique

2. ASEAN-US Watershed Project (AUWP)
   UPLB College of Forestry, College, Laguna

3. Buhi-Lalo Agroforestry Development Project (BLADP)
   Buhi, Camarines Sur

4. Bureau of Forest Development (BFD)
   MNR Building, Visayas Avenue, Quezon City

5. Bureau of Lands (BL)
   Insural Life Building, Plaza Cervantes, Binondo, Metro Manila

6. Bureau of Mines and Geo-Sciences (BMGS)
   L & S Building, North Avenue, Quezon City

7. Bureau of Soils (BS)
   Sunvesco Building, Taft Avenue, Manila

8. Central Visayas Rural Management Project (CVRMP)
   NEDA Building, Mandawe, Cebu

9. Environmental Center of the Philippines (ECP)
   Nayong Pilipino, MIA, Pasay City, Metro Manila

10. Farm Systems Development Corporation (FSDC)
    Rudgen Building, 17 Shaw Boulevard, Pasig, Metro Manila

11. Ministry of Agriculture and Food (MAF)
    MAF Building, Elliptical Road, Diliman, Quezon City

12. Ministry of Agrarian Reform (MAR)
    Philippine Tobacco Authority Building, Elliptical Road, Diliman,
       Quezon City

13. Ministry of Education, Culture and Sports (MECS)
    Aroceros Street, Metro Manila

14. Ministry of Health (MH)
    San Lazaro Compound, Rizal Avenue, Metro Manila

15. Ministry of Human Settlements (MHS)
    University of Life Complex, Pasig, Metro Manila

16. Ministry of Justice (MJ)
    Padre Paura Street, Metro Manila

17. Ministry of Local Government (MLG)
    New Quezon City Hall, Quezon City

18. Ministry of Muslim Affairs and Cultural Minorities (MMACM)
    Delta Building, Quezon Boulevard, Quezon City

19. Ministry of National Defense (MND)
    Camp Emilio Aguinaldo, Epifanio de los Santos Ave., Quezon City

20. Ministry of Natural Resources (MNR)
    MNR Building, Visayas Avenue, Quezon City

21. Ministry of Public Works and Highways (MPWH)
    Bonifacio Drive, Port Area, Manila

22. Ministry of Social Services and Development (MSSD)
    Batasan Complex, Constitution Hills, Quezon City

23. National Council on Integrated Area Development (NACIAD)
    2nd Floor, FBI Building, 60 Timog Avenue, Quezon City

24. National Economic and Development Authority (NEDA)
    NEDA Building, Amber Avenue, Ortigas Complex, Pasig,
        Metro Manila

25. National Electrification Administration (NEA)
    Dendro Thermal Development Office, Delta Building, Quezon
        Boulevard, Quezon City

26. National Environmental Protection Council (NEPC)
    6th Floor, Philippine Heart Center for Asia Building, East
        Avenue, Diliman, Quezon City

27. National Industrial Tree Corporation (NITC)
    c/o National Development Corporation, 6th Floor, Vernida IV,
        Alfaro Street, Salcedo Village, Makati

28. National Irrigation Administration (NIA)
    NIA Building, National Government Center, EDSA, Quezon City

29. National Power Corporation (NPC)
    BIR Road, Diliman, Quezon City

30. Natural Resources Management Center (NRMC)
    5th Floor, Triumph Building, 1610 Quezon Avenue, Diliman,
    Quezon City

31. Presidential Committee for Wood Industries Development (PCWID)
    Rosario Building, Scout Chautoco Street, Quezon City

32. Philippine-Australian Development Assistance Program (PADAP)
    Zamboanga del Sur Development Project, Pagadian City, Zam-
    boanga del Sur

33. Rainfed Resources Development Program (RRDP), Agroforestry
    Component
    MNR Building, Visayas Avenue, Diliman, Quezon City

B. Academic and Research Institutions

1. Agrarian Reform Institute (ARI)
   University of the Philippines at Los Banos, College, Laguna

2. Asian Institute of Management (AIM)
   Paseo de Roxas, Makati, Metro Manila

3. Central Luzon State University (CLSU)
   Munoz, Nueva Ecija

4. Central Mindanao University (CMU)
   Musuan, Bukidnon

5. Central Philippines University (CPU)
   Iloilo City

6. College of Forestry (CF)
   University of the Philippines at Los Banos, College, Laguna

7. College of Human Ecology (CHE)
   University of the Philippines at Los Banos, College, Laguna

8. College of Law (CL)
   University of the Philippines, Diliman, Quezon City

9. Cordillera Studies Center (CSC)
   University of the Philippines College at Baguio, Baguio City

10. De La Salle University Research Center (DLSU-RC)
    De La Salle University, 2401 Taft Avenue, Metro Manila

11. Development Academy of the Philippines (DAP)
    San Miguel Avenue, Pasig, Metro Manila

12. Don Mariano Marcos Memorial State University (DMMMSU)
    Bacnotan, La Union

13. Forest Products Research and Development Institute (FPRDI)
    UPLB College of Forestry Campus, College, Laguna

14. Forest Research Institute (FORI)
    UPLB College of Forestry Campus, College, Laguna

15. Forestry Development Center (FDC)
    UPLB College of Forestry, College, Laguna

16. Institute of Forest Conservation (IFC)
    UPLB College of Forestry, College, Laguna

17. Institute of Philippine Culture (IPC)
    Atneo de Manila University, Loyola Heights, Quezon City

18. Institute of Social Work and Community Development (ISWCD)
    University of the Philippines, Diliman, Quezon City

19. Isabela State University (ISU)
    Cabagan, Isabela

20. Highland Agriculture Research Center (HARC)
    Mountain State Agricultural College, La Trinidad, Benguet

21. Mindanao State University (MSU)
    Marawi City

22. Mountain State Agricultural College (MSAC)
    La Trinidad, Benguet

23. Notre Dame of Marbel College (NDMC)
    Koronadal, South Cotabato

24. Palawan National Agricultural College (PNAC)
    Aborlan, Palawan

25. Pampanga Agricultural College (PAC)
    Magalang, Pampanga

26. Panay Island Consortium on Research for Agricultural Development (PICRAD)
    c/o CPU, Iloilo City

27. Philippine Council for Agriculture and Resources Research and Development (PCARRD)
    Los Baños, Laguna

28. Philippine Institute for Development Studies (PIDS)
    106 Amorsolo St., Legaspi Village, Makati, M.M.

29. Program on Environmental Science and Management (PESAM)
    Univ. of the Phil. at Los Baños, College, Laguna

30. Research Institute for Mindanao Culture (RIMCU)
    Xavier University, Cagayan de Oro City

31. Silliman University Research Center (SURC)
    Dumaguete City, Negros Oriental

32. Visayas State College of Agriculture (VISCA)
    Baybay, Leyte

C. Non-Government Organizations

1. Agency for Community Educational Services (ACES) Foundation
   No. 12, 11th Avenue, Cubao, Quezon City

2. Allah Valley Muslim Association (ALMUSA)
   P.O. Box 8001, Marbel, South Cotabato

3. Asian Social Institute (ASI)
   1518 Leon Guinto Street, Malate, Manila

4. Cordillera Peoples' Alliance (CPA)
   U.P. College of Baguio, Baguio City

5. Ecology and Systematics Society (ESS)
   c/o Department of Life Sciences, UPLB-CAS, College, Laguna

6. Episcopal Commission on Tribal Filipinos (ECTF)
   CAP Building, 372 Cabildo Street, Intramuros, Manila

7. Farmers Assistance Board, Inc. (FABI)
   Room 210 Engalla Building, 1428 Taft Avenue, Ermita, Manila

8. Farmers Service Bureau (FSB)
   La Ignacia Apostolic Center, 2215 Pedro Gil, Sta. Ana, M.M.

9. Federation of Free Farmers (FFF)
   41 Highland Drive, Blue Ridge Subdivision, Quezon City

10. Haribon Foundation (HF)
    5th Floor, Marietta Apartments, 1200 Jorge Bocobo, Ermita, MM

11. Igorot Community Assistance Program (ICAP)
    Tabuk, Kalinga-Apayao or Church of the Resurrection
      Compound, Magsaysay, Baguio City

12. Kalahan Educational Foundation (KEF)
    Imugan, Santa Fe, Nueva Vizcaya

13. Mabuhay Agroforestry Industrial Tree Farming Program (MAITFP)
    Naawan, Initao, Misamis Oriental

14. Manila Seedling Bankd Foundation (MSBF)
    Quezon Boulevard Extension corner EDSA, Quezon City

15. Mindanao Baptist Rural Life Center (MBRLC)
    Dansalan, Davao del Sur or P.O. Box 94, Davao City

16. Mother Earth Ecosystem Society (MEES)
    Institute of Forestry, Gregorio Araneta University Foundation,
    Malabon, Metro Manila

17. Nasipit Lumber Company (NALCO)
    Nasipit, Agusan del Sur

18. National Congress of Farmers Organization (NCFO)
    41 Highland Drive, Blue Ridge Subdivision, Quezon City

19. National Council of Churches in the Philippines (NCCP)
    879 Epifanio de los Santos Avenue, Quezon City

20. Organization for Training, Research and Development (OTRADEV)
    4 Mahinhin Street, U.P. Village, Diliman, Q.C.

21. Pambansang Samahan ng mga Magsasaka sa Mataas na Lupa
      (PSMML)
    c/o PESAM, U.P. at Los Banos, College, Laguna

22. Philippine Action for Cultural Ties (PACT)
    c/o NCCP, 879 E. de los Santos Ave., Quezon City

23. Philippine Association for Intercultural Development (PAFID)
541 Retiro Street, Quezon City

24. Philippine Business for Social Progress (PBSP)
4th Floor, Yutivo Building, 270 Dasmarinas Street, Binondo, MM

25. Philippine Federation on Environmental Concern (PFEC)
P.O. Box 772, Manila

26. Philippine Wood Products Association (PWPA)
3rd Floor, LTA Building, 118 Perea Street,
Legaspi Village, Makati, Metro Manila

27. PICOP Agroforestry Development Project (PICOP-ADP)
Paper Industries Corporation of the Philippines Bislig, Surigao del
Sur

28. Program for Rural Agro-Industrial Services, Inc. (PRAISE)
P.O. Box 243, Pasay City, Metro Manila

29. Samahang Ekolohiya (SAMAEKO)
UPLB College of Forestry, College, Laguna

30. Share and Care Apostolic for Poor Settlers (SCAPS)
SAIDI Building, Intramuros, Metro Manila

D. International Institutions

1. Asia Foundation (AF)
Condominium #8, Corner Yakal and Mayapis Streets,
Makati, Metro Manila

2. Asian Development Bank (ADB)
ADB Building, 2330 Roxas Boulevard, Metro Manila

3. Ford Foundation (FF)
Rm. 601 Dona Narcisa Building, Paseo de Roxas, Makati, Metro
Manila

4. Foster Parents Plan, Inc. (FPPI)
Cacha Building, Del Pilar Street, Calapan, Or ntal Mindoro

5. Institute of Cultural Affairs (ICA)
11 Castilla Street, New Manila, Quezon City

6. International Development Research Center (IDRC)
Asia Regional Office, Tanglin P.O. Box 101, Singapore

7. International Institute of **Rural** Reconstruction (IIRR)
Silang, Cavite

8. International Rice Research Institute (IRRI)
Los Banos, Laguna

9. Resources Management International (RMI)
Ramon Magsaysay Building, Roxas Boulevard, Manila

10. Southeast Asian Regional Center for Graduate Study and
Research in Agriculture (SEARCA)
College, Laguna

11. Summer Institute of Linguistics (SIL)
P.O. Box 2270, 12 Horseshoe Drive, Quezon City

12. United States Agency for International Development (USAID)
15th Floor, Ramon Magsaysay Building, Roxas Boulevard, M.M.

13. United States Peace Corps Volunteer (USPCV)
2139 Agno Street, Malate, Manila

14. World Neighbors Soil and Water Conservation Project (WNSWCP)
P.O. Box 286, Cebu City

15. World Vision Philippines (WVP)
P.O. Box 527, MCC, Makati, Metro Manila

16. World Wildlife Fund International (WWF)
World Conservation Centre, 1196 Gland, Switzerland

APPENDIX 2: MATRIX OF DEVELOPMENT CONCERNS IN THE UPLANDS

Columns under **COMMODITY**: EC En Fo La Mi Wa Wi Wo AP AS CS — Columns under **SERVICES**: ER He ID In Li LT SP

NAME	EC	En	Fo	La	Mi	Wa	Wi	Wo	AP	AS	CS	ER	He	ID	In	Li	LT	SP
Gov't. Agencies & Projects																		
AUDP	+		*			+		+	*			*						
AUWP	+	+				+		+				*						
BDF	*	+	+	*		+	*	*	+						+			
BL				*												+		
BLADP	+		*			+		*	*			*				+		+
BMGS		+			*													
BS	+			+														*
CVRMP	+		+					+	*									*
ECP	*																	
FSDC		+	+					+	*				*					
MAF	+		*					+	*			+				*		
MAR				+														
MECS	+											+	*					
MH														*				
MHS	+	+	+					+	+			+			*		*	
MJ																+		*
MLG																+	*	
MMACM								+		*						+	+	
MND											*							
MNR	*	+	+	*	*	+		*	*					*			+	
MPWH												+						*
MSSD												*		*				
NACIAD																		
NEA		*				+		+						*				
NEDA																		
NEPC	*																	
NIA			+			*		+				+						
NITC								*										
NPC		*				+												
NRMC	*			+	+	+	+	+				*						
PADAP			+					+		+						+		
PCWID								*										
RRDP	+		+					+				+					+	
Foreign Institutions																		
ADB		+	+		+	+		+										
AF									+								+	+
FF	+		+			+		+				+						+
FPPI									+							+		
ICA											+							
IDRC	+	+	+					+	+			+						
IIRR									+			+				+		+
IRRI			+		+													
RMI												*						
SEARCA	+		+					+	+			+						
SIL											+							
USAID	+	+	+			+		+	+			+						
USPCV										+		+				+	+	+
WNSWCP	+		+	+		+		+	+									+
WVP									+							+		
WWF	+							+										

Academic and Research

NAME	EC	En	Fo	La	Mi	Wa	Wi	Wo	AP	AS	CS	ER	He	ID	In	Li	LT	SP
AIM												+						
ARI												*				+		
CF	+	+	+	+		+	+	+				*						
CHE	+											*						
CL												*						
CLSU	+							+				*						
CMU	+		+					+	+			*						
CPU								+				*						
CSC	+											*				+	+	
DAP	+											*	+			+	+	
DLSU-RC										+		*	+					+
DMMMSU	+		+					+	+			*						+
FDC	+			+			+	+	*			*						
FORI	+	+	+					*	+			*						
FPRDI								*				*						
IFC	*											*						
IPC										+		*				+	+	
ISU												*						
ISWCD												*						+
HARC	+		+					+	+			*						
MSAC	+		+					+	+			*						
MSU	+		+					+	+			*						
NDMC									+			*				+	+	
PAC			+					+	+			*						
PCARRD	+	+	+	+	+	+	+	+	*			*						
PESAM	+		+	+			+	+	+			*					+	
PICRAD									+			*						
PIDS	+		+					+				*						
PNAC	+		+					+	+			*						
RIMCU										+		*				+	+	
SURC	+		+					+	+	+		*						+
VISCA	+		+					+	+			*						

Non–Goverment Organizations

NAME	EC	En	Fo	La	Mi	Wa	Wi	Wo	AP	AS	CS	ER	He	ID	In	Li	LT	SP
ACES											*							+
ALMUSA									+							+		+
ASI										+								+
CPA									+							+		+
ECTF											+					+		+
ESS	+																	
FABI								+										
FFF								+			+					+		+
FSB								+										
HF	+						+											
ICAP								+		+						*		+
KEF								+		+						*		+
MAITFP		+					+											
MBRLC		+					+			+	+							
MEES	+																	
MSBF	+						*				+							
NALCO	+						*											
NCCP										+						+		+
NCFO								+										
OTRADEV																		+
PACT									+									+
PAFID										*	+					*		+
PBSP								+			+					+		+
PFEC	+							+										

PICOP-ADP	+	+		+	+			+	
PRAISE					*				
PSMML		+		+	*		+		+	+
PWPA				*					
SAMAEKO	*								
SCAPS						+			+

EXPLANATORY NOTES:

a. Column legend —
   1. Commodity: EC = Environmental Conservation, En = Energy, Fo = Food, La = Land, Mi = Mineral, Wa = Water, Wi = Wildlife, and Wo = Wood.
   2. Services: AP = Agricultural Productivity, AS = Armed Services, CS = Cultural/Sprititual, ER = Education/Research, He = Health, ID = Integrated Development, In = Information, Li = Livelihood, LT = Land Tenure, and SP = Social/Political.
b. Acronyms in rows: as previously defined in appendix 1
c. Symbols in cells: * = major concern, and + = minor concern.

APPENDIX 3 . READINGS

Aguilar, Filomeno Jr. V. 1982. Social Forestry for Upland Development: Lessons from Four Case Studies. Institute of Philippine Culture, Ateneo de Manila University, Loyola Heights,Quezon City.

Alvarez, Jesus B. 1984. Bridging the gap between research and policy formulation: The case of the BFD upland development working group. Paper presented at the meeting of the Working Group on Agroecosystem Research and Rural Resource Development Policy held on July 9-13, 1984 at the East-West Center, Honolulu, Hawaii.

Aquino, Rosemary M. 1982. Lessons from experiences in social forestry . Findings from nine selected case studies. Paper presented at the 11th Social Forestry Forum held November 4, 1982 at the Sulo Hotel, Quezon City

Bernales, B.C., et al. 1982. Social Forestry Projects in the Philippines: An Inventory and A Listing of Communal Forests and Pastures, 178pp. Integrated Research Center, De La Salle University, Taft Avenue, Manila.

Castro, Charles P. 1984. Uplands and uplanders. In search for new perspectives. Proceedings of the First National Conference on Research In the Uplands. C.P. Castro, ed. BFD Upland Development Program, Social Forestry Division, Bureau of Forest Development, Visayas Avenue, Diliman, Quezon City.

Cortes, Edmundo V. 1983. Management of natural forests. Proceedings of the First ASEAN Forestry Congress, October 10-15, 1983 at the Philippine International Convention Center, Manila. 205-209.

Del Castillo, Romulo A. 1980. Education and training needs in support of forestry for local community development. Proceedings of the FAO/SIDA Seminar on Forestry in Rural Community Development held on December 3-15, 1979 in Chiang Mai, Thailand. pp. 127-143.

Del Castillo, R.A. and E.S. Guiang. 1985. Rural employment through forestry activities. Paper for the IX World Forestry Congress held on July 1-10, 1985 in Mexico City.

Garilao, Ernesto D. and D.N. Morales. 1982. Non-government organizations in the uplands: Issues and prospects. Paper presented at the Participatory Approaches to Development Seminar Series, September 30, 1982 at De La Salle University, Taft Avenue, Manila.

Gibbs, Christopher and Jeff Romm. 1982. Institutional aspects of forestry development in Asia. Paper presented at the Conference on Forestry and Development in Asia held on April 19-23, 1982 in Bangalore, India under the sponsorship of Asia Society/USAID.

Korten, Frances F. 1982. Building national capacity to develop water users' Associations: Experience from the Philippines. World Bank Staff Working Papers Number 528. Agricultural and Rural Development Department, The World Bank, Washington, D.C., U.S.A.

Lambatlaya. 1983. Newsletter of the Network for Participatory Development (Inaugural Issue — Sept. to Oct. 1983) Inst. Center, Manila. pp. 647-659.

Makil, Perla Q. 1982. Toward a Social Forestry Oriented Policy: The Philippine Experience. Institute of Philippine Culture, Ateneo de Manila University, Loyola Heights, Quezon City.

Payuan, Edwin V. 1983. Social forestry in the Philippines. Proceedings of the First ASEAN Forestry Congress, October 10-15, 1983 at the Philippine International Convention Center, Manila. pp. 647-659.

PBSP. 1982. Directory of private voluntary organizations and cooperatives in the Philippines. A report prepared by the Philippine Business for Social Progress, Binondo, Metro Manila.

PESAM 1983. Directory of Farmer-Leaders, Researchers, Development Workers and Organizations Working Towards Upland Development.

2nd Edition, April 1983. UPLB Program on Environmental Science and Management, College, Laguna

Sajise, Percy E. 1982. Our critical uplands. . . And what we can do about them. Inaugural Issue of the Philippine Upland World, First Quarter 1982. BFD Upland Development Program, Visayas Avenue, Quezon City. pp.4-10.

12
PUTTING SOCIAL AND COMMUNITY FORESTRY IN PERSPECTIVE IN THE ASIA-PACIFIC REGION

James Kirchhofer and Evan Mercer

The purposes of this paper are: 1) to review briefly the concepts of community and social forestry, 2) to propose a typology of the various "social" approaches to forestry, and 3) to offer some comments on both the potential and the limitations of the community forestry concept.

Throughout the developing world, a myriad of projects and programs are being implemented under the titles "social forestry," "community forestry," "village woodlot forestry," and others. In concept, at least, these activities have, as a common denominator, a "social" approach to forestry — an approach that involves local participation in forestry-related activities to meet local needs.

Noronha (1982) suggests that the social approach to forestry differs from traditional forestry in three ways: 1) it is concerned with the non-monetized sector of the economy, 2) it involves direct participation of the beneficiaries, and 3) it requires that the forester change his role from that of "protector" of public forests to that of extension agent at the community level.

The forestry-related activities to be encouraged at the local level have been described by the FAO (1978) as:

> a spectrum of situations ranging from woodlots in areas which are short of wood and other forest products for local needs, through the growing of trees at the farm level to provide cash crops and the processing of forest products at the household, artisan or small-industry level to generate income, to the activities of forest-dwelling communities.

Specifically excluded are industry forestry or other forms of forestry which contribute to community development solely through employment of wages.

IMPETUS TO SOCIAL FORESTRY

Although a few nations (e.g., Korea, China, India) have utilized social or community forestry programs for several decades, widespread international interest in social approaches to forestry did not arise until the mid-1970s. Factors giving impetus to this interest included:

- a growing concern over the rapid rate of forest cutting and forest conversion in much of the tropical world, coupled with a not fully-substantiated belief that encroachment by people hungry for fuelwood and other forest products was a major cause of forest clearance (deforestation)
- a belated recognition of the critical dimensions of fuelwood shortages in parts of Africa and Asia, including the implications for nutrition and the diversion of dung from fertilizer to fuel use
- a realization that policies promulgating industrialization were not effectively attacking the problems of rural poverty
- the growth of an environmental ethic, giving rise to increased awareness of the ameliorative effects of forests on deteriorating or degraded environments
- a slowly dawning recognition by professional foresters that their custodial attempts to police the public forest domain were fruitless without local, public support and
- the fact that an increasing number of developing countries made the transition from wood-surplus exports to wood-shortage importers and felt impelled to make internal adjustments in their wood economies.

INSTITUTIONAL RESPONSE

This constellation of factors generated important institutional and organizational responses in the late 1970s. The World Bank, although not abandoning assistance to the industrial forestry sector, shifted its emphasis to encourage small-scale, local forestry activities designed to promote local development (Spears, 1978). In 1978, the FAO issued its position paper on "Forestry for Local Community Development" and, thereafter, has cooperated with the World Bank to actively promote the social forestry approach. In the same year, the Eighth World Forestry Conference in Jakarta adopted as its theme, "Forestry for People," and examined in detail the implications of social forestry for both the profession and its practitioners. Thus, encouraged by international and national donor agencies, many nations have now established social forestry programs. In fact, today some form of a social forestry program exists in every Asian nation. A research agenda for social forestry has been set forth by Romm (1982).

THE APPEAL OF SOCIAL FORESTRY

The widespread and rapid acceptance of the social approach to forestry is related, no doubt, to its broad appeal. Various proponents of social forestry have suggested that the establishment of village wood-lots, the afforestation of wastelands, and forest farming for cash crops would solve many different problems at once. Villagers with their own supply of fuelwood would have no need to collect fuelwood in national or state forests or biosphere reserves, which many nations find impossible to police. Village woodlots would solve the problem of fuelwood and fodder shortages, enable the creation of cottage industries, and supply building materials. The establishment of community forests would reduce erosion, local flooding frequency, downstream sedimentation in reservoirs, ameliorate village climates, and in many cases beautify the landscape. Social forestry would cost relatively little and would be self-perpetuating. Finally, the idea appealed to some Westerners because a sort of participatory democracy was to be established at the village level. It would be difficult for anyone to resist an approach with so many potential benefits.

CLASSIFICATION

From such beginnings, the social approach to forestry, now tempered by both successes and failures, continues to flourish and, indeed, has evolved into such a wide variety of programs at the international scale that a tentative classification scheme may be useful. One possible typology follows. The categories are by no means mutually exclusive. A major purpose in presenting the classification is to distinguish between the broad, all-encompassing category, "social forestry," and one of its components, "community forestry."

Social Forestry Programs

Sociological considerations dictate the division of social forestry into programs requiring collective action and programs requiring individual action. Programs requiring collective action include national campaigns, special interest group activities, and community forestry (village-based). Programs requiring individual action include trees for farmers, trees for residual areas, and contractual programs. Social scientists note that resource development innovations requiring collective adoption rather than individual adoption are more difficult to introduce due to the necessity for consensus and simultaneous action (West, 1978). Indeed, in most cases, social forestry programs at the individual level have been more successful than programs aimed at the community level. For example, in Gujarat, the "farm forestry" program that provides seedlings to individual landowners for ther own use has been resoundingly successful (Tawari, 1982), while the village self-help reforestation projects on communal forest lands have been disappointing (although some suggest that it has been too early for a fair evaluation).

The successful introduction of innovations through collective action requires a greater understanding of and sensitivity to the socioeconomic and sociocultural conditions of the community. For example, the designation of lands to be utilized for a collective social forestry program faces greater obstacles than for projects directed at the individual. In individual action situations, each farmer or contractor simply decides to use his/her own land or a designated plot for forestry purposes. When the inputs and benefits are to be shared by the community, however, an insensitive choice of sites can be disastrous.

For example, Eckholm (1975) described an example of a failed community reforestation project in an area of Ethiopia operating under a "quasi-feudal land tenure system." Since the land selected for the project belonged to a powerful landlord, the local people, believing they would not receive a fairshare of the benefits, sabotaged the project by planting seedlings upside down. In contrast to the Ethiopian situation and many villages in India, Noronha (1981) attributed the success of Korean village woodlot projects to the homogeneity of the Korean villages, unsegmented by caste, tribal affinity, or great disparities in wealth.

Beyond the distinction between collective and individual social forestry activities, classification becomes more difficult. Nonetheless, three general kinds of collective activity can be distinguished.

National Campaigns. Often symbolic, these campaigns are designed to raise the level of awareness of the benefits of tree planting. Included here are the Festival of Tree Planning *(Van Mahotsva)* in India (Srivastava, 1978) and national tree planting days in Gambia and Senegal (Honskins, 1980).

Special Interest Group Activities. Hoskins (1980) has described how urban women in Kenya solicited money for a tree planting campaign. Money collected by the women was used to pay rural women or handicapped individuals to plant and care for trees for a five-year period. In Senegal, Boy Scouts, and school groups are hired during the summer to plant trees. In India, Nature Lovers' Clubs, Friends of Trees Clubs, the YMCA, and many schools and universities are active in reforestation efforts (Dalvi, no date; Pant, 1980).

Community Forestry. Community programs are directed at the better management of village woodlots, the designation of parcels of public (state) forest as community forest, the reforestation of degraded public forest lands, the aforestation of wastelands, and the development of village level forest product, cottage, and artisanal industries to improve the living standards of the village. The primary characteristics of community forestry activities include collective decision making and action, and the sharing of benefits and costs by the community as a whole. As community forestry is the focus of this chapter, more detailed descriptions and analysis of these programs follow the discussion of the classification of social forestry programs.

Social forestry programs utilizing individual action also can be divided into three subclasses.

Trees for Farms. These programs encourage tree farming by individual landowners. In some cases, agroforestry systems (see following discussion of agroforestry) are encouraged, and in other cases, farmers with lands marginal for any kind of annual agriculture are encouraged to convert solely to tree crops. The Paper Industries Corporation of the Philippines (PICOP) joined with the Philippine Development Bank in a successful program that provided loans, technical assistance, and a guaranteed market to individual landowners to develop tree farms to produce pulpwood for a nearby pulp mill (Hyman, 1983b). Although the program was originally intended to encourage agroforestry, 93% of the participating farmers preferred to plant only trees because the pulpwood production was more profitable in the long run, and most of the participants were relatively well-off financially and able to wait for the financial returns (Hyman, 1983a). If poorer farmers are included in the program in the future, the restoration of the agroforestry approach would provide needed income during the eight years prior to tree harvesting.

Trees for Residual Areas. This version of social forestry encourages individuals or families to plant trees around homes and other private lands in both rural and urban areas. In India, through the "A Tree for Every Home and a Forest for Every Village" program, the Forest Department provides technical assistance and tree seedlings to individuals to plant around their homes. Both shade/aesthetic species and also economic species such as mango, guava, coconut, and curry leaf are provided under the assumption that, if every family (both rural and urban) planted one tree annually, at least 100 million trees would be planted every year. Another program in India entrusts the establishment and care of particular trees along avenues in residential areas to individual families (Krishna Murthy, 1982). Similar programs have existed in China since 1958, when Mao exhorted his people to "cover the country with green trees" (Rao, 1983). The Javanese homegardens provide an excellent example of a traditional agroforestry system developed spontaneously without a program. For centuries, gardens surrounding village homes in Java have included not only annual food crops but also a wide variety of trees for production of food, wood, fodder, and green manure (Widagda et al., 1984).

Contractual Programs. Many social forestry programs are based on contractual relationships between landless farmers and an outside entity (forest departments, private companies). In India, the "Social Security through Forest Plantations" program assigns 37.5 ha plots of degraded forest land to landless families who replant at a rate of 2.5 ha/yr for a 15-year period. The families receive a fixed salary, building materials, minor forest products, and 20% of the net profit from the harvest at the end of the 15-year period (Dalvi, no date). Many *taungya* agroforestry schemes that incorporate agricultural crops in the first few years of plantation development are also included under this class of social forestry. For example, the "prosperity approach" of the Indonesian State Forestry Corporation, Perum Perhutani, contracts farmers to carry out reforestation work in exchange for the use of inter-row spaces for crop cultivation (Atmosoedarjo and Banvard, 1978). As Rao (1983) stated, "In Indonesia it is now accepted that social forestry implies two-way traffic. e.g., foresters are allowed to go into the village to promote tree planting on agricultural lands, while villagers are encouraged to go to the forest to plant food crops."

AGROFORESTRY AND SOCIAL FORESTRY

Although agroforestry often is designated as a subdivision of social forestry, we view agroforestry schemes as production techniques suitable to a wide variety of contexts including but not limited to social forestry. Agroforestry can be defined as any system that combines (either spatially or sequentially) the production of woody perennials on the same unit of land as agricultural crops and/or livestock. The objectives usually include more stable, sustainable land use, soil conservation, increased net income, and/or risk minimization through diversification of production.

In this light, agroforestry techniques are compatible with (and indeed desirable for) several of our social forestry subdivisions. In community forest programs, taungya agroforestry techniques could be used in village woodlot establishment to provide returns to the community during the early years of the project prior to tree harvests. In community village woodlot projects in Khon Kaen, Thailand, silvopastoral techniques (agroforestry combinations of trees with grazing) were used to overcome the problem of the traditional use of village lands for grazing (Rathakette, 1983). Agroforestry possibilities under "individual action social forestry" are numerous and applicable to all three subdivisions as described above.

Furthermore, agroforestry techniques are not limited to social forestry programs but are also applicable to large industrial enterprises such as the Jari Forestral Project in Amapa, Brazil (Budowski, 1982). Thus, agroforestry is viewed as a production technique suitable to many social forestry schemes but not as a subdivision of social forestry.

COMMUNITY AND SOCIAL FORESTRY

The remainder of this paper is devoted to community village-based forestry programs requiring collective action — as a subset of the broader subject of social forestry. It should be recognized, however, that in some instances, it is difficult to make a clear distinction. In India, for example, there are two types of village woodlot programs: "supervised" and "self-help." Under the supervised system, the Forestry Department requests a village to set aside land (leased or owned by the community) and then undertakes the work of planting, maintenance, and protection. At harvest, the village receives 50% of the profit. The self-help woodlots, on the other hand, are managed entirely by the village with the Forestry Department providing free seedlings and technical advice. All profits go to the village. Clearly, the two schemes utilize significantly different levels of involvement by the villagers. However, since both programs involve collective decision making, action, and sharing in woodlot profits, both may be considered community forestry projects. Furthermore, in some village woodlots in West Africa, while trees are planted on communal land, the individual who planted each of the trees regards herself of himself as the owner of the identifiable plants.

The Potential of Community Forestry (or Why Community Forestry?)

Community forestry programs offer both the greatest challenge and the greatest opportunity for positive change in rural communities. Although individual action social forestry is easier to initiate and often less costly, the results too often fail to significantly aid the truly needy segments of the community and only enhance the financial situation of individual landowners who are already relatively well-off. Successful rural development schemes must include all strata of the community and ensure that benefits reach the poorest of the poor, the landless, and the unemployed. In individual action social forestry, however, the pattern of land ownership continues to determine who receives the benefits.

For example, one of the stated goals of the **PICOP** project in the Philippines was to improve the financial and employment situation of a substantial number of *kaingineros* (shifting cultivators). Yet, only land-owners were eligible to participate in the program (virtually eliminating the inclusion of kaingineros). In fact, the income of the eligible partici-pants averaged more than twice that of the Philippine national average (Hyman, 1983b). Furthermore, one of the reasons the farmers partici-pating in the PICOP project avoided the agroforestry approach was because it required greater labor inputs. Thus, although the project was successful, from PICOP's perspective, in reducing the cost and uncer-tainty of industrial raw materials and for the landowners who reduced labor costs and increased their profits, it did little to increase employ-ment or to improve the living standards for those in the community most in need.

In India, generating new employment opportunities and raising incomes of the rural poor (particularly landless agricultural laborers) are two major objectives of the social forestry program. Far too often, however, the programs directed at individuals achieve exactly the oppo-site effect. According to Shiva et al. (1982), conversion from food crops to *Eucalyptus* plantations results in a loss of 250 person days of employment per hectare per year, and despite the government's call for the "utilization of hitherto unutilized communal lands" for social forestry, 10,000 farmers in Gujarat have converted lands from food production to monocultures of Eucalyptus (which is usually sold direct-ly to the pulp and rayon industries). The resulting reduction in employ-ment opportunities and local food availability has led to increasing pressures on reserved forests by villagers who remove firewood not for their own use but for sale in urban and semiurban areas.

These are but two examples. Apparently, in individual action social forestry, market demand (primarily from large urban areas or industrial operations) dictates the patterns of land use, production techniques, and choice of tree species. Individuals, seeking to optimize their own financial situation, tend to reduce labor inputs and utilize lands and species that bring the highest price from national markets with little consideration of the local community's needs, desires, and ability to pay.

As Shiva et al. (1982) concluded:

> While afforestation can be taken up primarily for high commercial returns by individuals, if it is to lead to improvement in com-munity services, the better satisfaction of basic needs and a stable resource base, then the involvement of the community in plan-ning, raising, and using the forests becomes a practical necessity.

Thus, of the many varieties of social forestry, community forestry offers the best hope for broad-based rural development.

Limitations of Community Forestry

During the last decade, much has been learned about the extraordinary complexity that must be overcome in order to successfully implement community forestry projects. A concept appealing in its simplicity has turned out to be exceedingly difficult to put into practice. Scattered throughout the literature are many examples of lessons that have already been learned. Community forestry cannot be imposed from the top down — local residents must be involved at every stage of the planning process. The choice of which species of tree to plant is as much a cultural as an ecological decision. Means must be devised to permit consultation with the women of the community who have much to do with the choice of fuel for cooking or heating. The distribution of the benefits of the project must be spelled out in detail before the project commences. Governments must demonstrate a commitment to the project in terms of budget, human resources, and priority. Technical expertise must be available. Foresters must learn to assume new roles. Credit must be available. Short-term benefits must be offered in order to induce acceptance of a program with mainly long-term benefits. And, finally, the socioeconomic structure, culture, and local politics must be understood thoroughly.

Community Forestry in a Rural Development Context

To view the above lessons as conditions that must be met before implementing community forestry projects is probably an unrealistic goal. Yet their existence suggests another, perhaps more important, lesson: If community forestry is to succeed on a significant scale, it must be integrated into an overall, multifaceted rural development program.

The shortage of fuelwood in parts of the Third World is only a symptom of a serious population/environment imbalance, the causes of which are imperfectly understood. Certainly these include population growth, the breakdown of traditional values, unanticipated impacts of national policies, and growing inequities in the distribution of the benefits of modernization. Planting trees will not eradicate these problems. In Nepal's steeplands, much illegal fuel wood cutting is due not only to the search for scarce fuelwood, but also to a perceived need to convert more forestland into grazing or cropland (Bajracharya, 1981). This problem, too, cannot be rectified simply by planting more trees for fuel.

On a national scale, the only village woodlot program to reach its goals is that of the Republic of Korea. This program is notable not only for its scale, but also for the flexibility of its approach. The program began with a national survey of requirements and estimates of yield throughout the country. Fuelwood plantations were established on lands under varying types of ownership. The key to the program was the establishment of Village Forestry Associations (VFAs) that are integrated into a national hierarchy and that are intimately involved in local-level planning. In some cases, the VFAs are entitled to collect fuelwood and other forest products from "reserved forests" in return for forest protection activities. In other cases, the VFAs have entered into voluntary "yield sharing contracts" with private landowners (70%-75% of the forest land in Korea is privately owned), whereby the VFAs manage the land for the landowner and benefits are shared. "Trust management" of private lands also can be mandated by the government. In such cases, the VFAs take over management of the land, either through a yield sharing contract or by requiring reimbursement from the landowner for expenses incurred. Finally, there is a system for leasing "not to be reserved" forests to VFAs. If managed according to government standards, the forests are granted outright to the VFAs (Arnold, 1982).

Most significant, perhaps, is that the structure and function of the woodlot program complemented the existing rural development program already in place, called *Samaeul Udong* or "new community." This program was developed in order to reduce the income disparities between rural and urban dwellers. Thus, in 1973, when the government embarked on the national reforestation program, the village woodlot program was not viewed as a radical innovation but, instead, was seen simply as another element in the "new community" program. The experience of the Korean program suggests the advisability of coupling community forestry with other rural development programs.

Factionalism at the village level has been mentioned as a potential problem in community forestry projects. If an array of benefits are available to the various separate interest groups of a village through a rural development program, resistence or sabotage by any one group is less likely. Romm (1979) stated that since a village is a collection of people with different powers and interests:

> everyone whose cooperation is required for the success of a (collective) project must gain at least enough from it to sustain his or her commitment to it. The technology introduced must not simply be profitable for the village unit. It must be able to satisfy the minimal requirements of all necessary participants.

Finally, the experience of professionals outside the forestry discipline offers insights into the risks of myopic visions of community forestry. The plant geneticists who fathered the Green Revolution now know that agriculture cannot be divorced from its social, economic, and political context; higher yields do not necessarily signal the end of rural hunger. Sanitation engineers have learned the importance of literacy; civil engineers have been discomforted by the lessons of ecology; and public health workers, because of past mistakes, now often turn to the field of anthropology.

Putting Community Forestry in Perspective

Putting community forestry in perspective, then, begins with the understanding that the complex and multifaceted causes of rural poverty will not yield to any single remedy, no matter how well planned or brilliantly executed. In proper perspective, community forestry should be viewed as a valuable tool, but one subject to many limitations. Most importantly, these limitations need to be communicated to government planners and policymakers who, without guidance, may come to view community forestry as a panacea.

Putting community forestry in perspective also means that foresters should resist being placed in a position where they will fail. To the extent possible, this means insisting that community forestry projects be but one component of an overall rural development effort. Certainly there is value of expanding the scope of a forester's training to include exposure to economics and the social sciences, but the forester who is asked to leave his traditional habitat and venture into a village should seek the help and guidance of the economist, social scientist, and anthropologist. Indeed, the success of his own activities may largely depend upon his success in ensuring that representatives from these or related disciplines are there in the village with him. The fuelwood and fodder shortages plaguing parts of the Third World are symptoms of deep social problems far beyond the capacity of forestry alone to solve. Alleviation of these shortages will depend upon the extent that governments understand this fact. It is only within such a context that community forestry can succeed (see Rambo, 1983).

Finally, one should recognize that "rural development itself is like an old hat, shapeless and made to fit any head" (Rao, personal communication). The objectives, strategies, and target groups for rural development vary from government to government and region to region. Indeed, not all governments agree (and, in fact, many are opposed to the concept) that a community approach involving local people in decision-making and benefit-sharing, and addressing the collective needs of the rural poor, is essential for rural development.

Thus, although successful community forestry projects will necessarily be flexible and molded to fit the needs, abilities, and socioeconomic and cultural contexts of each community, a framework for an "ideal" community forestry program may be needed. A major thrust of this framework would include an analysis of the interactions and interrelationships betwen social systems and ecosystems and the implications for rural development and community forest management.

REFERENCES CITED

Arnold, J.E.M. 1982. Community forestry development in Korea. Paper presented at conference "Forestry and Development in Asia, " Bangalore, India, 19-23 April, under auspices of the Asia Society on behalf of the U.S. Agency for International Development.

Atmosoedarjo, S. and S.G. Banyard. 1978. The prosperity approach to forest community development in Java. Commonw. For. Rev. 57(2): 89-97.

Bajracharya, D. 1981. Implications of fuel and food needs for deforestation: an energy study in a hill village Panchayat in Eastern Nepal. Ph.D. dissertation. University of Sussex, Brighton, U.K.

Budowski, G. 1982. Applicability of agroforestry systems. In : Agroforestry in the African Humid Tropics. L.H. Macdonald ed., 13-16. Tokyo : United Nations University.

Dalvi, M.K. no date. Social forestry : An overview of the Indian experience. Unpublished manuscript.

Eckholm, E.P. 1975. The other energy crisis. Firewood American Forests 81(11): 12-13.

FAO. 1978. Forestry for local community development. FAO Forestry Paper No. 7. Rome.

Hoskins, M.W. 1980. Community forestry depends on women. Unasylva 32 (130): 27-32.

Hyman, E.L. 1983a. Pulpwood tree farming in the Philippines from the viewpoint of the small holder: An ex post evaluation of the PICOP project. Agric. Admin. 14(1983): 23-49.

1983b. Smallholder tree farming in the Philippines. Unasylva 35 (139) 25-31.

Krishna Murthy, A.V.R.G. 1982. Planning and execution of people's forestry and energy plantation programs in a developing country.

In: Tropical forests: Sources of Energy through Optimization and Diversification (proceedings of Int. Forestry Seminar, November 1980). Serdang, Selangor, Malaysia . Malaysia: Penerbit Universiti Pertanian. pp. 29-305.

Noronha, R. 1981. Why is it so difficult to grow fuelwood? Unasylva 33 (131): 4-12.

– 1982. Seeing people for the trees: Social issues in forestry. Paper presented at conference "Forestry and Development in Asia," Bangalore, India, 19-23 April, under auspices of the Asia Society on behalf of the U.S. Agency for International Development.

Pant, M.M. 1980. The impact of social forestry on the national economy of India. Int. Tree Crops Journ. 1 (1980) 69-92.

Rambo, A.T. 1983. Community forestry – the social view. Paper presented at "Regional Workshop on Community Forestry," Korat, Nakhon Ratchasima, N.E. Thailand, 22-29 August 1983.

Rao, Y.S. 1983. Social forestry and rural development. FAO Regional Office for Asia and the Pacific. Bangkok, Thailand.

Rathakette, P. 1983. Practical community forestry: Actual experiences of village woodlot project at Pra Youn District, Khon Kaen Province, Thailand. Paper presented at "Regional Workshop on Community Forestry," Korat, Nakhon Ratchasima, N.E. Thailand, 22-29 August 1983.

Romm, J. 1979. Local organizations for managing natural resources: The distribution of economic incentives and social shares. A & P Agricultural and Resources Staff Seminar, Ford Foundation Asia and Pacific, Yogyakarta, Indonesia, November 1979.

1982. A research agenda for social forestry. In. Tree Crops Journ. 2 (1) 25-59.

West, P.C. 1978. Some sociological aspects of forestry community development projects in developing countries. Paper presented at Eight World Forestry Congress, Jakarta, 17 October.

Widagda, L.C.O. Abdoellah, G. Marten, and J. Iskandar. 1984. Traditional agroforestry in West Java: The pekarangan (homegarden) AND Kebun-talun (perennial-annual rotation) cropping systems. East-West Environment and Policy Institute Working Paper, Honolulu.

13
CHANGE AND "DEVELOPMENT" IN THE UPLANDS: A SYNTHESIS OF LESSONS, UNRESOLVED ISSUES, AND IMPLICATIONS

Sam Fujisaka and Percy Sajise

This volume has dealt with change in the uplands. Change takes place and was examined in terms of ecosystemic, demographic, agricultural, cultural, social, economic, and policy variables. Somewhat naturally occurring change includes, for example, population growth, shifts from subsistence to cash economies, and increased privatization of lands within upland communities. Ecosystems change concomitant to human activities and practices includes deforestation and forest conversion and possible degradation. Change has also taken place in evolving laws, government policies, and institutional arrangements. Guided or introduced change — "upland development" — involves development of new or improved technologies for local adoption and strategies for related project implementation.

The chapters examined the interrelated topics of ecosystem, population, development strategies and technologies, the economics of such technologies, cultures and societies, land tenure and national laws, and agencies working towards upland development. Case studies compared several social forestry projects, discussed the dynamics of a pioneer migrant community, and outlined technology development in a land-reform community. Social and community forestry in the Asia-Pacific region were briefly considered. The contributions as a whole maintain that problems in the uplands start with ecosystem degradation and overpopulation and extend to losses of ancestral lands and poorly managed or coordinated upland development efforts. Problems were shown to be serious and worsening.

The essays gave evidence for several lessons or conclusions. While tropical forest ecosystems are productive, fragility and susceptibility to human use disturbances were said to have led to substantial losses and degradation. Population increases — especially if uplanders now represent almost one of every three Filipinos — were described as having exacerbated environmental problems while making development of possible solutions more difficult. Technologies being developed and tested to improve the productivity and sustainability of upland agroecosystems often offer only slight, long-term improvements while requiring significant immediate investments unavailable to most uplanders.

Upland societies and cultures are diverse and changing, with change often disruptive and associated with greater pressure on resources. Uplanders are a relatively powerless group within the national context and, as such, face substantial difficulties in obtaining secure rights over upland areas. Tribal or traditional groups are faced with continued losses of ancestral lands, while migrant uplanders are both displacing indigenous groups and are subject to similar land tenure insecurity. As upland development projects have been implemented, problems of inappropriate technologies, few real project benefits, low levels of participation, and poor or short-sighted planning have been apparent and were mentioned as problems. Research in specific upland communities reveals a range of social and institutional problems — especially concerning land access and competition — and demonstrates a need for careful, site-specific technology design, testing, and project implementation.

Some issues were and are unresolved. More information is needed to understand what have been thought to be problems, the severity of such problems, dynamic interactions among sets of factors comprising problems, and possible problem solutions. For example, there is a lack of consensus as to the nature and magnitudes of problems affecting tropical forest ecosystems. For the Philippines, data on rates of deforestation, watershed degradation, soil erosion losses, and upland population magnitudes and increases are incomplete. Problems encountered in the uplands and in carrying out upland development are being studied, and the increasing concern and effort being turned towards the uplands and the uplanders are encouraging. Going beyond individual component topics, however, this chapter attempts a systemic synthesis of findings, evaluates problems facing the uplands, uplanders, and upland development in the light of lessons learned and other research evidence, and offers a simple (eco)systems model-diagram of change and development.

THE TROPICAL RAINFOREST ECOSYSTEM AND HUMAN PRACTICES

The conceptual starting point of the book was the interaction between tropical rainforest, the predominant upland ecosystem, and human activities within that ecosystem. The tropical rainforest was characterized as a dynamic and previously stable system in which human settlement, exploitation, and — relatively recently — overuse have led to degradation and destruction. While the world's tropical forests have existed for 60 million years, deforestation has been extensive in about the last two hundred. Given poor basic information, however, the magnitude of the problem with estimates of global tropical forest deforestation ranging from 7 million to 20 million ha per year is controversial.

The Philippines was reported to be predominantly forested until the turn of this century. Today, forests are estimated to constitute from 30% to 50% of the country's 30 million ha land area. Estimates of deforestation rates vary, although a government figure of 200,000 ha in 1970 has been considered a reasonable estimate. Of the 16.7 million ha classified as public forest (56% of the land area of the country), about 5 million ha were described by the government as "open, denuded, and unproductive" as of 1982. Deforestation has resulted from logging and wasteful logging methods, conversion of uplands for agriculture, and exploitation for fuelwood.

In Sajise's structural-functional description, the three interdependent productive, protective, and regenerative functions are key to understanding the tropical forest ecosystem. Biomass production is high. Representing one-third of the earth's forests and 4% of its surface, the tropical rainforest produces four-fifth, of the earth's vegetation and fixes at least 25% of its terrestrial carbon. Productivity of the stable system is enhanced through high water-holding capacity, high cation exchange capacity, and stable and porous soil structure.

The canopied layers of the tropical forest protect soils from high solar radiation, heavy and intense rainfall and strong winds. The ecosystem maintains required high humidity, carbon dioxide, and a diverse gene pool, and exists on a tight nutrient budget with continual, almost closed cycling between biomass and a thin topsoil and humus layer. Important mineral nutrients are held in the biomass, although most nitrogen and phosphorous reserves are in the top soil. Nutrients are transferred through decomposition of litterfall, timberfall, and root materials, and are taken up by plants and recycled. The total stock of nutrients is small, but recycling is rapid and efficient. Without significant disturbances, tropical rainforests have self-sustaining and regenerating capacities.

Stability depends upon disturbances not exceeding certain threshold values and upon relatively constant environmental conditions. Human-generated disturbances, such as logging and agriculture, have disrupted and changed forest ecosystems. Effects are said to include altered hydrological cycles, reduced upland watershed capacities, increased downstream flooding, increased soil erosion and decreased land productivity, and losses of genetic resources. Logging, for example, opens the upper canopy, exposes the soil, and can result in permanent forest loss.

Shifting cultivation is a major form of land use by settlers and most native peoples in the tropical forest. The practice, especially when combined with increasing populations and decreasing fallow periods, was also said to contribute to soil impoverishment, leaching, erosion, and expansion of grasslands.

There is general agreement about the basic structure and functioning of the tropical forest ecosystem and about the effects of disturbances in causing systems change or conversion and loss of genetic resources. There is less agreement as to the extent of reduced watershed capacities, increased downstream flooding, increased soil erosion, decreased land productivity, and the harmful effects of shifting cultivation.

Hamilton (1983, 1985) detailed evidence showing that assumptions about the magnitude of negative effects of tropical forest cutting and transition to grasslands have not been proven. Grasslands, if not overused and over-grazed, can actually be superior to forest in terms of water yield and soil erosion control. Sherman (1980) also compared grassland and forest — grasslands have more water run-off, less leaching, and only twice the sediment loss of primary forest. The *Imperata cylindrica* grassland root system improves soils, and because farmers are waiting for biomass regeneration and not fertility restoration, grassland shifting cultivation has naturally developed as a viable alternative to other types of swidden.

The benefits of burning for the shifting cultivator includes ash rich in minerals (K, P, Ca) and the neutralization of acidic soils (due to K_2CO_3). Some nitrogen and sulfur is lost, however, and swiddens do not produce much humus and are subject to declining soil fertility, weeds, insects and diseases (Parfitt, 1976). Productivity and sustainability of shifting cultivation depends upon regeneration, which, of course, depends upon adequate fallows. Intensification of root crop swiddens in New Guinea created a chain of events in which fallows decreased with greater population; more sweet potato rather than yams and taro were produced; root crop diversity decreased; and root crop pests increased (Gagney, 1981). In Belize, intensification led to a decline in soil fertility and crop yields, with phosphorous becoming the limiting factor, although plot abandonment was a response more to weeds than declining fertility (Arnason, 1982).

Plot yields decrease due to deterioration of soil nutrient status and physical conditions, changes in soil flora and fauna, and increases of weeds, pests, and diseases. Regeneration after abandonment is favored by few and selective weedings, short cropping periods, tree planting, and fire prevention. On the other hand, cash cropping, population increases, and over-use of fire have led to degradation (Clarke, 1976). Comparing rainforest and gardens in New Guinea, Wood (1979) found that with cultivation there were reductions in available P, exchangeable Ca and Mg, cation exchange capacity, and percent C and N. Decline in soil fertility was due largely to reduced soil organic matter.

Chin and Chua (1984) examined human impacts on a Malaysian tropical forest. Frequency data on trees and species in primary and secondary/fallow plots suggested that diversity was higher in the secondary forest. *Macaranga* was "the genus par excellence of pioneer trees" (Whitmore, 1975: 71, cited by Chin and Chua, 1984); common Dipterocarps used for timber were missing from the secondary forest there were more fruit trees in the secondary forest. On the other hand, Kartawinata et al. (1981) studied logging and shifting and permanent cultivation in East Kalimantan in which "genetic erosion" combined with few benefits to the people were the main effects.

From research in Mindanao (Kellman, 1969), softwood tree fallows recycled the greatest amount of nutrients and showed rapid increases of soil carbon, but original soil qualities could not be regained under most shifting cultivation systems, with nitrogen as the main limiting factor. Considering erosion, runoff, and different regeneration stages, prolonged cultivation is found to be the main culprit in degradation. Woody fallow, a stage in effective land protection second only to forest cover, depends upon not cropping, weeding, and burning too many times.

Furtado (1978) carefully examined the dynamics of Southeast Asian tropical moist forests and agreed that cultivation reduced species diversity, exposed soils to surface runoff and erosion, and too-heavy use led to conversion to grassland. On the other hand, Zambales Negrito root crop swiddens — located on grasslands **intentionally** maintained by burning — are productive sustainable, even though located in an area that is otherwise ". . . the ideal setting for extreme rates of sheet erosion and soil loss" (Brosius, 1982).

In all, much is being learned about the dynamics of the interactions between man and the tropical forest ecosystem. Some issues are unresolved and need further investigation:

1. There is general agreement as to structural-functional characteristics of the tropical rainforest ecosystem, and that human practices have disrupted and changed forest eco-

systems. However, assumptions about the magnitudes of negative effects of tropical forest conversion have not been proven. While deforestation from logging has been extensive, more field research is needed concerning forest conversion, regeneration, soil losses, hydrological changes and downstream flooding, and characteristics of grassland ecosystems. Shifting cultivation — even on grasslands — can be productive and sustainable, depending on regeneration through adequate fallows. Field research must continue to examine the human practices of clearing, burning, cropping, weeding, and shifting on ecosystems in terms of soil characteristics and nutrient status, soil flora and fauna, weeds, pests, diseases, water run-off and nutrient losses, regeneration, and "downslope" effects. Evidence shows that plot shifting and fallows allow regeneration, thus maintaining agroecosystems, while the increased cropping, weeding, and burning associated with many systems of land use intensification lead to systems degradation.

INCREASING POPULATION

Population increases in the uplands can exacerbate environmental problems in that the "adequate fallows" of shifting cultivation require available lands for plot shifting, which in turn, depend upon sufficiently low population densities. Among others, Cruz discussed migration to the uplands, population pressure, and resource use, presented data and calculations to estimate the current upland population, and gave evidence for continued heavy population movements to the uplands.

Estimates — mostly "armchair" deductions — of the upland population range from less than half a million people in the mid-1960s to 7.5 million in 1983. Cruz used actual available number and derived a sample estimate using 486 municipalities, 1,179 barrios, 1970 NCSO census figures, and a listing of 1983 BFD social forestry project communities. The result was a startling 11 million people in the uplands as of 1970, then representing about 29% of the total population and over one-half of the total 1960-70 migrant population of the country. If the estimate is accurate, one of the author's conclusions is understated. "It is on the basis of these estimates that the problem of population pressure in the uplands appears to be much greater than the current literature leads one to expect."

The method of estimation and resulting population figure will probably stir up some needed discussion, however. The relationship between reported municipality and barrio totals and estimated numbers of uplanders may be subject to further discussion. It was somewhat unclear if all residents of a barrio or municipality in which there was a social forestry project were included as "uplanders". In our experience, the municipalities seeking to administratively claim Calminoe, now a social forestry project community, have populations some 5 to 10 times that of the local community, and populations of these municipalities are **not** considered "upland". Additional discussion of calculations, assumptions on which they were based, and possible direction of bias arising from the assumptions would be helpful. In brief:

    2. Population is a key variable in the sustainability of shifting cultivation in tropical forest ecosystems since "adequate fallows' depend upon a sufficiently low man-land ratio. With increased population, land use is intensified and competition for land increases — often with damaging effects. The man-land ratio has been increasing in the uplands of the Philippines, but as in other tropical forest countries, the numbers of people — like rates of deforestation — have been largely subject to speculation. The considerably more rigorous approach taken by Cruz led to a figure of 11 million uplanders in 1970, but this figure is likely to remain open to that further work on the numbers and methods used in arriving at them.

"UPLAND DEVELOPMENT" TECHNOLOGIES, ADOPTION, AND INTRODUCED VS. IN-PLACE TECHNOLOGIES

Samson said that technologies for upland development must be based upon solid, basic research, must attempt to rehabilitate, maintain, or improve productivity and sustainability of upland agroecosystems, must consider the inherent fragility of the environment, and must be appropriate to resources available to potential adopters. That is, not only must introduced technologies be effective, but, to be adopted, must be beneficial to potential adopters in light of the resources required for adoption and comparing benefits of introduced vs. in-place practices.

Soil erosion was identified as the most critical problem stemming from human resource use and contributing to instability and low-sustainability of upland agroecosystems. Engineering methods to counter soil erosion include bench and ridge terraces, diversion canals, dams, dikes, and channels. Uplanders without indigenous terracing can design and construct terraces using the A-frame for laying out contours. Adoption of terracing in social forestry projects has been minimal due to the labor required and uncertainties as to who would benefit in the long-term. Contour methods — contour ditching, contour planting, strip cropping, and contour strips that terrace naturally over time — generally reduce soil erosion, but usually require significant labor investments and have been inconsistent in terms of improving crop yields.

Vegetative soil erosion control methods and soil-conserving tillage methods require less labor than terracing. Zero and minimum tillage combined with mulching and weeding are promising for reducing soil erosion. Labor for weeding, however, may be a constraint to adoption. Soil-improving and improved humus production technologies, still largely at the research stage, include production and use of organic fertilizer, composting and mulching, use of nitrogen fixing plants and inoculant microorganisms, use of microorganisms to speed decomposition, green manuring, and use of mycorrhizal relationships.

Agricultural intensification for greater production includes increasing lands cultivated, cropping more often, and increasing inputs. With limited land and inputs, multiple cropping systems are being developed. These include sequential, inter-, and relay cropping. Potential benefits are higher yields per unit area, greater yield stability over time, more even production over time, less susceptibility to pests and diseases, more efficient use of sunlight, moisture, and nutrients, reduced risk of total crop failure, and decreased soil erosion. However (again, the big "however"), introduced upland multiple cropping systems have often required relatively expensive (for the upland farmer) inputs. Adoption has been low, indicating, among others, a need for more research and development on farmers' fields.

Agroforestry, the intercropping of trees and food crops, can mimic the tropical forest in providing a protective, layered canopy and a nutrient-cycling system of high species diversity. Many traditional swidden agricultural systems — including "cyclical" and *Taungya* — were productive and stable, but depended upon alternating forest fallows. "Integral" agroforestry, the **permanent** cropping of trees and food crops, has been characterized as suitable for hillside farming, acceptable to target beneficiaries, inexpensive, productive, sustainable, efficient, and ecologically sound. While traditional integral agroforestry exists scattered throughout the Philippines, adoption of the introduced technology remains limited, and more development on this ideal technology is needed.

Aguilar considered technology adoption in the social forestry projects. Technologies promoted and subsidized included soil conservation measures, soil fertility improvement, reforestation, agroforestry, ipil-ipil planting, and forest fire control. Results varied but were not encouraging, given limited adoption even with subsidiaries and other inducements.

According to Samson, technology effectiveness requires research for understanding specific problems addressed by particular technologies and for appropriate use and application. The case studies also indicate that technology design and testing must take place not only at research stations or universities, but also in upland communities and on farmers' fields. Technology development must be an interactive process between researchers and intended adopters that considers both specific local, technical appropriateness and the economics of the introduction.

Torres and Raintree illustrated the use of community and site-specific preliminary "diagnosis and design" research for upland technology (agroforestry) development. Research in Tabango, Leyte, examined local circumstances, diagnosed problems, screened possible technologies or interventions for testing, and designed farming systems experiments to test, develop, and to then possibly extend "best bet" technologies.

Land-use includes coconut-perennial intercropping with some livestock and permanent upland field crop cultivation. A few larger coconut-based enterprises were successful, but the many small tenant farmers producing annuals under perennials and coconut on lands received through "Operation Land Transfer" were less successful. Since tenants could buy only lands tenanted, parcels were too small to support a family through tree cropping patterns viable for large holders. The upland field crop alternative was "only marginally productive and subject to severe upland degradation." Research sought technologies to make such small-holder upland farming sustainable. Target farmers had land of one ha to two ha, and produced rice and maize, cash crops of peanut and mung bean, root crops and banana as supplements, and ipil-ipil for cash (leaf meal for firewood).

Problems were inadequate cash income and low efficiency of livestock (especially pig) enterprises. Low yields were due to low and declining soil fertility, lack of soil regeneration through fallowing, lack of mineral fertilizers, and soil erosion.

Technologies for screening included reduced tillage, contour plowing, maintenance of organic matter on fields (in place of burning), new cash crops, and improved pig management. Technologies had to have low capital needs, be labor efficient, be land efficient, not substantially increase pest problems, and consider local market potentials.

Non-forestry technologies were examined. However, inorganic fertilizers faced cash, credit, and infrastructure constraints. Composting was constrained by lack of knowledge and labor. Green manuring faced land, labor, and direct profitability constraints. Incorporation of crop residues required a new plow technology and stronger traction. No-tillage methods were constrained by knowledge, equipment, and herbicide needs. Bench terraces required high labor investments. Perennial crops would compete with food crops. Cattle would require fodder production and would compete with food and cash crops.

Contour intercropping of nitrogen-fixing tree hedgerows and food/ cash crops was promising and could provide organic fertilizer, form vegetative barriers to soil erosion, and would naturally terrace over time. Plowing would have to follow contours, further reducing erosion. An ex ante evaluation of the most appropriate tree/shrub system component led to the choice of giant ipil-ipil strips. Cropping patterns research on farmers' fields, at a local field station, and at a local college was designed, and components, management practices, treatment variables, and parameters to be monitored were stated.

At best, technology development and testing is slow and painstaking, and must be done both at the research station and along with potential "beneficiaries" if adoption is to result.

Costs required and benefits forthcoming to potential adopters must be considered. As Capistrano and Fujisaka (1984) concluded in a discussion about upland development technologies and productivity, costs to both projects and to potential adopters are usually high, benefits have been minimal or long-term in realization, and adoption has been limited.

In examining a range of the economic factors involved in upland development, Segura-de los Angeles first precisely identified factors influencing the uplanders' use of resources: 1) "open access to uplands and upland resources, 2) differential effects of upland cultivation on the uplands and on the lowlands and between present and future users, 3) lack of more permanent alternative sources of livelihood for uplanders, and 4) a high preference for present consumption over future consumption."

The economics of in-place upland cultivation practices must be understood in considering possible changes in different upland resource use patterns. The goals of introduced upland development expressed by the government social forestry program are either individual (improved standard of living) or societal (stable forest occupancy, prevention of further forest destruction, rehabilitation of degraded forest areas, and minimization of soil erosion). Participation in projects meeting individual goals has been stronger than in projects addressing societal goals.

Many ex-ante economic feasibility studies have been over-optimistic in assuming cooperation of intended "beneficiaries" and ex-post studies have offered a mixed bag of findings. Prior land use influenced participation of assisted farmers in the **PICOP** project, while non-assisted

farmers were influenced by income, education, and degree of risk
aversion. Land types and water resources affected adoption of new
cropping patterns in Antique. Segura-de los Angeles found that initial
higher organic matter, larger farm size, higher income, less knowledge
of conservation, and less household labor implied less adoption of
terraces. Her analysis of projects in Daidi, Nueva Viscaya, and in Pantabangan, Nueva Ecija, had problems attributing income increases to
project activities, and another study provided the unsurprising result
that, to the farmer, traditional *kaingin* farming ". . . is as advantageous
as the modified systems."

Overall, the findings of economists complement the examination of
upland development technologies: technologies are being developed and
introduced; adoption has been limited; where adopted, usually on a
limited project basis, both (agro)ecosystems and income or welfare
benefits have apparently been limited or difficult to determine. In
short:

3. "Upland development" includes development of technologies and implementation of projects intended to maintain
   or improve the productivity and sustainability of upland
   agroecosystems and associated farmer practices. Introduced
   technologies have so far offered potential adopters limited,
   long-term, or societal benefits and substantial costs. As
   such, in-place or indigenous practices have necessarily been
   preferred by most uplanders; and adoption of new technologies and project participation has been limited. Technology development is generally a slow, painstaking activity
   that involves research carried out both at the university or
   research station and on farmers' fields, alongside the potential "beneficiaries" or adopters. More research is needed —
   on the technologies, on costs and benefits for potential
   adopters, and on impacts on ecosystems, agroecosystems,
   and society of projects and related policies. "Upland
   development" must consider and balance productivity,
   equity and individual welfare, systems sustainability, and
   off-site, macro-concerns.

THE UPLANDERS: DIVERSITY AND SOCIAL AND CULTURAL CHANGE

Russell analyzed a wealth of ethnographic materials dealing with
diverse, changing Filipino upland cultures and societies, and Fujisaka
and Capistrano examined a single, rapidly evolving and internally hetero-
genous pioneer upland community. Uplanders, far from being "change
resistant", have long been a part of dynamic processes of change and
are diverse in terms of subsistence strategies, social and economic orga-

nization, beliefs and perceptions, languages, and — somewhat as a result of such differences — respective responses to changing conditions.

Russell described the diversity of upland groups in the Philippines and, among a range of characteristics indicating such diversity, examined different subsistence strategies (swidden, wet-rice, and hunting and gathering), the exploitation of different ecosystems by single groups, and differing socioeconomic organizations. Detailed and compared across groups were different systems of property access and control, inheritance, kinship organization, political organization and social stratification, agricultural labor patterns, water management systems, and ritual regulation.

The spontaneous transition from subsistence to market economy has been increasing and disruptive for many upland groups in the last several decades. Indebtedness, changes in land access patterns, greater inequality within groups, land losses to outsiders, decline in reciprocal arrangements, and vulnerability to outside market forces have been among the results.

While indigenous technologies are best-bet starting points for trying to introduce "improved technologies", evidence supports neither that all indigenous resource use practices are somehow ecologically sound nor that all lowlanders settling in the uplands utilize destructive practices in the uplands. Increased population pressure on land resources, increased dependency on a commercial economy, and demand for cash incomes have brought environmentally destructive changes in indigenous resource use practices.

The rapidly evolving pioneer migrant community of Calminoe was examined in terms of interactions among agroecosystem, farmer practices and patterns of resource use, and emergent social organization structuring resource access and use. Settlers established diverse home gardens and farm plots in the logged-over secondary forest. In-migration, commercial cropping, and small-scale logging combined to increase resource competition and demands on the ecosystem, and resource-competing factions formed. More than local circumstances were relevant: spacing the settlers and the national, regional, and local government agencies were unaware of who held legal jurisdiction over the land. Ambiguity contributed to strife. Different factions allied themselves with different municipal and provincial-level claimants, and claims and counter-claims concerning the lands further polarized residents.

The UPLB-Program on Environmental Science and Management and the BFD's Integrated Social Forestry Program are implementing a research and development project in Calminoe. After basic research, initial efforts involved less the design and testing of appropriate technologies, and more the settling of the status of the lands, of jurisdictions of local municipal and provincial governments, and of local conflicts through clarification of resource rights and jurisdictions.

In brief:

4. Upland cultures and societies — indigenous and migrant — are heterogeneous and dynamic. Diversity and change can be seen in resource use strategies, related social and economic organization, and interactive effects of those strategies on local ecosystems. While in-place or indigenous practices are starting points for technology development, neither indigenous peoples nor migrants have been shown to be better or worse in terms of practices leading to upland degradation. Rather, dynamic factors such as increasing populations and transitions to cash economies have been accompanied by changes in resource access and use practices and related social-economic organization. Such change can be socially and culturally disruptive, and often involves greater pressures on resources and damage to upland ecosystems. Not only local, but national and institutional factors are a part of the context of change and development.

LAND TENURE

Essays by Lynch, Russell, and the case study chapters demonstrated that access to land resources is perhaps the major issue facing uplanders. Lynch detailed the (national) legal situation — the colonial foundation of current laws, current jurisdictions over so-called public lands, increasing control by the executive branch of public lands, property rights of upland occupants, and traditional or customary land laws.

The Regalian Doctrine said that all lands belonged to Spain other than where contrary rights were recognized. The subsequent three hundred years of Spanish occupation, however, saw several laws protective of indigenous property rights. The U.S. administrators of the Philippines returned to the Regalian Doctrine, and the Philippines later followed U.S. public land and resource laws. As of 1984, about three-fifths of the land of the country was "public". Lynch concluded that possibilities for upland residents to acquire recognition of private property rights are decreasing, and that "displacement of indigenous and migrant occupants is now widespread and increasing."

As in Calminoe, land access depends upon classification. Laws of the early 1900s said that public agricultural lands could be privately owned. Cultivators could apply for titles unless lands were shown to be not agricultural. The policy was inverted after 1940, and public lands were presumed to be forest unless classified as agricultural. The executive branch and the forest bureaucracy gained power over public lands, and individuals faced greater difficulties in gaining ownership. Executive powers have increased since declaration of martial law in 1972, while the judiciary's power to recognize private rights has decreased.

In 1982 the private property rights of long-term occupants were reconfirmed. Unfortunately, getting public agricultural lands classified as alienable and disposable has been "virtually impossible" for most upland residents.

Again as demonstrated in Calminoe, Philippine public lands are currently (overlappingly) claimed by the Bureau of Forest Development, the Bureau of Lands, the Bureau of Mines, the Ministry of Human Settlements, the Ministry of Agrarian Reform, the National Development Corporation, the National Power Corporation, the National Irrigation Administration, and other government agencies and institutions.

Several laws protect the land rights of uplanders, but have generally been ineffective due to current executive powers, lack of application of laws favoring claimants, and procedures required of uplanders to have rights recognized. Laws protect native rights over ancestral lands owned since "time immemorial". Migrants continuously occupying and cultivating public agricultural lands since mid-1945 are also entitled to ownership. Tribal and Muslim migrants occupying lands since mid-1955 are similarly protected. Migrants settling forest lands prior to mid-1975 were not to be prosecuted, and later, migrants settling in forest lands prior to the end of 1981 became eligible to receive 25-year renewable leases over plots of up to 7 ha through the Integrated Social Forestry Program of the BFD. Overall, laws recognizing or allowing uplanders access to land resources are on the books, except that implementation of such laws in the face of government agencies actually holding power over the lands remains limited. In short:

5. Land access is a major problem facing uplanders and efforts at upland development. Unfortunately, legal and institutional factors have tended to complicate matters. Land tenure laws and policies have developed over time. Several protect the rights of certain categories of upland occupants, but these have not been effectively implemented. Rational, coherent, and integrated policies for the use and manage-

ment of upland forest resources in the Philippines do not yet exist. Executive branch powers over land have increased, and different national agencies are competing for control of lands and forest resources. Ambiguities concerning current jurisdictions over land and other resources have contributed to local problems in the uplands.

INTRODUCED CHANGE: "UPLAND DEVELOPMENT" PROJECTS

In spite of problems, difficulties, and unresolved issues, needed research continues; technologies are being developed; agencies are working on projects intended to bring development to the uplands.

Aguilar examined several "social forestry" projects that attempted to address environmental problems, improve living conditions, increase income, and/or provide participants with access to land. Each project had its own goals, policies, and problems of implementation. Associations of "beneficiaries" generally had low rates of participation and were "largely ineffectual." Most participants felt they had little project decision-making power. Participants could name benefits, but these generally fell short of expectations. In some projects participants acquired land access, but land tenure insecurity appeared to continue. Some projects provided employment, but income improvements were otherwise limited. Aguilar concluded: "The available data indicate that few respondents saw positive returns to their involvement in the projects."

Del Castillo and Castro described agencies and institutions active in upland development work in the Philippines. These were classified as commodity-oriented, service-delivery organizations, community organizations, research and educational institutions, and funding institutions. It was recommended that: 1) more coordination in the planning and implementation of upland development projects is needed, 2) research needs to be speeded up, 3) interdisciplinary and multisectoral research approaches are desirable, 4) research should be linked to action, and 5) the "working group" may be an effective model for interagency collaboration.

Philippine upland development efforts currently subsume "social" approaches — from the use of social sciences to projects using "participatory approaches". Different social approaches to forestry in the Asia-Pacific region are concerned with the non-monetized sector of the economy, seek participation, and require a change of the forester's role from forest protector to extension agent. Starting in the mid-1970s, such approaches were a response to increasing tropical forest destruction combined with, among others, the idea that resource use by poor forest occupants was not the major cause of deforestation. Other contributing factors named by Kirchhofer and Mercer were continued high fuelwood needs, continued rural poverty in spite of policies promoting industrialization, growth of an environmental ethic, recognition of the futility of trying to bar forest settlement and use, and the transition of many developing countries from wood exporters to importers.

Social forestry programs require either collective or individual action. Collective action programs include, among others, village based community forestry. These are difficult to implement since they require consensus and coordinated action. Individual action social forestry programs — trees for farms, trees for residual areas, and contractual programs — have been more successful: "The successful introduction of innovations through collective action requires a greater understanding of and sensitivity to the socioeconomic and sociocultural conditions of the community."

Community forestry — a social forestry program requiring collective action — is a response to several needs. Development needs to reach all strata in a community, including the landless and unemployed, rather than further benefit only the relatively better-off. In some projects, benefits have accrued to the better-off, and individual activities have been detrimental to community well-being. The PICOP project (Philippines) benefited land owners, but not the intended poor *kaingineros* or unemployed. Individual Indian farmers planted *Eucalyptus* in place of food crops, decreasing employment opportunities and food availability.

Community forestry projects are difficult to implement. They "cannot be imposed from the top down" and need the involvement of women, prior specification of the distribution of benefits, government commitment, technical experience, foresters willing to assume new roles, credit, short-term benefits and long-term goals, and understanding of local economic, cultural, and political structures. "If community forestry is to succeed on a significant scale, it must be integrated into an overall, multifaceted rural development program."

We can also mention that development activities necessarily differ in the uplands and lowlands. Compared to the lowlands, the uplands are more heterogeneous in terms of physical factors — slope, slope aspects, vegetation, water and soil resources — and in terms of peoples and cultures. As one result, a mosaic of "micro-patches" must be considered in developing appropriate improved upland systems of land use. As has been demonstrated, there is a relative dearth of baseline data that can be used for planning such introduced change in the uplands. Because of such diversity and lack of basic information, a "menu" of technologies for site testing is often more appropriate for the uplands rather than a technology "package" of the type used in the lowlands (table 1). The "menu" technologies are currently usually based upon in-place or indigenous technologies, which are often identified through ethnographic methods (e.g., "rapid rural appraisal"), and are heavily subject to farmer selection and testing. On the other hand, improved lowland technologies can be, to a much greater extent, generated at research stations.

Table 1. Upland and lowland technology generation compared

CHARACTERISTIC	LOWLANDS	UPLANDS
1. Variability	Relatively homogeneous, some micro-environmental variation	Very heterogeneous
2. Baseline Data	Considerable	Very little
3. Obtaining Information for Design of Appropriate Technologies	Standard survey, extension agents, research stations	Adapted ethnographic methods, "rapid rural appraisal"
4. Technology Generation	Heavily based on research station work	Heavily based on indigenous practices, knowledge
5. Technology	"Packaged"	"Menu"

6. "Upland development" now includes activities ranging from research in upland communities to research, development, and testing of technologies, and to community projects that attempt to extend technologies or otherwise introduce change. Guided and introduced change in the form of social forestry projects seek "participation" combined with technology adoption. Success has been limited. Popular participation is considered desirable but has been difficult to achieve. Problems include determining indicators of participatory project success and forms of participation that improve possibilities for such success and balancing "participation" in terms of individual vs. group and societal benefits. Agencies and institutions concerned with upland development in the Philippines now include government and private institutions, national and international funding sources, and research, policy-generating, and project-implementing agencies. More coordination between agencies

is needed. Consideration of social and community forestry at the regional (Asia-Pacific) level indicates a need for programs requiring collective rather than individual action. While more difficult to implement, such programs are more successful in reaching and being applicable to "the poorest of the poor". The generation of appropriate technologies differs for the lowlands and uplands due to the greater heterogeneity and lack of baseline data characteristic of the latter.

LESSONS, IMPLICATIONS AND AN (ECO)SYSTEMS MODEL OF CHANGE AND DEVELOPMENT

The following statements reiterate lessons learned and areas in which continued work is needed, relate these to upland development policies of the Philippines' social forestry program, and synthesize findings into a working (ecosystems) model-diagram. Key lessons and areas in which work is needed:

- The tropical forest ecosystem is productive and stable given few or low disturbances, but is fragile in the face of some human practices.

- Logging, a major human-generated disturbance, has led to forest degradation, destruction. More data on magnitudes are needed. Rational national policy is needed.

- Shifting cultivation, another major human practice in the tropical forest, combined with low populations can be relatively sustainable.

- Population has been increasing in the uplands, in large part due to migration.

- Population increases in communities practicing shifting cultivation are linked to agroecosystem degradation via increased demand for land, shorter fallows, more cropping, weeding, burning, and less effective regeneration.

- More data are needed on the magnitudes of other supposed effects such as significantly increased water run-off, siltation, downstream flooding, and on the supposed negative effects of forest conversion to grassland.

- Cultures and societies in the uplands are diverse and changing. Some changes are "disruptive" to peoples and ways of life, and damaging to local ecosystems. Others are more "adaptive" in that new dynamic sets of interrelationships allow systems to continue.

— Farming and other resource use systems are naturally changing in many upland communities as populations increase and resources become scarcer. These changes need to be carefully observed for respective lessons.

— Technologies for improved resource use and management and strategies for introducing change to the uplands are being developed and tested. Continued work is needed.

— Technology adoption has so far been limited by lack of (economic) superiority over in-place technologies — especially in meeting short-term needs, the additional investments required of potential adopters, and the long-term and social rather than individual pay-offs.

— Land tenure laws and policies seem to protect the rights of various groups of uplanders — both indigenous and migrant — but have not been effectively implemented. Obviously, "political will" is needed.

Research and development of upland technologies and field projects are being carried out by many agencies and institutions. The Integrated Social Forestry Program of Bureau of Forest Development is the major project-implementing program in the Philippines. Specific upland development policies of the social forestry program of the Bureau of Forest Development should be evaluated in the light of lessons learned and research and development needs. Alternative policies, where appropriate, are suggested.

— Individual project participants receive 25-year renewable leases or "Stewardship Contracts" for up to seven ha of land. Because land tenure security is desirable, some individuals and communities may be induced to participate. Most uplanders, however, feel that they already "own " their land, regardless of national laws or policies. This appears to be as true for long-established peoples of the Cordillera as for pioneer settlers in Calminoe.

— Seven ha may be adequate for a Mountain Province vegetable farmer, but may be inadequate for the cultivator of more marginal land. Policy flexibility to meet local conditions may be impossible, but is suggested.

— Project participants receive contracts for a single parcel that they must permanently crop. This chapter has discussed the importance of shifting and fallowing fields for regeneration under existing farming systems. However, project policy expressly and effectively attempts to do away with shifting cultivation.

— Permanent agroforestry cropping is to be encouraged and is an attractive idea, but evidence shows that such technologies for marginal uplands and necessary and equitable structures for credit, marketing, inputs, information access, and labor are not yet available. Shifts are naturally being made by some upland groups to more permanent commercial cropping; however, many of the social, cultural, economic, and ecological impacts have not been favorable. Examples of needed research include observation of spontaneous shifts to vegetable farming in Benguet, commercial shifting tomato cultivation in Calminoe, and the careful, location-specific technology design and testing in Tabango.

— Tree planting and other forms of reforestation are encouraged. The ecosystemic results of conversion to grasslands, secondary forests, and cropped areas need further investigation and site-specific comparison to forest ecosystems. Where trees have little or no advantage for ecosystem or farmer, they can be omitted as a technology.

— For social forestry projects in migrant communities, participation is supposed to be limited to those settling in project areas prior to 1981. The impossibility of establishing arrival dates in pioneer communities means that the program can encourage migration to the uplands and that conflicting claims, especially for pioneer shifting plots, cannot be easily settled — both with net negative effects.

— Policy prohibits settlement in, and use of "critical" watershed areas. The negative effects of cultivation compared to forest cover in terms of watershed degradation have not been established. More research is needed if another unenforceable, unpopular, unnecessary, and possibly expensive policy is to be avoided.

— Project strategy includes "participatory approaches" and "community organization". Long-established communities in the uplands and elsewhere have their own functioning social and economic organization. In such cases, the difficult-to-implement strategy of working within the existing social structure while trying to ensure that project benefits are equitably distributed must be recommended. In heterogeneous pioneer communities,

FIGURE 1. A SIMPLE (ECO) SYSTEMS MODEL
FOR UPLAND CHANGE AND DEVELOPMENT

ECOSYSTEMS DEGRADATION	TROPICAL FOREST-BASED AGROECO-SYSTEMS	ECOSYSTEMS SUSTAINABILITY
migration to uplands greater population increased land demand shortened fallows more cropping/plot more weeding/plot more burning/plot	DEMOGRAPHIC, HUMAN PRACTICES	regeneration adequate fallows plot shifting current shifting cultivation decreased migration to uplands
increased timber demand destructive logging practices more logging/deforestation forest conversion some commercial cropping systems	ECONOMIC, HUMAN PRACTICES	alternatives in the lowlands decreased timber demand less logging/less destructive practices
shifts to some commercial cropping systems breakdown of traditional systems greater stratification factions, resource strife greater land, resource privitization greater demand on land	SOCIAL CHANGE	emergence of new, adaptive social-economic structures
use/adoption of inappropriate technologies	CULTURAL CHANGE	innovative/experimentation indigenous technical knowledge adoption of new, adopted re-source use/farming practices
confused/ambiguous resource jurisdiction lack of political will to enforced "good" policy	INSTITUTIONAL LEGAL POLICY	implementation of policies wholistic/"enlightened" policy settle judistic, ambiguities
inappropriate introduced technologies inappropriate strategies inappropriate strategies for working with local communities	"UPLAND DEVELOPMENT"	appropriate strategies to work with uplanders appropriate technologies technology testing technology development continued research

some characterized by factionalism, a formal organization established by the project may either be superfluous or can be monopolized by a single group — leading obviously to further problems. Appropriate forms of "participation" must be determined on a case-to-case basis.

The current dynamic situation in the uplands involves several interacting factors: 1) an ecosystem or perhaps various agroecosystems in interaction with human groups, 2) increasing populations, 3) different, naturally changing cultures, societies, and technologies, and 4) efforts to develop improved technologies and to introduce or guide change (i.e., upland development). On the one hand, change may be leading to continued ecological damage, resource depletion, and eventual systems collapse combined with severe negative downstream or downslope effects. On the other hand, responses to systems degradation and increasing population may include emerging systems-adaptive upland resource use practices. Such adaptive practices include both naturally occuring agricultural intensification and research-based development of "productive, stable, sustainable, and equitable" technologies and strategies. Successful application of such technologies and strategies, of course, requires associated institutional change, rationalization of national resource-use policies, adequate "political will", and economic change in the lowlands that would decrease migration to the upland. These factors are presented in a simple systems model-diagram (fig 1).

Hopefully, this volume will contribute to adaptive change in the uplands by bringing together and holistically synthesizing research and "development" leassons learned, by indicating areas of needed continued work, by considering the efforts of scientists of various disciplines and applied fields in terms of basic relevant parameters, ongoing "natural" change processes, and guided change, and by providing a better understanding of the risks involved in **not** adequately addressing the problems and potentials at hand.

REFERENCES CITED

Arnason, T., J.D.H. Lambert, J. Gale, J. Cál, and H. Vernon. 1982. Decline of soil fertility due to intensification of land use by shifting agriculturalists in Belize, Central America. Agroecosystems 8:27-37.

Bernstein, R. and R.W. Herdt. 1977. Towards an understanding of Milpa agriculture: The Belize Case. Journal of Developing Areas 11:3:373-392.

Brosius, Peter. 1982. The analysis of swidden systems: perspectives from succession theory. EAPI-Working Paper. Honolulu: East-West Center.

Capistrano, A. Doris and S. Fujisaka. 1984. Tenure, technology, and productivity of agroforestry schemes. Philippine Institute of Development Studies 84-06.

Chapman, E.C. 1974. Shifting cultivation in tropical forest areas of South East Asia. Paper, Regional Meeting of the I.U.C.N. at Bandung.

Chin, S.C. and T.H. Chua. 1984. The impact of man on a southeast Asian tropical forest. Malayan Nature Journal 36:253-269.

Clarke, W. 1976. Maintenance of agriculture and human habitats within the tropical forest ecosystem. Human Ecology 4:3:247-259.

Eder, James. 1977. Agricultural intensification and the returns to labor in the Philippine swidden system. Pacific Viewpoint 18:1:1-21.

Furtado, J.I. 1978. The status and future of the tropical moist forest in Southeast Asia. In: Developing Economies in Southeast Asia and the Environment. C. MacAndrews and L.S. Chia, eds. Singapore: McGraw-Hill.

Gagney, Wayne. 1981. The transformation and intensification of shifting agriculture: past and present conservation practices. In: Traditional Conservation in Papua New Guinea: Implications for Today. Morauta, Pernetta, and Haney, eds. Boroko, PNG: Institute for Applied Social and Economic Research (IASER) 16.

Hamilton, Lawrence. 1985. Overcoming myths about soil and water impacts. In: Soil Erosion and Conservation. S.A. El-Swaify, W.C. Moldenhauer, and A. Lo, eds. Ankeny, Iowa: Soil Conservation Society of America.

1983. Some water/soil consequences of modifying tropical rain forests. International Symposium on the Future of Tropical Rainforests in SE Asia. Kepong, Malaysia: Forest Research Institute.

Jordan, Carl and Rafael Herrera. 1981. Tropical rainforests: Are nutrients really critical? American Naturalist 117:2:167-180.

Kartawinata, K.H. Soedjito, T. Jessup, A.P. Vayda, and C.J.P. Colfer. 1984. The impact of development on interactions between people and forests in East Kalimantan: A comparison of two areas of Kenya

Dayak settlement. In: Traditional Life Styles, Conservation, and Rural Development. J. Hanks, ed. IUCN (EAPI Reprint 76).

Kartawinata, K., S. Adisoemarto, S. Riswan, and A.P. Vayda. 1981. The impact of man on a tropical forest in Indonesia. Ambio 10:2, 3:115-119.

Kellman, M.C. 1969. Some environmental components of shifting agriculture in Upland Mindanao. Journal of Tropical Geography. 28:40-56.

Margolis, Maxine. 1977. Historical perspectives on frontier agriculture as an adaptive strategy. American Ethnologist 4:1:42-64.

Parfitt, Roger. 1976. Shifting cultivation — how it affects the soil environment. Harvest 3:2:63-67.

Sherman, George. 1980. What "Green Desert"? The Ecology of Batak Grassland Farming. Indonesia 29:113-148.

Wood, A.W. 1979. The effects of shifting cultivation on soil properties: An example from the Karinui and Bomai Plateaux, Simbu Province, Papua New Guinea. PNG Agricultural Journal 30 (1-3).

THE AUTHORS

Filomeno V. Aguilar, Jr.: MS, Social Planning, University of Wales; was Research Associate, Institute of Philippine Culture, Ateneo de Manila University; now with the research group of the Ramon Magsaysay Award Foundation; upland development, development planning and policy.

A. Doris Capistrano: Economist; MS, Economics, University of the Philippines, Diliman: was Instructor of Economics at the University of the Philippines at Los Baños and staff member of the Program on Environmental Science and Management; now working on PhD, resource economics at University of Florida; resource economics, natural resources policy and management, economic evaluation of upland development technologies.

Charles P. Castro: Forester; BS, Forestry, University of the Philippines at Los Baños; Forester, Social Forestry Division, Bureau of Forest Development, Ministry of Natural Resources, Philippines; social forestry and forestry extension.

Romulo A. Del Castillo: Forester; PhD, Forest Biometry, North Carolina State University; Professor, Forest Resources Management, University of the Philippines at Los Baños; Consultant, Ford Foundation, as facilitator of Foundation assisted Upland Development Program of the Bureau of Forest Development, Philippines; extensive work in the Philippines; additional work in Vietnam and Indonesia; forestry education and training, forestry statistics, upland and forest resources management and development.

M. Concepcion J. Cruz: Ecologist and Water Resources Management; PhD, Development Studies, University of Wisconsin-Madison; Assistant Professor, College of Development Economics and Management, University of the Philippines, Los Baños; Chairperson of the UPLB Masters of Science, Environmental Studies Program; river basin development, irrigation policy, water rights and upland water management, natural resources policy, upland development, environmental science education.

Sam Fujisaka: Ecological-Agricultural Anthropologist; PhD, Cultural Anthropology, University of Oregon; Visiting Associate Professor, Program on Environmental Science and Management, University of the Philippines, Los Baños; work in the Philippines and Bolivia; cultural ecology in upland-fragile lands, farming systems development, pioneer slash-and-burn agriculture, changing human-ecosystem interactions, RRA methods.

James Kirchhofer: Geographer; PhD, Geography, University of Hawaii, Manoa; was graduate student, University of Hawaii; work in Thailand; biogeography, natural resources management and conservation, plant ecology, soils, climatology, Southeast Asian studies, social forestry.

Owen J. Lynch, Jr.: Attorney; Juris Doctor (J.D.), Catholic University of America; Master of Laws (LL.M.), Yale University; was Visiting Professor, University of the Philippines College of Law, Diliman, College of Law; Philippine legal history and ethnography, national and customary laws pertaining to ancestral lands.

D. Evan Mercer: Forest Ecologist; MS, Natural Resources, University of Michigan, School of Natural Resources; was Research Intern, Environment and Policy Institute, East-West Center, Hawaii; is PhD student, Resource Economics, School of Forestry and Environmental Studies, Duke University; work in the US, Indonesia, Costa Rica; forest ecology and management, tropical forestry and agroforestry.

John B. Raintree: Ecological Anthropologist; PhD, Applied Ecological Anthropology, University of Hawaii; Project Leader for the Diagnosis and Design Project and the Tree Tenure Project, International Council for Research in Agroforestry (ICRAF); research on agroforestry systems in Kenya, the Philippines, Indonesia, Nigeria, Sierra Leone, and the US; socioeconomic aspects of agroforestry, cultural ecology, farming systems diagnosis, bio-economic modeling, village-level technology assessment, cross-cultural extension methods, knowledge engineering.

Susan Russell: Economic Anthropologist; PhD, Anthropology, University of Illinois at Urbana-Champaign; was Visiting Professor, School of Economics, University of the Philippines, Diliman; is Assistant Professor, Department of Anthropology and Center for Southeast Asian Studies, Northern Illinois University, DeKalb; extensive work with indigenous tribal groups of the Luzon Highlands, Philippines; cooperatives, marketing, urban food markets, highland and rural development, socioeconomic analysis methods.

Percy E. Sajise: Plant Ecologist; PhD, Plant Ecology, Cornell University; Director, Program on Environmental Science and Management, University of the Philippines at Los Baños (UPLB), Professor, UPLB; extensive work in the Philippines; additional work in Indonesia; tropical and grassland ecology, upland development, upland development technologies, natural resources managementt policy, agroecosystems analysis, environmental science and education.

Benjamin K. Samson: Botanist; MS, Botany, University of the Philippines at Los Baños; working on PhD in botany/tropical ecology at the

University of Florida; was instructor, Institute of Biological Sciences and staff member of the Program on Environmental Science and Management at the University of the Philippines at Los Baños; water relations, tropical ecology and botany, upland development, agroforestry technologies.

Marian Segura-De Los Angeles: Economist; PhD, Economics, University of the Philippines, Diliman; was Senior Lecturer, School of Economics, University of the Philippines, Diliman; is Research Fellow, Philippine Institute for Development Studies; economics of upland development, natural resources policy, economic and social impact analysis.

Filemon Torres: Range Management and Livestock Production Specialist; PhD, Animal Nutrition, Cornell University; Coordinator of the Collaborative and Special Projects Programme and Project Leader for Review on Silvopastoraliam, International Council for Research in Agroforestry (ICRAF), Kenya; extensive work in Argentina; beef production systems with special reference to plant-animal relationships.